Toxicants of Plant Origin

Volume II
Glycosides

Editor

Peter R. Cheeke

Professor of Comparative Nutrition
Department of Animal Science
Oregon State University
Corvallis, Oregon

CRC Press
Taylor & Francis Group
Boca Raton London New York

CRC Press is an imprint of the
Taylor & Francis Group, an **informa** business

CRC Press
Taylor & Francis Group
6000 Broken Sound Parkway NW, Suite 300
Boca Raton, FL 33487-2742

© 1989 by Taylor & Francis Group, LLC
CRC Press is an imprint of Taylor & Francis Group, an Informa business

First issued in paperback 2019

No claim to original U.S. Government works

ISBN 13: 978-0-367-45103-5 (pbk)
ISBN 13: 978-0-8493-6991-9 (hbk)

Visit the Taylor & Francis Web site at
http://www.taylorandfrancis.com

and the CRC Press Web site at
http://www.crcpress.com

FOREWORD

Natural toxicants in plants or "nature's pesticides" are increasingly recognized as significant items of the diet of humans and domestic animals. These diverse compounds function as chemical defenses of plants to deter herbivory. Selection of forage and crop plants for resistance against insects and other pests often increases their content of chemical defenses, thus inadvertently adversely affection their nutritional value. With an increasing emphasis on alternatives to agricultural chemicals in control of pests in crop production, natural plant toxicants will assume even more significance. It is appreciated that animal products such as meat, milk, and eggs may also be a means of exposure of humans to plant toxins. Thus knowledge of the metabolic fate of plant toxins in livestock is very important, both in terms of efficient animal production and the safety of animal products in the human diet.

These four volumes are intended to provide a comprehensive, up-to-date treatment of major plant toxins. The authors are authorities in their respective disciplines, and have provided fresh viewpoints and an integration of existing knowledge. These state-of-the-art treatments also draw attention to major areas where additional research is needed.

THE EDITOR

Peter R. Cheeke, Ph.D., is Professor of Comparative Nutrition, Department of Animal Science, Oregon State University, Corvallis. He received his B.S. and M.S. degrees in animal science at the University of British Columbia, Vancouver, in 1963 and 1965, respectively, and his Ph.D. in animal nutrition at Oregon State University in 1969. His research program has reflected a broad interest in livestock production and the components of feeds which adversely affect animal performance, including pyrrolizidine and quinolizidine alkaloids, saponins, protease inhibitors, and glucosinolates. His achievements in this area have resulted in extensive participation in international symposia and conferences. His research has resulted in over 100 technical publications and several books. In addition to work on plant toxins, he has pioneered research on the domestic rabbit as a new livestock species for use as a meat source in developing countries and has extensively studied the relationships between secondary compounds in plants and feeding and digestive strategies in the rabbit. He is recognized internationally for his work in rabbit nutrition, and has received several awards including the Mignini International Award for rabbit research and the outstanding paper award of the American Association of Laboratory Animal Science.

Dr. Cheeke is a member of several scientific organizations including the American Institute of Nutrition and the American Society of Animal Science. His current research emphasizes pyrrolizidine and quinolizidine alkaloid metabolism and the use of the rabbit as an herbivore model for the study of metabolism of plant toxicants.

CONTRIBUTORS

Richard J. Cole, Ph.D.
Location Coordinator
National Peanut Research Laboratory
U.S. Department of Agriculture
 Agricultural Research Service
Dawson, Georgia

Horace G. Cutler, Ph.D.
Plant Physiologist and Research Leader
Plant Physiology Unit
U.S. Department of Agriculture
 Agricultural Research Service
Athens, Georgia

G. Roger Fenwick, Ph.D.
Group Leader
Department of Chemistry and
 Biochemistry
AFRC Institute of Food Research
Norwich, England

**Robert K. Heaney, Higher National
 Certificate**
Research Scientist
Department of Chemistry and
 Biochemistry
AFRC Institute of Food Research
Norwich, England

Iwao Hirono, M.D.
Professor
Department of Pathology
Fujita-Gakuen Health University School
 of Medicine
Toyoake, Aichi, Japan

Eustace A. Iyayi
Research Officer
Rivers State University of Technology
Port Harcourt, Nigeria

J. P. J. Joubert, B.V.Sc.
State Veterinarian
Department of Agriculture: Veterinary
 Service
Regional Veterinary Laboratory
Middelburg, Cape Province, South Africa

Walter Majak, Ph.D.
Research Scientist
Research Station
Agriculture Canada
Kamloops, British Columbia, Canada

Ronald R. Marquardt, Ph.D.
Professor
Department of Animal Science
University of Manitoba
Winnipeg, Manitoba, Canada

Rodney Mawson, Ph.D.
Research Scientist
Unilever Research Laboratory
Sharnbrook, Bedford, England

David G. Oakenfull, Ph.D.
Principal Research Scientist
Division of Food Research
CSIRO
North Ryde, New South Wales, Australia

Michael A. Pass, Ph.D.
Reader in Physiology
University of Queensland
St. Lucia, Queensland, Australia

G. S. Sidhu, Ph.D.
Principal Research Scientist
Department of Food Research
CSIRO
North Ryde, New South Wales, Australia

Barry P. Stuart, Ph.D.
Manager and Veterinary Pathologist
Department of Pathology Services
Mobay Corporation
Stilwell, Kansas

Olumide O. Tewe, Ph.D.
Associate Professor
Department of Animal Science
University of Ibadan
Ibadan, Nigeria

Martin Weissenberg, Ph.D.
Senior Scientist
Department of Chemistry of Natural
  Products and Pesticides
Agricultural Research Organization
The Volcani Center
Bet Dagan, Israel

TOXICANTS OF PLANT ORIGIN

Volume I

ALKALOIDS

Volume II

GLYCOSIDES

Volume III

PROTEINS AND AMINO ACIDS

Volume IV

PHENOLICS

TABLE OF CONTENTS

Chapter 1

GLUCOSINOLATES

G. R. Fenwick, R. K. Heaney, and R. Mawson

TABLE OF CONTENTS

I. INTRODUCTION

The history of glucosinolate research is now well into its second century. Three main phases may be seen within this period. The first, stretching from 1831, when the first crystalline glucosinolate, sinalbin, was isolated from the seed of white mustard, until about 1950, represented a period of sporadic research, centered on the detailed structural studies of Gadamer at the end of the last century.[1] The second period, from 1950 until 1970, was characterized by a tremendous increase in knowledge about the chemistry of glucosinolates, centered primarily on the research groups of Schulz and Kjaer.[2] This period also saw the correction of the glucosinolate structure proposed by Gadamer and its ultimate confirmation by synthesis and X-ray studies. The final period, from 1970 onward has seen the emphasis on glucosinolate research shift from chemical to biological, prompted initially by the emergence of rapeseed as an oilseed of commerce and the ready availability of the defatted meal as a protein source of potentially great value for farm animals.[3] Current research is also fueled by the legitimate concerns over the effects of glucosinolates and their products in the human diet.[4] It is, thus, appropriate to consider both aspects of glucosinolates in this chapter.

In common with other natural toxicants, the effects of glucosinolates in animals are readily documented; in comparison those in man are much less easy to identify, being the consequences of long-term, low-level exposure (although, as will be seen, in this context "low" may be something of a misnomer). In addition, there has been considerable interest shown in the anticarcinogenic and enzyme-inducing properties of the products of certain glucosinolates which may be present in the human diet.[5] Thus, although it is impossible yet to provide any measure of risk associated with glucosinolate intake in man, it is apparent that this will reflect an overall balance of deleterious and beneficial properties.

In recent years a number of authoritative reviews of glucosinolates in foods and feedstuffs have been published, notably by research groups from the U.S. Department of Agriculture[6,7] and, in the U.K., the Agriculture and Food Research Council.[1,8] Many of the areas covered in these reviews, notably biosynthesis, enzymic breakdown, and relationship to flavor, have not advanced greatly in the intervening period, and, so in general, earlier references will be cited, the topics being dealt with here only in sufficient depth as to facilitate understanding of the remainder of the chapter. In other areas, notably those of human intake and the relationships of glucosinolates and their products to physiological disturbances in animals and man, the opportunity has been taken to critically assess recent findings.

II. DISTRIBUTION OF GLUCOSINOLATES

It is generally held that glucosinolates are limited to certain families of dicotyledonous angiosperms, predominantly within the order Capparales, *sensu* Cronquist or Taktajan, embracing the Capparaceae, Cruciferae, Moringaceae, Resedaceae, and Tovariaceae.[9] In the limited context of cultivated human foods and animal feedstuffs, it is members of the family Cruciferae which are most important, including oilseeds and forage crops, condiments, relishes, and vegetables[10] (Table 1). At the time of writing no authenticated member of the Cruciferae has been found to be devoid of gucosinolates or, when examined, the associated enzyme, myrosinase (see Section IV.A), and their presence has been suggested to be an important chemotaxonomic criterion for classification within this family.

Recently, claims for the presence of glucosinolates in such botanically diverse species as onion,[11] cocoa,[12] and mushroom[13] have been made, largely on the basis of the presence of trace amounts of known glucosinolate degradation products. Detailed reexaminations of the presence of glucosinolates in onion[14] and cocoa[15] have been conducted, and no evidence for their occurrence has been obtained. At this moment, then, there is little reason to modify the above boundaries of glucosinolate occurrence.

III. CHEMICAL ASPECTS OF GLUCOSINOLATES

A. Structure

The common skeletal structure for glucosinolates which is currently accepted is shown in Figure 1 (I). Based upon the work of Ettlinger and Lundeen,[16] the structure has been confirmed by X-ray studies[17] and direct synthesis.[18] The significant features of the structure are a sulfonated oxime grouping, which has been shown to be *anti* with respect to the side chain, R, and *syn* with respect to a thioglycosidic moiety. In almost all cases the sugar is (or, rather has been assumed to be) D-glucose, although a glucosinolate isolated from radish seed was found to possess a 6-sinapoyl-β-D-glucose group.[19] Danish workers have suggested that "bound" glucosinolates, containing sinapic, malic, and caffeic acids esterified to the sugar moiety, also occur naturally, although no evidence has yet been published to support these claims.[20]

Table 1
ECONOMICALLY IMPORTANT
GLUCOSINOLATE-CONTAINING PLANTS

White mustard	*Sinapis alba*
Brown mustard	*Brassica juncea*
Abyssinian mustard	*B. carinata*
Horseradish	*Armoracia lapathifolia*
Wasabi	*Wasabi japonica*
Radish	*Raphanus sativus*
Cress	*Lepidium sativum*
Indian cress	*Tropaeolum majus*
Water cress	*Nasturtium officinalis*
	B. oleracea L.
Kohlrabi	var. *gongylodes*
Cabbage (red, white)	var. *capitata*
Cabbage (savoy)	var. *sabauda*
Brussels sprouts	var. *gemmifera*
Cauliflower	var. *botrytis* subvar. *cauli-flora*
Sprouting broccoli, calabrese	var. *botrytis* subvar. *cymosa*
Kale	var. *acephela*
Pe-tsai	*B. pekinensis*
Pak-choi	*B. chinensis* var. *chinensis*
Turnip	*B. campestris* L. spp. *rapifera*
Turnip rape	*B. campestris* L. spp. *oleifera*
Swede, rutabaga	*B. napus* L. var. *napobrassica*
Rapeseed	*B. napus* L. var. *napus*

The glucosinolate side chain may comprise aliphatic (saturated and unsaturated), aromatic, or heteroaromatic groupings. Common substituents include hydroxyl groups (which may occasionally be glycosylated), terminal methylthio groups (and oxidized analogues), esters, and ketones.[1] As will be seen, the side chain determines the chemical nature of the products of enzyme hydrolysis and, thereby, their biological effects and potencies. A number of important glucosinolates in plants consumed by animals and man are listed in Table 2, together with the trivial nomenclature which, although discouraged, is still used.

B. Separation and Isolation of Pure Glucosinolates

The vast majority of the 100 different glucosinolates which are now known have not been isolated in the pure state. In recent years there has been a need for pure glucosinolates, initially as chromatographic standards and, thereafter, for assessing their antinutritional and toxicological properties. While methods for the synthesis of glucosinolates have been described, in general, isolation rather than synthesis has been employed. In some cases (for example, benzyl and 4-hydroxybenzyl glucosinolates in *Lepidium sativum* and *Sinapis alba* seed, respectively), botanical sources contain essentially only a single glucosinolate, but generally, complex mixtures occur, so that their separation and isolation/purification is a challenge.

In most cases, seed meal (defatted, ground seed) is the preferred source, but whether this or other plant material is to be used, extreme care must be taken to ensure that glucosinolates are not decomposed (enzymically or chemically) during extraction and isolation. Commonly, plant material is extracted with boiling aqueous alcohol, the glucosinolates being separated as a class by anionotropic alumina column chromatography. Separation of individual glucosinolates has been achieved by passage through Sephadex® G-10 or A-25,[21] but recently Peterka and Fenwick have found flash chromatography on reverse-phase bonded octadecyl silane to be effective.[22] Separation and isolation may also be possible using preparative

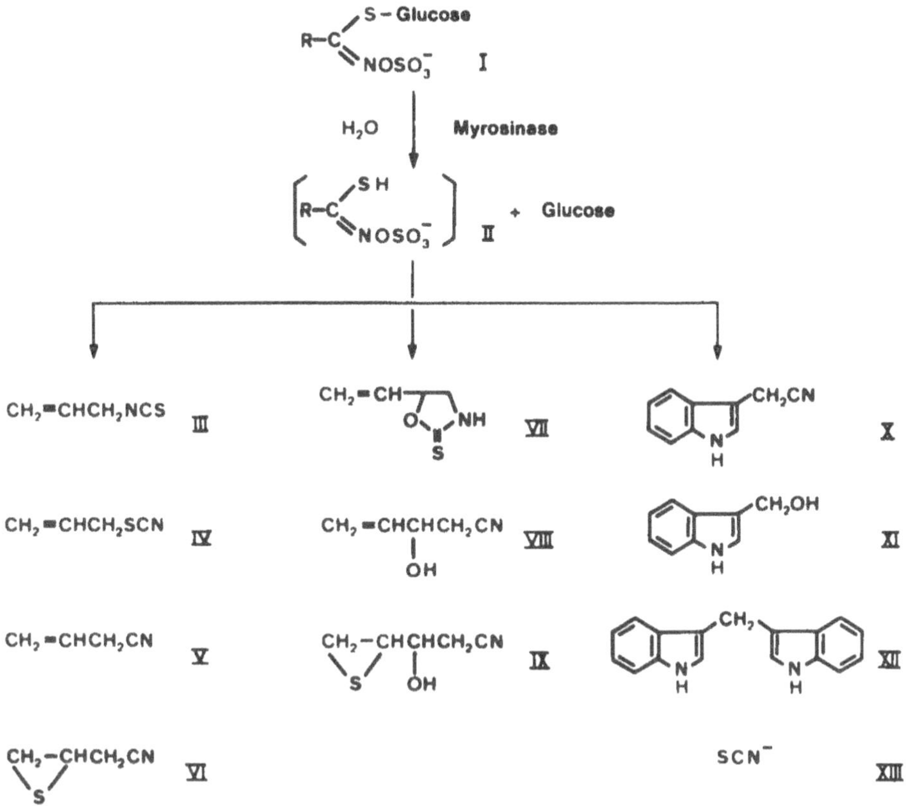

FIGURE 1. Enzymatic breakdown of glucosinolates. The products, which are characteristic of the three classes of glucosinolates referred to in the text, include prop-2-enyl isothiocyanate (III), prop-2-enyl thiocyanate (IV), 1-cyanoprop-2-ene (V), 1-cyano-2,3-epithiopropane (VI), 5-vinyloxazolidine-2-thione (VII), 1-cyano-2-hydroxybut-3-ene (VIII), 1-cyano-2-hydroxy-3,4-epithiobutane (IX), 3-indolylacetonitrile (X), indole-3-carbinol (XI), 3,3'-diindolylmethane (XII), and thiocyanate ion (XIII). (From Watson, D. H., *Natural Toxicants in Food, Progress and Prospects*, Ellis Horwood, Chichester, 1987, 78. With permission.)

reverse-phase high-performance liquid chromatography (HPLC) and volatile buffers, but details are not yet available.[23] It is considered that complete structural elucidation should now include both the side chain, R (Figure 1, I), and the sugar moiety. To assist in the direct structural elucidation, or confirmation of glucosinolate identity, detailed rationalization of mass spectrometry (MS) (EI, CI, and FAB) fragmentation processes[24] and nuclear magnetic resonance (NMR) spectra[25] have been published.

C. Chemical Stability

The effects of chemical treatments on the breakdown of a variety of glucosinolates are indicated in Table 3. In general, nitriles are produced although, as will be seen, glucosinolates possessing a β-hydroxy group yield atypical products such as thionamides and oxazolidinones. Oxazolidine-2-thiones, the important products of enzymic breakdown (see later), have not been identified. The effect of ferrous ion, illustrated in Table 3, is not produced by other ions, such as ferric, cobaltous, nickel, or stannous.[29] The result of conventional and microwave heating on prop-2-enyl glucosinolate in aqueous solution or model systems (soya flour) have been studied,[31] although the products of breakdown were not identified. Decomposition in the soya model system was a function of both the initial moisture level and the period of microwave heating. MacLeod[32] has attributed the adverse flavor of cooked cabbage and other brassicas to chemical, rather than enzymic, degradation products of glucosinolates, especially nitriles.

Table 2
TRIVIAL NOMENCLATURE AND STRUCTURES OF MAIN GLUCOSINOLATES OCCURRING IN EDIBLE PLANTS

Prop-2-enyl	Sinigrin
But-3-enyl	Gluconapin
Pent-4-enyl	Glucobrassicanapin
2-Hydroxybut-3-enyl	Progoitrin, *epi*-Progoitrin
2-Hydroxypent-4-enyl	Gluconapoleiferin
3-Methylthiopropyl	Glucoiberverin
4-Methylthiobutyl	Glucoerucin
5-Methylthiopentyl	Glucoberteroin
3-Methylsulfinylpropyl	Glucoiberin
4-Methylsulfinylbutyl	Glucoraphanin
4-Methylsulfinylbut-3-enyl	Glucoraphenin
3-Methylsulfonylpropyl	Glucocheirolin
4-Methylsulfonylbutyl	Glucoerysolin
Benzyl	Glucotropaeolin
2-Phenylethyl	Gluconasturtiin
4-Hydroxybenzyl	Glucosinalbin
3-Indolylmethyl	Glucobrassicin
1-Methoxy-3-indolylmethyl	Neoglucobrassicin
4-Methoxy-3-indolylmethyl	—
1-Hydroxy-3-indolylmethyl	—

Table 3
PRODUCTS OF CHEMICAL BREAKDOWN OF GLUCOSINOLATES

Glucosinolate	Conditions	Product(s)	Ref.
Prop-2-enyl	H_2O, 100°, 5 h	1-Cyanoprop-2-ene	26
	119°, 2 h	1-Cyanoprop-2-ene	26
	175°	1-Cyanoprop-2-ene	27
	Fe^{2+}, 100°, 15 min	1-Cyanoprop-2-ene; but-3-ene thioamide	28
Benzyl	200°	Benzyl nitrile; benzyl isothiocyanate	27
	119°, 2 h	Benzyl nitrile	26
	Fe^{2+}, H_2O	Benzyl nitrile	29
2-Phenylethyl	150°	2-Phenylethyl isothiocyanate; 1-cyano-2-phenylethane	27
(*R*) and (*S*)-2-hydroxybut-3-enyl	200°	1-Cyano-2-hydroxybut-3-ene	27
	H_2O, 100°, 5 h	1-Cyano-2-hydroxybut-3-ene	26
	119°, 2 h	1-Cyano-2-hydroxybut-3-ene	26
	Fe^{2+}	1-Cyano-2-hydroxybut-3-ene; 3-hydroxypent-4-ene-thioamide	29
	borate, pH 8—12	5-Vinyloxazolidinone	30
2-Hydroxy-2-phenylethyl	Fe^{2+}	1-Cyano-2-phenyl-2-hydroxy butane; 3-hydroxy-3-phenyl propanethionamide	29

IV. BIOCHEMICAL ASPECTS OF GLUCOSINOLATES

A. Glucosinolate-Degrading Enzymes

Disruption of plant tissue results in a loss of glucosinolates due to hydrolysis by the enzyme myrosinase (thioglucoside glucohydrolase, EC 3.2.3.1) present in all glucosinolate-containing plants. Although found in specialized "myrosin" cells, myrosinase activity appears not to be restricted to those cells.[33] Myrosin cells are differentiated early in leaf development and are, in effect, diluted with leaf expansion. Myrosinase occurs in a number of isoenzymic forms, but the number and activity of these isoenzymes seems not to be correlated with particular glucosinolates.[34] Myrosinase shows activity over a wide pH range[35] and is stable at temperatures up to 60°C, with an optimum activity at around 50°C. The reaction with the glucosinolate substrate releases a stoichiometric amount of glucose, and the resulting aglucone (II) undergoes a spontaneous Lossen rearrangement with elimination of bisulfate (Figure 1). The nature of the breakdown products is affected by a number of cofactors which are discussed below. Autolysis or hydrolysis due to endogenous enzyme can occur in a somewhat random manner, but the reaction, which is the basis for a number of quantitative analytical methods, can be controlled by adding myrosinase from an exogenous source (usually *S. alba*) to an inactivated extract. Glucosinolates are also susceptible to hydrolysis by sulfatase (notably that from the edible snail *Helix pomatia*), releasing sulfate and yielding the corresponding desulfoglucosinolate. This reaction is commonly used as a highly specific cleanup step in the analysis of glucosinolates.[24]

B. Products of Enzyme Hydrolysis

Myrosinase-induced hydrolysis of glucosinolates yields a wide variety of products, the exact nature of which is determined by a number of factors including pH, the presence of certain cofactors, and, most importantly, the structure of the parent glucosinolate.[1,6] Most glucosinolates can be conveniently divided into three groups according to the products of their hydrolysis (Figure 1). By far the largest group are those glucosinolates generally having either an alkyl or alkenyl side chain which on hydrolysis with myrosinase at pH 5 to 7 yields glucose and, following a loss of bisulfate, primarily isothiocyanates (III). Glucosinolates possessing a β-hydroxyl substituent form unstable hydroxyisothiocyanates which spontaneously cyclize to oxazolidinethiones (such as VII). The third group having an indole nucleus are hydrolyzed via an unstable isothiocyanate to the 3-carbinol (XI) and thiocyanate ion (XIII); the former can react further to yield diindolylmethane (XII). In more acid conditions of hydrolysis (pH 3), all three groups produce increasing amounts of nitriles (V, VIII, X). Significant amounts of such products are also produced during autolysis of cruciferous plant tissue, even when the pH would be expected to favor isothiocyanate formation. Recently, Uda et al.[36,37] have offered explanations for this effect. While ferrous ion markedly depressed myrosinase-induced isothiocyanate formation under acid conditions, it had little effect at pH 7.5. The authors suggested that the ferrous ion was acting on the aglycone rather than on the parent glucosinolate, since rate of release of glucose was unaffected by ion concentration. Further support was forthcoming from the finding that, in combination, ferrous ion and thiols inhibited isothiocyanate formation at neutral pH. Experiments using silver sinigrate as an aglucone model led the authors to propose a mechanism whereby ferrous ion and thiols together facilitate desulfuration of the aglucone (yielding nitriles) at the expense of the isothiocyanate-producing Lossen rearrangement. The involvement of thiols may also, at least in part, explain the reduction in nitrile production which is evident after heating and which has been attributed to destruction of a thermolabile "nitrile-forming factor."[6]

Ferrous ion is also a necessary cofactor, together with an inactive epithiospecifier protein[6,7,38] for the myrosinase-induced addition of sulfur across a terminal double bond (e.g., the formation of the episulfides VI or IX). Epithiospecifier protein does not occur in all cru-

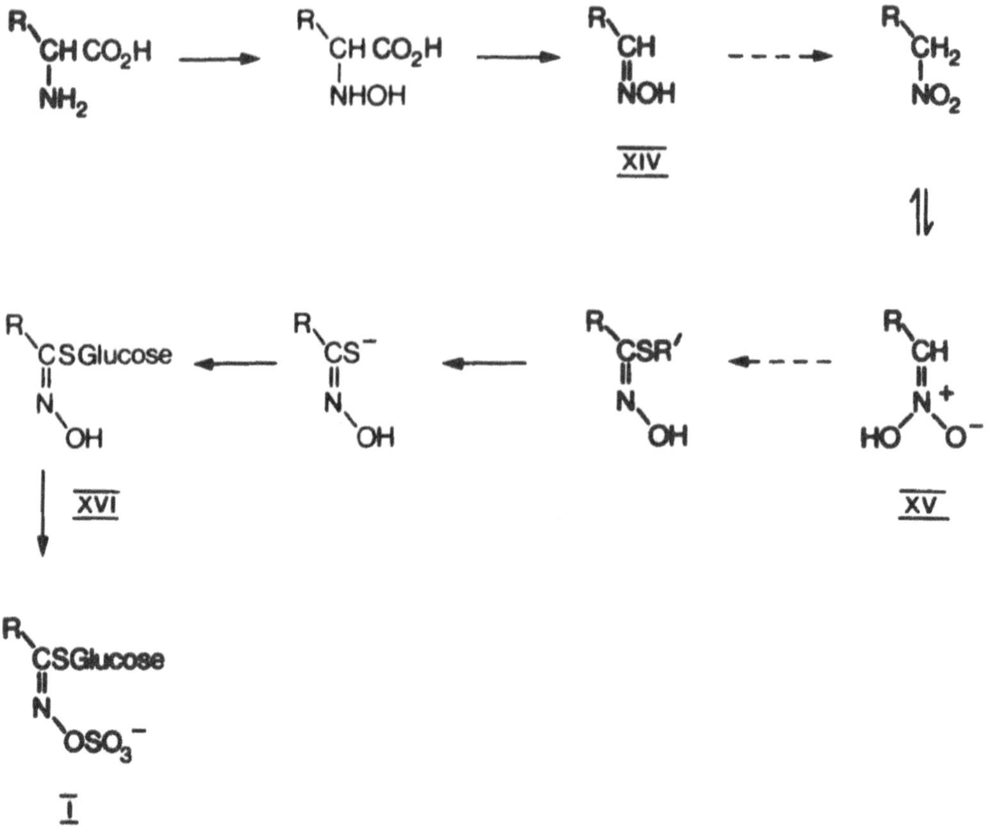

FIGURE 2.   The biosynthesis of glucosinolates. (From Bell, E. A. and Charlwood, B. V., Eds., *Encyclopaedia of Plant Physiology*, Vol. 8, Springer-Verlag, Heidelberg, 1980, 503. With permission.)

ciferous plants, and recently MacLeod and Rossiter have speculated that it is absent from those species which do not contain glucosinolates possessing terminally-unsaturated side chains.[39] The presence of only a trace of such a glucosinolate, however, appears to be necessary for the presence of this cofactor.

Various mechanisms have been offered to explain the formation of episulfides.[38-41] Kinetic evidence suggests that epithiospecifier protein acts as a noncompetitive inhibitor of myrosinase[38] while the sulfur atom is transferred to the terminal site by a substantially intramolecular process.[41] In marked contrast, the factors responsible for the unusual formation of organic thiocyanates (IV), rather than isothiocyanates, in certain species are as yet little understood.

C. Biosynthesis

Many studies, conducted over a decade ago, have led to the view that most glucosinolates are derived from amino acids, with most lying on a common biosynthetic pathway, indole glucosinolates being apparent exceptions. A comprehensive review has been presented by Underhill and Wetter,[42] and more recent advances have been discussed by Kjaer[43] and Underhill.[44]

The currently accepted pathway (illustrated in Figure 2) involves *N*-hydroxylation and oxidative decarboxylation to yield an aldoxime (XIV) which is a common intermediate for the biosynthesis of glucosinolates and cyanogenetic glycosides. It is only recently, however, that in *Carica papaya* an example has been found of the co-occurrence of these classes of secondary metabolites. It has been suggested that for glucosinolate biosynthesis this aldoxime undergoes oxidation to a nitro compound, the *aci*-tautomer of which (XV) may be envisaged as the site for introduction of the thioglycoside-*S*. In studies reported by Underhill and Wetter,[42] cysteine-*S* was most readily incorporated to produce a thiohydroxamic acid. UDP-Glucose-mediated *S*-glucosylation yields the penultimate desulfoglucosinolate (XVI), the final product being obtained following reaction with PAPS, 3'-phosphoadenosine-5'-phosphosulfate. While some glucosinolates, for example, those possessing methyl, isopropyl, or 4-hydroxybenzyl side chains, are clearly derived from corresponding protein amino acids (alanine, valine, and tyrosine), others require elaboration, as can be seen from Table 2, most commonly homologation, elimination or addition, and hydroxylation. There is ample evidence that such processes can occur at various points along the biosynthetic pathway. In a few cases both modified amino acids and glucosinolates containing corresponding side chains have been identified.[42] A large family of glucosinolates contain homologous series possessing a terminal methylthio grouping (or oxidized analogues, methylsulfinyl, methylsulfonyl). These are considered to be derived from methionine by acetate-type chain elongation. Terminal elimination of the elements of methylthiol affords another series of glucosinolates possessing a terminal ω-double bond. In view of their current importance in rapeseed breeding, it is surprising that more research has not been carried out on the biosynthesis of indole glucosinolates, their interconversion, and, most importantly, the point where their biosynthesis deviates from that of most other glucosinolates.

Not all of the common amino acids have side chains which are found in corresponding glucosinolates, but, as Kjaer has pointed out,[43] in view of the well-defined and ordered processes involved, the future identification of these "missing" compounds should occasion little surprise.

D. Metabolism

Metabolic studies on glucosinolates are relatively few. The breakdown of 2-hydroxybut-3-enyl glucosinolate to (in part) 5-vinyloxazolidine-2-thione has been reported in both man and the rat.[45] However, in retrospect both the purity of the chemicals and the analytical methods employed may be questioned. Recently various glucosinolates have been shown to be broken down when incubated with human fecal extract, although the products have not as yet been identified.[46] Little or no breakdown occurred under conditions (e.g., pH, digestive enzymes) simulating those of the stomach. The breakdown of 2-hydroxybut-3-enyl glucosinolate to 5-vinyloxazolidine-2-thione has been suggested to occur in rats[47] and poultry[48] on the basis of the known antithyroid and enzyme-inhibiting properties of the latter, but Smith and Campbell[49] observed greater amounts of nitriles in the crops of fowl fed rapeseed meal. Various studies have indicated that intact glucosinolates in the rat are probably broken down, but the nature of the products is generally unknown. Because of the nature of the resulting physiological effects the assumption is often made that these products are similar, if not identical to, those obtained following treatment with plant myrosinase.

Recently Macholz and co-workers[50] have found traces (<1%) of benzyl glucosinolate in urine and feces following the feeding of this glucosinolate to rats. Benzyl nitrile and benzyl isothiocyanate were found in the feces, the former alone in the thyroid, but no quantitative information was given in the preliminary description of this work. When 2-hydroxybut-3-enyl glucosinolate was fed to germ-free rats, larger amounts were excreted than was the case for conventional animals. Products "other than 5-vinyloxazolidine-2-thione" were observed, one of which may have been the 4-hydroxy metabolite reported earlier.[1] Independently, Nugon-Baudon and Szylit[51] have found that rapeseed meal is of equivalent nutritional value to soya meal only when fed to germ-free rats or chickens. Rat commensal microflora was responsible for the goitrogenic effect, hypertrophy of liver and kidneys, and growth retardation. In the fowl the commensal microflora showed a dramatic goitrogenic influence but had only a marginal effect on growth. When germ-free rats or chickens were inoculated with the total fecal flora of the alien species, the goitrogenic effect was maintained, but the depressive influence of the rat microflora on growth was lost when implanted into chickens.

V. ANALYSIS

The analysis of glucosinolates has been a fruitful area of research for many years. Indeed, methods for the analysis of glucosinolates and their products have been described for over a century, and there have been few major analytical techniques which have not been applied to these compounds. There are many methods currently used for glucosinolate analysis, having been developed in response to the specific, and sometimes mutually exclusive, requirements of industry, researcher, and legislator. The development of methods for the analysis of glucosinolates and their products has been discussed by McGregor et al.[24] In this section it is only relevant to review those methods currently widely used, and, in this context, it is appropriate to consider the analysis of total and individual glucosinolates separately.

A. Total Glucosinolates

Measurement of the total glucosinolate content is particularly useful for plant breeders and legislators, when concerns about the nutritional value (and, hence, complement of individual glucosinolates) are not paramount. Most methods are based upon the measurement of a product obtained following the controlled enzymic hydrolysis with myrosinase. Usually glucose is measured, spectrophotometrically or by a chromatographic procedure. Interfering substances may be removed by ion exchange[52] or microcolumn methods.[53] The measurement of enzymically released sulfate has been proposed but has been little used. Recently, Møller et al.[54] have adapted a method, based upon the observed color reaction between glucosinolates and tetrachloro(IV)palladate, and have automated this for the screening of spring-grown rapeseed. Specific concerns over the availability of a reliable, rapid method suitable for the crushing industry have been addressed recently by Thies[55] and Smith and Dacombe.[56] There would be obvious advantages if a nondestructive method of analysis could be developed, and a number of groups have explored the potential of NIR, so far with little success.[57] A recent paper[58] has described the use of X-ray fluorescence, the time of analysis being less than 1 min. The cost of the equipment may be the major factor mitigating against the wider, industrial use of this method.

The specific requirements of the plant breeder — speed, inexpensiveness, simplicity, and the ability to analyze as little as half a cotyledon — have been addressed by glucose-release methods, where the level of glucose (and, hence, glucosinolates) is revealed by change of color on a test paper or stick.

The current international standard method[59] for glucosinolate content of rapeseed has a number of limitations. Two procedures are required: measuring isothiocyanates (by gas chromatography [GC]) and oxazolidine-2-thiones (spectrophotometrically). The "total" figure arrived at by summing these contributions underestimates the real total because indole glucosinolates (which may, however, be estimated independently via the thiocyanate ion formed upon myrosinase hydrolysis[24]) are not included. Moreover, this underestimate is more serious in low (improved) glucosinolate seed. A matter of considerable additional confusion to the layman is that isothiocyanate levels are specified in terms of prop-2-enyl (allyl) isothiocyanate. This compound is absent in rapeseed, and legislation exists in the U.K. and elsewhere prohibiting the use as animal feedstuffs of seed meals containing (precursors of) this substance. It seems likely that this method, which does give some indication about the individual glucosinolate composition, will shortly be replaced by a chromatographic method allowing separation and analysis of individual glucosinolates.

B. Individual Glucosinolates

Individual glucosinolates may be separated and quantified by GC or HPLC. Methods based upon the separation of desulfoglucosinolates, produced enzymatically, using temperature-programmed GC conditions are now in official use in Canada and Europe.[60] While certain indole glucosinolates can be separated and quantified by this method, problems with 4-hydroxy-3-indolylmethyl glucosinolate, which only recently has been found to be present in rapeseed, have meant that attention has been turned to HPLC. Methods based upon both intact glucosinolates and desulfoglucosinolates have been described,[24] with the latter having received more support.

Glucosinolate breakdown products may be determined by a variety of techniques including HPLC (oxazolidine-2-thiones, indoles), GC (nitriles, isothiocyanates), and spectrophotometry (thiocyanate ion).[24] While these methods have proved effective for analysis of plant material, there have been few applications to biological samples, although oxazolidine-2-thiones have been monitored in milk by HPLC[61] and in plasma and urine by capillary GC.[62]

VI. LEVELS OF GLUCOSINOLATES IN FOODS AND FEEDSTUFFS

While glucosinolates have been studied in a wide variety of plants, quantitative studies have been confined mainly to cruciferous plants having economic importance as human foods and animal feedstuffs. Brassicas are widely consumed in the human diet, fresh, cooked, or othrwise processed and constitute the major source of glucosinolates.[1,4] While 10 to 20 individual glucosinolates are usually present in brassica species, only a few predominate. There is a large variation in the absolute amounts of glucosinolates both in individual cultivars of the same species and between species (Table 4). Factors which affect total glucosinolate contents are discussed in the following section.

In general, it has been found that the same glucosinolates usually occur in a particular subspecies irrespective of genetic origin. There is, however, a marked variation between cultivars in the absolute amounts of glucosinolates. This is reflected in the wide ranges in the levels of individual glucosinolates in the species listed in Table 5. It should be emphasized that this table lists major glucosinolates, and for a consideration of the total complement the individual references should be examined. Investigations conducted on North American cultivars employed methods based upon analysis of hydrolysis products so that distinction between the individual indole glucosinolates was impossible (all contributing to the figure obtained by thiocyanate ion release).

The relative distribution of glucosinolates has been shown in the case of brussels sprouts to be characteristic of a particular variety affording a possible chemotaxonomic tool. Such cultivar specificity is not, however, exhibited by other important species such as oilseed rape which poses particular problems in defining distinctness for establishing breeders' rights.

A. Foods

There have been many studies of the nature and amount of glucosinolates in crops commonly consumed by man, and an indication of the average levels and range of these compounds in some of the more important crops is given in Table 5. It should, however, be noted that, in general, these studies used representative samples of material as marketed, and breeding lines may be significantly different. Although only a minor component of the Western diet, radishes are an important feature of the Japanese diet[62a] and are the only vegetable in which 4-methylsulfinylbut-3-enyl glucosinolate occurs.[72] Cruciferous seeds contain high amounts of glucosinolates, and, although only a small proportion of the Western diet, mustard (*S. alba* and *Brassica juncea*) contains large amounts of 4-hydroxybenzyl and but-3-enyl glucosinolates, respectively, which in the form of their enzymic breakdown products are responsible for the desirable pungency and flavor of these condiments.

Table 4
TOTAL GLUCOSINOLATE CONTENTS µg/g OF AGRICULTURALLY IMPORTANT PLANTS[a]

| Species | Glucosinolate content | | |
	Range	Mean	Ref.
Cabbage			
White	260—1,060	530	63
	420—1,560		64
Red	410—1,090	760	63
Savoy	470—1,240	770	63
	1,210—2,960		64
Chinese cabbage	170—1,360	540	65
Brussels sprouts	600—3,900	2,000	66
	1,070—2,760	1,770[b]	67
Cauliflower	610—1,140		68
	100—180	161[b]	67
Calabrese	420—950	620	69
Swede (peeled)	200—1,090	550	70
	1,130—2,310		71
Turnip (peeled)	210—600	420	70
	970—2,270		71
Radish			
Oil	920—1,120	1,010	72
European	340—570	450	72
European-American			
Red	420—1,170	680	72
White	570—1,190	770	72
Black	1,240		72
Japanese daikon	660—2,530	1,390	72
Korean	704—1,650	1,090	72
Horseradish	33,200—35,400		73
Mustard			
White	22,000—52,000		74
Brown	<44,000		74
Black	18,000—45,000		74
	33,000—60,000		75
Rapeseed			
High glucosinolate	>42,000	(equivalent 100 µmol/g defatted meal)	
Intermediate glucosinolate	25,000	(60 µmol/g)	
Low glucosinolate	13,000	(30 µmol/g)	
Canola	20,000[c]		
Spring (Danish)	5,000—10,000	(10—20 µmol/g)	

[a] Additional comprehensive data on the glucosinolate content of U.S.-grown broccoli, brussels sprouts, cauliflower, collards, kale, mustard greens, kohlrabi, and turnips has recently been published.[76,77]
[b] Calculated from data on cooked vegetables.
[c] Specification 13,000 µg/g (30 µmol/g defatted meal) excluding indole glucosinolates.

B. Feedstuffs

Brassicas find widespread use in animal feedstuffs, both as forage crops, such as fodder rape, kale, cabbage, swedes, and turnips, and as oilseed crops such as rape or mustard which, after removal of the oil, provide a meal rich in protein. In general the glucosinolates of forage crops reflect those of vegetable crops grown for human consumption, and, until recently, problems of thyroid disorder in animals grazing such crops were addressed by

Table 5
CONTENT OF MAJOR INDIVIDUAL GLUCOSINOLATES (µg/g) IN AGRICULTURALLY IMPORTANT CROPS[a]

	A	B	C	D	E	F	G	H	I	J
White cabbage		35—590 74—414				46—270 23—890		0—144		
Red cabbage		6—102 0—160	16—53			22—143		150—390		
Savoy cabbage		125—646	19—64 30—500			70—420 334—1290				
Brussels sprouts		11—1560			0—295					
Cauliflower		10—627				6—419		151—270		
Calabrese						0—71				
Swede	0—240						37—380			4—330
Turnip	0—180		4—501	4—250			9—137			24—220
Radish, oil									594—768	
European									178—343	
European-American									292—1,070	
Japanese									498—2,230	
Korean									494—1,193	
Chinese cabbage			0—250	13—275						
Horseradish		27,000—29,000								
Mustard										
White			33,000—69,000	0—320						
Brown		320—8,000								
Black		18,000—45,000								
Rapeseed										
High		8,800—9,600		2,200—2,700						
Low		500—700		200—300						

	K	L	M	N	O	P	Q	R	S
White cabbage						51—510[b] 8—83			

Table 5 (continued)
CONTENT OF MAJOR INDIVIDUAL GLUCOSINOLATES (µg/g) IN AGRICULTURALLY IMPORTANT CROPS[a]

	K	L	M	N	O	P	Q	R	S
Red cabbage						155—330[b]			
Savoy cabbage						300—526[b]			
Brussels sprouts						24—126			
Cauliflower						40—990			
Calabrese						0—101			
Swede	0—110	24—82	0—190		5—470	189—357	265—491		
Turnip		220—670	0—450		28—553	60—560[b]	70—570[b]		
Radish, oil									
European									
European-American									
Japanese									
Korean									
Chinese cabbage	4—257	3—194			22—320				
Horseradish					4,200—7,200			97—547[b]	
Mustard									
White									
Brown									
Black				22,000—52,000					
Rapeseed (Danish Spring)									
High		26,000—28,000	1,300—1,600			<100			1,600—1,900
Low		1,200—1,700	<100			<200			1,200—1,600

[a] Data from References 63 to 75. Key to column heads as follows: A — 1-Methylpropyl glucosinolate; B — Prop-2-enyl glucosinolate; C — But-3-enyl glucosinolate; D — Pent-4-enyl glucosinolate; E — 3-Methylthiopropyl glucosinolate; F — 3-Methylsulfinylpropyl glucosinolate; G — 4-Methylthiobutyl glucosinolate; H — 4-Methylsulfinylbutyl glucosinolate; I — 4-Methylsulfinylbut-3-enyl glucosinolate; J — 5-Methylthiopentyl glucosinolate; K — 5-Methylsulfinylpentyl glucosinolate; L — 2-Hydroxybut-3-enyl glucosinolate; M — 2-Hydroxypent-4-enyl glucosinolate; N — 4-Hydroxybenzyl glucosinolate; O — 2-Phenylethyl glucosinolate; P — 3-Indolylmethyl glucosinolate; Q — 1-Methoxy-3-indolylmethyl glucosinolate; R — 4-Methoxy-3-indolylmethyl glucosinolate; S — 4-Hydroxy-3-indolylmethyl glucosinolate.

[b] Indole glucosinolates assayed as a group.

Table 6
MEAN DAILY INTAKES OF INTACT
GLUCOSINOLATES

(mg/person/day)

	U.K.[83]	Canada[84]	U.S.[a]	Holland[67]
Cabbage	19.4 (14.0)[b]	3.4	8.5[c]	nd
Brussels sprout	17.2 (9.4)	1.5	0.6[d]	(6.5)
Cauliflower	6.4 (4.4)	1.2	2.7[e]	(1.2)
Broccoli		1.4	3.3[f]	nd
Swede/turnip	3.1 (1.6)	5.8		nd
Total	46.1 (29.4)	13.5	15.1	(7.7)

[a] Calculated.
[b] Figures in brackets refer to calculations based upon cooked vegetable data.
[c] Including 1.0 mg from canned sauerkraut.
[d] Including 0.5 mg from frozen brussels sprouts.
[e] Including 0.6 mg from frozen cauliflower.
[f] Including 1.4 mg from frozen broccoli.

assaying for thiocyanate ion content. Recent studies have shown, however, that rape and radicole, unlike kale, contain significant amounts of 2-hydroxybut-3-enyl glucosinolate, the precursor of the potent goitrogen 5-vinyloxazolidine-2-thione, and it was concluded that its level should be closely monitored in future breeding programs.[78,79] Glucosinolate levels in cruciferous seeds are generally much higher than in the corresponding vegetative material, and the high content of glucosinolates in rapeseed is one of the main factors limiting its use as a protein supplement in animal feeds,[3] as is described in Section VIII. However, intensive breeding programs have produced much improved rapeseed varieties having much lower levels of glucosinolates.[75] The glucosinolate content of high and low rapeseed cultivars is shown in Table 5.

C. Food Chain

Glucosinolates, or more accurately their breakdown products, can pass indirectly into the human diet in foods derived from animals feeding on cruciferous material. Little is known about the metabolic fate of glucosinolates in the digestive tract, but it has been suggested that in Finland endemic goiter may be related to goitrogens in milk.[80] VanEtten et al.[81] have conducted a detailed study into the possible occurrence of glucosinolate hydrolysis products in the organs and meat of cattle fed rations containing crambe seed meals, but no such products could be detected. Low levels of 1-cyano-2-hydroxybut-3-ene (1 to 3 mg l^{-1}) have been detected[82] in milk following the feeding of rapeseed meal to dairy cattle; given the toxicological properties of this compound further studies using more modern analytical techniques would appear justified. Recently, 5-vinyloxazolidine-2-thione levels of up to 1 mg l^{-1} have been found in milk following the feeding of rapeseed meal to cattle,[61] and under certain circumstances the levels may be even higher.

D. Glucosinolate Intake

Information on glucosinolate intake is shown in Table 6. The most comprehensive study has been conducted in the U.K. where a mean daily intake of 46 mg has been calculated.[83] This is certainly an underestimate since the calculation takes no account of processed or home-grown vegetables. Seasonal dietary trends and, in particular, the consumption of large amounts of fresh glucosinolate-rich brussels sprouts result in a much higher-mean glucosinolate intake in the winter months (58.8 mg vs. 32.9 mg in the summer). When this information, together with social and agronomic factors, is taken into account, it is calculated that 5% of the U.K. population (95th percentile) consume >300 mg daily. It is pertinent to contemplate whether consumers and legislators would be so sanguine about the daily consumption of 300, 30, or even 3 mg of a synthetic chemical ("additive") for which equally little biological information was available.

In 1978, Mullin and Sahasrabudhe[84] reported a mean daily intake of 7.9 mg glucosinolates on the basis of Canadian dietary data. The analytical techniques available at that time mean that this is an underestimate. Using data obtained by various workers on North American- and (where the former is unavailable)* U.K.-grown vegetables, the present authors have calculated a mean daily intake of 13.5 mg glucosinolates. When the same procedure is applied to American dietary tables, the mean daily intake is calculated to be 15.1 mg. Recently Gramberg and colleagues[67] have reported mean daily glucosinolate intakes from brussels sprouts and cauliflower in the Netherlands to be 6.5 and 1.2 mg, respectively. These values are calculated on the basis of cooked vegetables, and, given earlier findings, the combined intake (based upon fresh produce) would likely be rather higher, perhaps 12 to 14 mg.

From these comparisons it may be seen that the U.K. has the highest intake of glucosinolates, thus justifying the continuing programs of research on dietary glucosinolates in that country. However, populations in certain Oriental countries consume much higher amounts of crucifers and so, at least potentially, are exposed to much higher levels of glucosinolates. In Japan,[85] for example, the average daily consumption (1975 figures) of radish, cabbage, and Chinese cabbage is 75.5 g and of fermented root and leaf vegetables (some of which will be crucifers) 37.2 g. In contrast, the total daily crucifer intake in the U.K., from which the figures shown in Table 6 were obtained is 45.2 g. Information on the glucosinolate content of Japanese vegetables is, unfortunately, limited, but recently Carlson et al.[72] have examined various types of radish, including Japanese daikon (Table 4). The mean daily glucosinolate intake in Japan *due to daikon alone* may thus be calculated as 46 mg. Differences in food preparation practices may reduce the actual intake of glucosinolates significantly (although a proportion of this "loss" may be revealed as an increase in the intake of glucosinolate breakdown products).[1] Thus, boiling has been shown to reduce glucosinolate levels in brassica vegetables such that the U.K. mean daily intake, calculated on a cooked-vegetable basis, is reduced to 29.4 mg.[83] Intact glucosinolates are readily broken down under fermentation conditions.[86]

VII. FACTORS AFFECTING GLUCOSINOLATE LEVELS

Although it has been shown that the relative amount of glucosinolates in a particular subspecies is comparatively stable, absolute levels are subject to a wide variety of influences. These have been reviewed in detail by Fenwick et al.[1] and consequently will be discussed only briefly here.

A. Genetic and Botanical

Glucosinolate levels in cruciferous seeds have been manipulated to advantage by plant breeders. Mustard quality has been enhanced by increasing the levels of glucosinolates, whereas breeding from selected lines of oilseed rape has reduced the glucosinolate levels in the seed to around 10% of those in high-glucosinolate varieties. In rapeseed, only alkenyl glucosinolates are reduced, and indole glucosinolates, which remain unaffected, assume greater significance.[87] In cruciferous vegetables, the flavor is due to certain glucosinolates,

See comment at foot of Table 4.

and manipulation of the relative amounts in order to reduce levels of undesirable precursors such as 2-hydroxybut-3-enyl glucosinolate may be a more desirable goal. In this context the finding that F_1 hybrid brussels sprouts have a glucosinolate content close to the average of both parent lines may be significant.[88] Differences have been noted between vegetables grown in North America and the U.K. Both individual and total glucosinolates differ, suggesting both genetic and agronomic factors are involved.[64]

Absolute and relative amounts of glucosinolates are determined by the part of the plant examined, with leaves, stems, and roots showing wide differences.[1] Apical leaves of brussels sprout plants have been found to contain up to fourfold more 3-indolylmethyl glucosinolate than mature leaves, and the glucosinolate content of fully expanded leaves decreases rapidly with the onset of senescence. VanEtten et al.[89] have examined the variation of individual glucosinolates within the pith, leaves, and cambial cortex of red, white, and savoy cabbage. Glucosinolates occurred in relatively low abundance in the older, outer leaves; within the cabbage head proper, the highest concentration was found in the cambial cortex, being approximately double that in the pith and leaves. When seeds of rape were germinated, isothiocyanates were decreased, and nitriles increased in comparison to the behavior of the autolyzed seed.[90] This is consistent with changes in the microenvironment (pH, ion concentration) during germination. Highest glucosinolate concentrations have been found in seeds from the top siliques of the lowest branch of *B. campestris*,[91] and, overall, higher levels were found in siliques from upper, rather than basal, portions of the branch. Such variability has obvious implications for the screening of plant-breeding programs.

B. Agronomic

The use of sulfur-based fertilizers has been shown to increase glucosinolate levels, and it has been suggested that nitrate application results in a lowering of values. Stress factors such as drought and close plant spacing are also known to increase levels.[1] The effect of sowing time on glucosinolate content of harvested rapeseed has recently been emphasized.[92] Harvesting techniques have been suggested to be responsible for the observed instability of glucosinolates in some low-glucosinolate cultivars of oilseed rape,[93] with harvested seed sometimes containing twice the glucosinolate content of sown seed. This problem may have important implications in the context of any legislation governing the levels of glucosinolates in internationally traded seed.

C. Processing

Although some cruciferous crops for human consumption are eaten fresh, most are processed in some way. Such processes include cutting or chopping, cooking and blanching, freezing, pickling, and fermenting. Any process which causes disruption of cellular integrity results in some hydrolysis occurring with a loss of glucosinolates. Although blanching and cooking inactivate the myrosinase, some loss of glucosinolates occurs through leaching into the cooking water, and it has been shown that boiling of brassica vegetables in the normal manner reduces glucosinolates by up to 50%.[83] Enzymic reactions continue during storage; thus, a loss of glucosinolates in sauerkraut is accompanied by the formation of nitriles.[94] The dehydration of *Wasabia japonica* by air drying or freeze drying has been found by Kojima et al.[95] to be associated with only minor losses of glucosinolates.

The processing of rapeseed includes steps which inactivate the myrosinase, but further processing of the rapeseed meal to reduce glucosinolate content (a goal which does not necessarily equate with detoxification) has been superceded to some extent by the success of plant breeding. However, processing remains an option to achieve greater utilization of this valuable protein source.[96]

VIII. ANTINUTRITIONAL EFFECTS OF GLUCOSINOLATES IN CRUCIFEROUS FOODS AND FEEDSTUFFS

A. Animals

There have been many studies concerned with the antinutritional effects of rapeseed and, to a lesser extent, mustard and crambe seeds. Such effects are generally attributed to the glucosinolate content and composition of these products. The purpose of the present chapter is to supplement earlier reviews,[97-100] critically considering the range of effects produced by glucosinolates and their hydrolysis products. Particular attention will be paid to threshold dietary levels* their action, and effects on target organs.

1. Growth and Performance

The characteristic symptoms following the ingestion of substantial amounts of glucosinolates include reduced performance, enlarged thyroid glands, and reduced levels of circulating thyroid hormones. A number of studies have related glucosinolate intake, via comparisons of high- and low-glucosinolate rapeseed, to nutritional value defined in terms of animal growth. Changes in the latter do not, however, necessarily reflect differences in nutrient digestibility and utilization but may result from direct effects on thyroid function and intake control. A similar argument may apply to other criteria of nutritional value, such as protein efficiency ratio (PER) and net protein utilization (NPU), which include an intake component. There is similarly no unambiguous evidence to relate glucosinolate intake to metabolizable (ME) or digestible energy (DE) values, even though low-glucosinolate rapeseed meal does tend to have higher ME and DE values and improved nitrogen digestibility. Glucosinolate-related effects on palatability and thyroid function are considered elsewhere, but are clearly integral to the growth responses described here.

a. Swine

Glucosinolate-related effects are especially severe in young pigs (<20 kg) which exhibit reduced growth and intake even with diets containing low-glucosinolate rapeseed.[101,102] Table 7 relates observed growth response to calculated dietary levels and intakes of glucosinolates.

In practical terms the results suggest that no deleterious effects were observed with total glucosinolate and oxazolidine-2-thione levels below 360 μg g^{-1} and 80 μg g^{-1}, respectively.

Experiments to compare the effects of high- and low-glucosinolate-rapeseed meals in older pigs have demonstrated the superiority of the latter.[105,109] It is generally difficult to identify the effects due to glucosinolates beyond the gross comparisons described above. Many experiments have examined the nutritional value of rapeseed meals, and thus diets were not nutritionally balanced, confounding dietary levels of glucosinolates and nutrients. Recent experiments in Sweden [110-112] and Finland [113] have examined the growth response of pigs to increasing levels of low glucosinolate rapeseed meal. There is a clear trend emerging from these results, with growth depression being associated with dietary glucosinolate levels in excess of 2.7 mg g^{-1} diet.

b. Cattle and Sheep

Growth responses of young calves to dietary glucosinolates are dependent upon the age and weight of the animal.[114] Growing steers were, thus, unaffected by diets containing low-glucosinolate rapeseed meal[115] and appeared to tolerate diets containing 12 to 16% high-glucosinolate rapeseed meal (3.6 to 6.3 mg glucosinolate g^{-1} diet) without detriment to growth or feed conversion.[116] In general, feeding experiments with young calves indicate growth to be depressed by high-glucosinolate rapeseed meal, but to be unaffected by the

---

* For purposes of comparison, the average molecular weight of rapeseed glucosinolates is taken as 420.

Table 7
EFFECTS OF GLUCOSINOLATES (IN RAPESEED MEAL) ON GROWTH
AND PERFORMANCE OF YOUNG PIGS

Growth response	Glucosinolate		Oxazolidine-2-thione		Ref.
	Level (mg/g diet)	Intake (g/d)	Level (mg/g diet)	Intake (g/d)	
No significant effects	0.36[a] (0.1—0.8)	0.14 (0.08—0.26)	0.08 (0.03—0.2)	0.03 (0.01—0.08)	103—107
Tendency for reduced gain, intake, or conversion	0.9 (0.8—1.2)	0.37 (0.34—0.48)	0.18 (0.15—0.21)	0.07 (0.06—0.09)	104,106,108
Significant depression in performance	2.1 (0.6—5.8)	0.7 (0.29—1.3)	0.3 (0.11—0.54)	0.11 (0.04—0.21)	103,104,107,108

[a] Quoted values are means, the range of values being given in parenthesis.

inclusion of improved rapeseed varieties.[117] Calculation suggests that growth of young calves is not affected by dietary glucosinolate levels of 0.3 to 2.7 mg g^{-1}.

Early Canadian feeding trials with cows led to the recommendation that high-glucosinolate rapeseed meal be restricted to less than 10% in dairy rations. In more recent work, high-glucosinolate rapeseed meal has been fed at levels ranging from 8 to 25%, corresponding to dietary glucosinolate concentrations of 2.7 to 9.8 mg g^{-1} diet. While levels of 2.7 to 6.3 mg g^{-1} had no significant effect on concentrate intake, milk yield, or composition, higher levels (9.8 mg g^{-1}) caused reductions in both intake and milk yield.[118,119] In contrast, low-glucosinolate rapeseed meals have been fed at up to 34% in the diet without adverse effects,[120-122] these inclusions corresponding to a maximum glucosinolate intake of <1.8 mg g^{-1} diet. The finding by Fisher and Walsh[123] that inclusion of 34% low-glucosinolate rapeseed meal (corresponding to 1.5 mg glucosinolates g^{-1} diet) depressed intake and butterfat yield may, in retrospect, be associated with the high residual oil and hexane (20 to 30 ppm) contents of the particular meal used. The effects of glucosinolates on body weight change in dairy cattle fed rapeseed meal have not been widely studied, although there has been a suggestion that such feeding is associated with a loss of body weight. Growth of store and finishing lambs is unaffected by moderate levels (12%) of high-glucosinolate rapeseed meal,[125] but higher levels produce poorer performance.[126] The effect may be attributed to glucosinolates since low-glucosinolate rapeseed meals have been shown to produce good performance in lambs.[115] Dogan and Yucelen[127] have suggested that problems associated with glucosinolates may be overcome by pelleting diets containing rapeseed. The performance of lambs grazed on rape is superior to that for kale or fodder radish,[128] although fodder rape supplemented with barley gives even further improvement.[129]

c. Poultry

Detrimental effects of high glucosinolate intake on growth performance have been observed in broilers, layers, ducks, geese, quail, and turkeys following the feeding of rapeseed,[97,130] cabbage,[131] mustard products,[132] and meadowfoam (*Limnanthes alba*).[133] Glucosinolates are clearly implicated since their removal or reduction markedly improved performance.[133-135] Supplementation of glucosinolate-containing diets with potassium iodide[136] or iodinated casein[137] gave a small improvement in layer performance, but had no effect on growth. Problems in broilers appear to be less severe than in laying hens or turkeys; however, weight gain decreases linearly with increasing levels of dietary glucosinolates.[138] Inspection of the available literature suggests that (1) dietary glucosinolate levels above 4.5 mg g^{-1} will cause significant growth depression in broilers, (2) growth depression of about 10% will follow consumption of 3.1 to 4.5 mg glucosinolates g^{-1} diet, and (3) a trend towards reduced growth is observed at dietary glucosinolate levels of 0.9 to 1.8 mg g^{-1}. The commercial significance of this latter point merits consideration.

While only a few studies have considered the effects of glucosinolate products, a marked growth depression due to a reduction in feed intake has been associated with the presence of glucosinolate-derived nitriles.[139,140] Brak and Henkel[140] have suggested several explanations for this finding, including toxic effects per se, formation of cyanide and anoxic effects on nervous control of feeding, and impairment of iodine metabolism by thiocyanate formation. Clearly, the role of nitriles merits further attention, particularly in relation to rapeseed processing.

The inclusion of high-glucosinolate rapeseed meal in layer rations has been associated with increased mortality and decreased egg production together with small effects on egg size and egg Haugh units.[97] Such findings led to the recommendation that a maximum of 5% of such rapeseed meals be included in layer rations. These effects are all reduced if low-glucosinolate rapeseed meal is fed, for which an upper limit of 10% has been recommended. The inclusion of low-glucosinolate rapeseed meal in layer rations is, however, associated with a reduction in egg weight,[110,141] especially during early lay;[142] but whether this is a result of the low energy value of the rapeseed or of extreme sensitivity to dietary glucosinolate-derived goitrogens is unclear.

d. Rats and Mice

Antinutritional effects in rats and mice have been shown with intact glucosinolates, hydrolysis products, and plant products. Severe depression of both intake and growth has been observed on feeding high-glucosinolate rapeseed[143,144] and weed-seed meals.[145] These effects were much reduced or removed when low-glucosinolate rapeseed was fed, or when glucosinolates were reduced by processing.[146] Food intake and weight gain were depressed when dietary glucosinolate content exceeded 1.8 mg g^{-1} diet[147] or when 2-hydroxybut-3-enyl glucosinolate (3 mg g^{-1} diet, in the presence of myrosinase)[144] or high levels of p-hydroxybenzyl glucosinolate[148] were fed. In contrast, there was no effect on these parameters when a range of glucosinolates was fed individually to rats at levels equivalent to those found in low-glucosinolate rapeseed.[149] At higher levels (typically 4.5 mg g^{-1}) effects on growth and food intake were noted.[150] Inclusion of 3 mg 5-vinyloxazolidine-2-thione g^{-1} diet depressed rat growth by 85%[47] while autolyzed crambe or rapeseed meals, rich in nitriles, produced poor growth and increased mortality in conjunction with severe liver lesions,[47,151] discussed elsewhere.

e. Other Species

Rapeseed flours supplying 40% of the dietary protein have been fed to beagle dogs without impairment of growth.[152] In short- and long-term feeding trials with beagles and malamutes dietary glucosinolate levels of up to 4.5 mg g^{-1} had no adverse effect on growth or thyroid function.[153] In contrast growing, furring mink were adversely affected by levels of 0.5 mg g^{-1}.[154] No significant effects on performance of rainbow trout were observed with glucosinolate levels of <1.6 mg g^{-1} diet.[155] Similarly, Higgs et al. have recommended that low-glucosinolate rapeseed meal can provide 25% of the dietary protein for chinook salmon (<1.2 mg glucosinolate g^{-1} diet) without adversely affecting growth or feed conversion.[156]

2. Reproduction

a. Swine

Hill[157] reported no ill effects of high-glucosinolate rapeseed on breeding sows, but referred to possible problems in gilts. Studies in this area are summarized in Table 8 and indicate that no adverse responses occurred at a dietary level of <1.8 mg glucosinolate g^{-1} diet.

Table 8
REPORTED PHYSIOLOGICAL RESPONSES OF
PIGS TO DIETARY GLUCOSINOLATES IN
RAPESEED

| | Glucosinolate level | | |
| | Dietary level (mg/g diet) | Daily intake (g/pig) | Ref. |
Effect			
Lower prebreeding weight gain	2.7	7.1	158
Delayed estrus	5.2	14.1	158
Incidence of subestrus	4.0	8.1	159
Increased services per conception	2.7	6.2	160
Reduced litter size, weight	3.6	7.1	161
Poor survival to weaning	9.8	19.7	162
No significant effect	1.3	3.5	158
	1.1	2.4	160
	1.8	3.5	162
	0.9	1.9	162

Recently Fiems and Buysse[100] have graphically illustrated similar responses to dietary inclusion of high- and low-glucosinolate rapeseed meals. There is no clear evidence to fully describe the mechanisms involved in glucosinolate-related effects in swine reproduction. It has been suggested that the main effect of goitrogens is delayed sexual maturity,[160] and observations of lack of estrus from immature reproductive tracts in gilts,[160] poor conception rates,[158] incidence of subestrus,[162] and generally poor reproductive performance in gilts fed high-glucosinolate rapeseed meal support this view. Detrimental effects of rapeseed during gestation feeding have been attributed to its goitrogenic activity,[158] and this is supported by studies in which thiouracil was fed to pigs.[163]

b. Cattle and Sheep

For a decade, 10% inclusions of high-glucosinolate rapeseed meal have been used commercially in the U.K. without any adverse effects.[164] Higher levels (25 to 32%) of such meals have been fed to yearling heifers without detriment to ovarian cyclic activity or behavioral estrus.[165] Preliminary results of a long-term study have revealed a significant increase in days from calving to conception, related to reduced conception rates and increased services for conception with low-glucosinolate rapeseed diets (cv. Topas, glucosinolate content 12.5 mg g^{-1} defatted meal).[166] Daily glucosinolate intakes in these studies were approximately 31 g. Recent studies with pregnant ewes[167-169] have shown no effect of high-glucosinolate rapeseed meal on reproduction.

c. Poultry

Reproductive efficiency in poultry is usually defined in terms of fertility, egg hatchability, chick mortality, and performance, and most studies indicate that rapeseed, irrespective of glucosinolate content, has no significant effect on these criteria.

3. Goitrogenicity

The earliest report of glucosinolate-related goitrogenicity resulted from feeding high levels of cabbage to rabbits.[170] The condition, termed cabbage goiter, was not very reproducible, nor were the benefits claimed following iodine treatment.[171] It was later established that thiocyanate ion was the causative agent[7,8,173] and that its effect was dependent upon low dietary iodine status. The ingestion of seeds of cabbage and other crucifers produced a thyroid condition, brassica seed goiter, which was less amenable to iodine therapy.[174] The active principle, isolated from turnips and other cruciferous seeds, has been shown to be 5-vinyloxazolidine-2-thione.[175]

A great deal of information has been obtained on the goitrogenic properties of glucosinolates following feeding trials involving laboratory and domestic animals. These have been comprehensively reviewed[1,6-8,45,176] by various authors and so will not be dealt with in detail here. As would be expected from the above, the goitrogenicity of rapeseed is dependent upon the levels and nature of glucosinolates present in the diet and upon the intake.[97-100] Thus, thyroid responses to low-glucosinolate rapeseed are observed, but are greatly reduced compared with that resulting from feeding high-glucosinolate rations.[97-100,110] Similar trends were noted with rapeseed samples processed to reduce or remove glucosinolates.[177,178]

The histopathology of glucosinolate-induced thyroid dysfunction has been described for various species,[179,180] the severity being influenced by the duration of feeding and glucosinolate intake. Thus, low levels of intake initiate increases in follicular epithelial cell height and foamy cytoplasm; thyroid hypertrophy results from continued exposure, with thyroid weights being increased 200 to 500%; reduced circulating levels of T_3 and T_4,[110,144,179] and reduced secretory responses to administration of TSH are also observed.[181] Loss of follicular colloid and development of hyperplasia are typical of prolonged, high glucosinolate intake.

Species effects are observed, with pigs,[98,,100] poultry,[97,130] and salmonids[155,156] appearing more sensitive than growing steers,[116] dairy cattle,[100] or carp.[102] Marks[183] has suggested that resistance to goitrogenic effects can develop with selective breeding in quails.

It is difficult to establish a clear threshold for glucosinolate-related antithyroid effects; Hardy and Sullivan[155] recommended a maximum content of 1.2 mg glucosinolate g^{-1} diet for salmonids, and inspection of the literature for other species suggests that glucosinolate-related thyroid responses are apparent at levels of approximately 1 mg g^{-1}. In the latter case effects on growth and production were not always evident. This is probably a reflection of the compensatory role of thyroid responses to intake of goitrogens whereby animals establish new physiological equilibrium and growth is unaffected. Performance effects, described elsewhere, thus, become significant only when the intake of glucosinolates exceeds the capacity of the thyroid to adjust its hormone output.

In addition, attempts to specify ''no-effect levels'' in terms of glucosinolates are confounded by the plethora of glucosinolate-derived products which may induce hyperthyroidism by a variety of mechanisms and to different degrees. Responses to individual glucosinolates have been described,[144,149,150] and for 2-hydroxybut-3-enyl glucosinolate these were related to intake and potentiated by addition of exogenous myrosinase.[144,150] Oxazolidine-2-thiones are probably of major significance because they interfere with the synthesis of thyroid hormones (by blocking tyrosine iodination and the coupling reactions necessary for T_4 formation), effects which are not alleviated by iodine supplementation. Significant increases in rat thyroid weights, coupled with suppression of radioiodine uptake, followed the daily administration of 20 to 80 µg for 20 d.[184] Elfving[80] later demonstrated thyroxin synthesis in rats to be inhibited by a dose of 1 µg kg^{-1} body weight; at five times this dose goiter resulted.

Thiocyanate ion, which may be derived directly from dietary indole glucosinolates[185] or indirectly via the metabolism of isothiocyanates and nitriles,[45,176] is much less potent and acts by blocking or reducing iodine capture by the thyroid. Thus, iodine supplementation of the diet has proved an effective solution. Increased thiocyanate ion concentrations in cows' milk,[186] plasma and urine of pigs,[103,187,188] and urine of rats[189] has been observed following the feeding of rapeseed meal. Both thiocyanate excretion and depression of T_3/T_4 secretion were found to be linearly related to the intakes of rapeseed[188] and kale,[190] respectively. Jahreis et al.[188] have recently observed hypothyroidism, reduced growth, and reduced circulating levels of somatostatin in pigs fed high levels of nitrate and glucosinolates.

Chicks from hens fed rapeseed meal exhibited thyroid hypertrophy, the extent of which was related to glucosinolate intake.[97] When the chicks were subsequently reared on a control ration, the condition disappeared within 28 d.[191] Iodine transfer to the egg is reduced on feeding rapeseed meal,[192] with thiocyanate ion blocking active transfer mechanisms for iodine in the ovary.[193]

Milk from cattle fed diets containing up to 35% high-glucosinolate rapeseed meal caused thyroid enlargement when fed to rats on a low-iodine diet,[194] an effect attributed to markedly reduced milk-iodine levels.[186,194]

4. Liver Abnormalities

Published data on the hepatotoxic effects of glucosinolate-containing feedstuffs are dominated by studies of liver hemorrhage in laying hens. The relative unimportance of this problem in other species may reflect (1) variable species susceptibility, (2) the nature and extent of glucosinolate-derived products formed *in vivo*, or (3) the efficiency of detoxification pathways.

a. Liver Weight

i. Swine

Significant increases in the liver weights of early weaned piglets, growing and finishing pigs, and sows have been observed following the feeding of rapeseed meals; overall, the enlargement varied from 6 to 25% over that of the controls and was more severe in pigs receiving high-glucosinolate meals.[112,195,196] Examination of the data from these and other experiments implicates glucosinolates as the major causative factors and suggests that dietary levels of up to 1 mg and above 2 mg glucosinolates g^{-1} diet will cause small (10%) and large (20%) weight increases over controls, respectively. The histological appearance of the affected livers appeared to be normal with no evidence of gross abnormalities.

ii. Cattle and Sheep

There are no reports of gross liver enlargement in dairy cattle, although Witowski et al.[197] found calves fed rapeseed meal to have livers weighing 11 to 38% more than those of controls. These changes did not appear to be related to glucosinolate intake, however, and were partially offset by administration of L-thyroxine. Significantly higher liver weights have been observed in steers fed kale,[198] but conflicting results have been published for lambs fed marrow stem kale/rape[199] (liver weight doubled) and kale[200] (reduced liver weight).

iii. Broilers

The feeding of cabbage,[131] meadowfoam (*L. alba*),[133] and rapeseed[138,201] to broilers has resulted in enlarged livers. The incidence of liver enlargement was related to glucosinolate content and was much reduced when low-glucosinolate rapeseed meals were fed.[202] Total glucosinolate levels of <0.7 mg g^{-1} diet appeared to have a minimal effect. Despite liver enlargement of 33 to 41% over controls, no liver lesions have been observed.[131,203]

iv. Rats

Liver enlargement occurs in the presence of glucosinolates or their hydrolysis products,[45,149,150,204] and severity is related to dietary concentration. Thus, while most samples of rapeseed tested caused some degree of enlargement, in the majority of cases the severity was reduced if low-glucosinolate rapeseed products were fed or if the glucosinolates were first removed.[205] Addition of myrosinase to pure glucosinolates exacerbated liver enlargement.[144,150]

Gross morphological and histological abnormalities have been observed following the feeding of rapeseed, *Crambe abyssinica*,[45] glucosinolates,[45] and their hydrolysis products.[45,206] Such lesions were characterized by disruption of the normal lobular architecture, bile duct proliferation, fibrosis, and megalocytosis of hepatocytes.

Rats fed with 2-hydroxybut-3-enyl glucosinolate (1 to 5 mg g^{-1} diet), 2-hydroxy-2-phenylethyl glucosinolate (1 mg g^{-1}), but-3-enyl glucosinolate (1 mg g^{-1}), and prop-2-enyl glucosinolate (1 mg g^{-1}) all exhibited enlarged livers, but it seems improbable that it is the intact glucosinolates which are the active species. Nishie and Daxenbichler[107] found that 2-hydroxybut-3-enyl glucosinolate had no effect on the liver weights of pregnant rats when administered subcutaneously, and the response to intact glucosinolates increased following treatment with myrosinase. Furthermore, liver enlargement associated with severe incidence of lesions has been found following the feeding of epithionitriles (330 μg g^{-1} diet). The latter compounds are considered to be detoxified by nucleophilic opening of the epithio group, conjugation with glutathione, and excretion as the mercapturic acid.[208]

b. Liver Hemorrhage

Inclusion of rapeseed products in diets for certain strains of laying hens is associated with progressive liver damage culminating in hemorrhage and increased mortality.[1,209] The problem occurs mainly in adult female poultry; susceptibility varies between strains and may be genetically controlled.[210]

The lesions associated with hemorrhagic liver have been well characterized.[211] Degenerative changes in hepatocytes of laying hens have been described including cytoplasmic vacuolation, swelling of the mitochondria, and distortion of the rough endoplasmic reticulum. These changes result in multifocal lysis and necrosis; sinusoids in the necrotic areas are also damaged, and abnormalities occur in bile canaliculi which may be associated with cholestasis.[212] The lesions leading to hemorrhage are controversial, but Martland et al.[213] have observed that reticulolysis was significantly worse in hens fed rapeseed meal and that it appeared to be related to dietary glucosinolate content. Liver hemorrhage occurs with feeding all types of rapeseed, although it is generally less severe, or absent, with low glucosinolate varieties.[214,215] The causative agents have not yet been defined, although epithionitriles have been suggested, these being hepatotoxic in other animals.[206] There was no association with hemorrhagic liver on feeding diets containing varying amounts of intact glucosinolates, 5-vinyloxazolidine-2-thione, isothiocyanates, and 1-cyano-2-hydroxybut-3-ene,[216,217] but intact glucosinolate and the latter nitrile were associated with increased hemorrhage when fed together.

5. Skeletal Abnormalities

Leg abnormality (perosis) has been a problem in the poultry industry for many years, resulting in poor growth, mortality, condemnation, and down-grading of carcasses with resultant financial loss to the producer. The incidence of perosis increased with the feeding of high-glucosinolate rapeseed[218] but was reduced when low-glucosinolate or extracted rapeseed meal was fed.[219,220] Examination of the published data suggests that the level of perosis was not significantly different from controls with glucosinolate levels of 0 to 1.8 mg g^{-1} diet, but that higher levels (3.6 to 16.1 mg g^{-1}) caused severe leg problems. Recently, Mawson[221] has observed a high incidence of transitory leg weakness in 4-week-old broilers fed a diet containing 0.6 mg 1-cyano-2-hydroxybut-3-ene g^{-1}. According to Timms,[222] rapeseed-induced perosis was increased in birds infected by reovirus.

B. Humans

Antinutritional effects of glucosinolates or their products in man are difficult to identify, being the result of long-term exposure. The determination of hazard, if any, associated with such intake requires toxicological information which, in the main, is not yet available. Concerns over the possible goitrogenicity in man have led to a number of investigations with the objective of establishing levels of 5-vinyloxazolidine-2-thione in milk. Recently, levels of up to 1 mg l^{-1} have been found following the feeding of rapeseed meal to dairy cattle.[61] In man a single dose of 50 to 200 mg of this compound caused inhibition of radioiodine uptake for 24 h.[80] According to Greer,[45] natural oxazolidine-2-thiones exhibit 33% more activity in man than propylthiouracil (PTU). It seems improbable that large amounts of goitrogenic 5-vinyloxazolidine-2-thione will result from cooked vegetables. In a recent study volunteers consuming each day an amount of cooked sprouts having a glucosinolate content equivalent to 40 mg 5-vinyloxazolidine-2-thione showed no disturbance of normal thyroid function after 4 weeks.[223] Dahlberg et al.[224] have found a similar lack of effect following the prolonged exposure of normal subjects to thiocyanate ion (8 mg daily for 96 d), but it is possible, especially in the case of certain subgroups, that intakes may at least potentially exceed this figure. On the basis of earlier studies, it would seem that very high levels (0.5 to 1 g) of thiocyanate ion are necessary to disturb normal thyroid function.[1] Despite these findings, glucosinolate hydrolysis products cannot yet be excluded as factors contributing to endemic goiter in certain parts of the world; moreover, any effect may be aggravated if dietary iodide levels are limiting.

IX. EFFECTS ON PALATABILITY AND FLAVOR

Sulfur compounds are often associated with strong and distinctive flavors. A variety of descriptions have been used for the products of individual glucosinolates including hot, pungent, cabbagy, garlicky, acrid, biting, and bitter.[1] While certain of these qualities may contribute to the overall desirable quality of vegetables, relishes, and condiments, they can be detrimental to the acceptance by animals of cruciferous oilseed meals and forages.

A. Animals

Palatability may be defined as initial acceptance (determined by preference tests or rate-of-eating trials) or as intake measured over a longer time span. Diet acceptability varies with dietary components, feeding situation, and the species of animal under test (see below). While initial acceptability may be influenced by flavor and taste, longer term intake is governed by the animal's drive to realize its genetic potential for growth and production, the nutritional quality of the diet, and the presence of antinutritional factors. It is, thus, difficult to separate the effects of glucosinolate-derived products on intake which arise directly (e.g., via a reduction in palatability) and indirectly (as a consequence, for example, of interference with normal thyroid function). In addition, rapeseed and other cruciferous oilseeds contain other unpalatable and antinutritional components, notably sinapine and tannins.[97] In the great majority of cases no attempt has been made to quantify their effects and relative importance *vis a vis* glucosinolates.

1. Swine

A major limitation to the inclusion of rapeseed products, especially those containing high levels of glucosinolates, in diets for swine is their apparent low palatability. Young pigs are especially sensitive and exhibit reduced intake with high rapeseed inclusions, even if this is of the low-glucosinolate type (Table 4).[101,102] The overall consensus of published information is that at lower levels of inclusion, low-glucosinolate rapeseed will replace soybean meal without detriment to diet palatability.[225,226] Nevertheless, if offered a choice, pigs will select soybean diets or those containing the lowest levels of low-glucosinolate rapeseed.[102] The finding that diets containing *B. campestris* are consumed more readily than those containing *B. napus*[227] is probably a result of their respective individual glucosinolate contents, the latter being much richer in glucosinolates yielding bitter oxazolidine-2-thiones. This is consistent with the findings of Lee et al.[139] that 2-hydroxybut-3-enyl glucosinolate at dietary levels below 1 mg g^{-1} diet had little effect, while palatability started to decrease between 1 to 2 mg g^{-1} diet and was significantly depressed thereafter.

2. Cattle and Sheep

Palatability is particularly important in high-yielding dairy cows and rapidly growing young animals. Dairy cattle have been found in both long-[228] and short-term[76,229] experiments to tolerate high levels of rapeseed without adverse effect on feed intake. In such experiments rapeseed (both high and low varieties) was included at levels up to 25%, with daily intakes of oxazolidine-2-thiones varying between 0 to 6.5 g per cow. While this clearly suggests that dairy cattle are insensitive to rapeseed glucosinolates, when offered a choice they will select against diets containing mustard meal.[230] This effect is presumably a result of the presence of the prop-2-enyl isothiocyanate precursor, since caustic soda treatment, known to remove this compound, markedly improved palatability. While the overwhelming balance of opinion indicates few problems in dairy cattle, it should be noted that in most trials no specific observations on rate of eating and incidence of feed refusal were recorded.

In contrast to the above, rather contradictory results have been reported for calves. Thus, when high- and low-glucosinolate rapeseed meals were compared, only the former depressed intakes, and then, only at high levels of incorporation (20 to 25%).[76,117] In other experiments no effect on intake was reported. In two such cases[231,232] animals over 8 weeks of age were involved, and it is clear from other work[115] that older animals (>100 kg) can tolerate both low- and high-glucosinolate diets (12 to 16%).

An examination of available data[76,233] suggests that young calves can tolerate dietary oxazolidine-2-thione and total glucosinolate at 0.7 mg and 3.3 mg g^{-1} diet, respectively, while intake is depressed at twice these levels. Gorrill et al.[234] found that the inclusion of rapeseed flour in calf-milk replacer diets (7.1 mg glucosinolates g^{-1} diet) depressed intake while half this level was without effect.

The limited data available on young lambs[115,235] suggest that no effects are found with low-glucosinolate rapeseed meals and that grazing intakes of stubble turnip and other crucifers, including forage rape, are unaffected by variety.[236]

3. Poultry

While rations containing high-glucosinolate rapeseed meal have a low palatability for pigs and young ruminants, it has generally been assumed that this does not apply to poultry, since it was thought that their senses of taste and smell were not as well developed.[157] Intake depression does occur with the feeding of high levels of high-glucosinolate rapeseed, but there is little or no evidence to suggest that this is associated with palatability. Thus, in the majority of cases where this has been observed for broilers, turkeys, or laying hens, the lower consumption has been attributed to reduced growth and/or other problems such as perosis, liver hemorrhage, or enlargement of thyroid or liver.

B. Humans

It is commonly held that glucosinolate breakdown products make a major contribution to the flavor of cruciferous vegetables, relishes, and condiments. While the latter is certainly true, with the volatile products produced by enzymic hydrolysis being the major contributors to the flavor profiles of mustard, radish, horseradish, and watercress, the flavor of vegetables is more complex.[1] There is a great deal of information available on the composition of flavor volatiles in vegetables, but the contribution of individual components to the overall, balanced flavor is generally little understood. Early work in this area has been reviewed by MacLeod.[32] A number of interesting findings were reported which would certainly repay further investigation using the more sophisticated techniques and equipment now available.

Off-odors associated with over-cooked cabbage have been suggested to be due to nitriles, arising by chemical, rather than enzymatic, processes.[32] This is clearly important in the case of cruciferous foods which have been thermally processed. An objectionable odor found by Srisangnam et al.[237] in rehydrated cabbage samples was attributed to the same source.

Increased bitterness in the heart of leafy brassica vegetables is well known, and MacLeod[32] has suggested this is due to increased levels of prop-2-enyl isothiocyanate. Recent work[238] has shown that this compound lacks bitterness and has linked the bitterness to the parent prop-2-enyl glucosinolate and 5-vinyloxazoldine-2-thione. As a result of this work, it has been possible to relate perceived bitterness in brussels sprouts and other brassicas to glucosinolate profile. This work has incidentally also shown that, in the absence of any source of exogenous myrosinase, glucosinolates are not broken down in the mouth.

Factors which affect glucosinolate content would be expected to affect flavor, and studies have confirmed this. Thus, the sulfate status of the growing medium has been related to flavor strength in watercress, cabbage, radish, and mustard. Various plant stresses, e.g., close plant spacing or reduced moisture, have been shown to increase glucosinolate content and strengthen overall flavor.[1]

The alteration of flavor by processing is clearly highly complex. Not only will cooking and blanching destroy most, if not all, the myrosinase, but intact glucosinolates will be leached out, in varying amounts, into the cooking liquor, Volatile-flavor compounds formed during processing may be lost, undergo hydrolysis or other secondary reactions which will greatly change, or destroy completely, their flavor characteristics. Similar changes may occur with fresh-processed brassicas (pickled cabbage, sauerkraut, coleslaw), and it would be expected that flavor release would be related to degree of tissue disruption (e.g., coarse or fine shredding). Recent work[239] has shown the flavor volatiles of pickled crucifers to be formed by autolysis, but as in the case of processed Japanese radish,[240] considerable changes in the composition and content of flavor volatiles would be expected with time. In addition, great variation would occur between batches.

X. OTHER EFFECTS

Glucosinolates or, more specifically, their hydrolysis products show a wide range of biological activities. These have been discussed in detail[1] in a recent review and so will be considered only briefly here.

A. Insects

A great deal of work has been carried out into the allelochemical properties of glucosinolates and their products.[1] A variety of glucosinolates have been found to stimulate the feeding of the diamondback moth (*Plutella maculipennis*) and the large white butterfly (*Pieris brassicae*). Prop-2-enyl glucosinolate is, however, acutely toxic to the black swallowtail butterfly (*Papilio polyxenes*). Ovipositing by the adult cabbage maggot (*Hylemya brassicae*) is induced by a mixture of prop-2-enyl glucosinolate and its isothiocyanate whereas, alone, the latter serves only as an attractant. Hardman and Ellis[241] have found the ovipositing preference of the cabbage root fly (*Delia brassicae*) for radish plants to be correlated with 4-methylthiobut-3-enyl isothiocyanate, levels of which increase with plant age. Isothiocyanates have been used in agriculture to reduce pest damage, with both synthetic *t*-butyl- and *o*-methoxyphenyl isothiocyanates being more effective than prop-2-enyl isothiocyanate. 2-Phenylethyl isothiocyanate has effective insecticidal activity against house- and vinegar flies and confused flour beetles and was the most toxic of various compounds tested against eggs of the brassica specialist, *Dasineuria brassicae*.[242] The insecticidal effects of the crude extracts of other cruciferous plants have also been attributed to their isothiocyanate content.

B. Microorganisms

Isothiocyanates possess wide-ranging fungistatic and bacteriostatic effects.[1] Of a variety of isothiocyanates tested against *Aspergillus niger* 11/13, *Penicillium cycloprium* 11/17, and *Rhizopus oryzae* 5/1, the most active were those possessing a 5-methylthiopentyl or aromatic grouping. Significant increases in the activity of the latter would be achieved by aromatic substitution. The presence of high levels of glucosinolates is considered to be associated with resistance of brassicas to cabbage downy mildew, ringspot, and stem canker. Recent studies[243,244] have shown a number of glucosinolates to produce, after treatment with myrosinase, products which inhibit the growth of *Leptosphaeria malculans*, the stem canker organism, and levels of alkenyl glucosinolates in leaves were inversely correlated with lesion extension.

Isothiocyanates are considered to exert their effect by binding to biologically-active sulfhydryl centers, a mechanism which also served to explain the activity of the sulfur-containing allicins in onion, garlic, and related species.[245]

C. Plants

The possibility of a link between glucosinolate content and clubroot resistance was first suggested over 50 years ago. It seems highly probable that the biological and biochemical factors associated with this disease are complex, but recently Butcher et al.[246] have demonstrated that the development of clubbing is associated with a large increase in the synthesis and degradation of indole glucosinolates. In support of this, a recent analysis of 145 commercial and breeding cabbage cultivars showed those possessing clubroot resistance to have low levels of indole glucosinolates.[247]

Glucosinolate-derived indoles and isothiocyanates have been shown by various workers to inhibit germination,[1] an effect which may, in retrospect, explain the phytotoxic effect of woad which was known in the Middle Ages. The effects of these compounds in the soil may not be long lasting, being reduced by leaching or microbial conversion to less toxic species.

D. Taints

Off-flavors have been detected in milk following the feeding of white mustard cover crops or rapeseed contaminated with stinkweed (*Thlaspi arvense*). The hedgerow biennial *Alliaria petiolata* (garlic mustard) smells of garlic when bruised and reputedly causes an onion flavor in milk when fed in quantity to lactating cows. In the latter two cases, volatile thiocyanates rather than isothiocyanates are responsible.[1] An economically significant taint problem in Australia associated with the consumption of land cress (*Coronopus didymus*) is associated with the breakdown of benzyl glucosinolate to a variety of products, of which benzyl mercaptan and benzyl methyl sulfide are apparently most offensive.[248] Moss[230] has reported that off-flavors result from the feeding of raw mustard meal to dairy cattle. The effect, attributed to prop-2-enyl isothiocyanate could be prevented by treatment with 3% caustic soda. According to Fredricksen et al.[249] inclusions of rapeseed expeller cake, containing high levels of glucosinolates, were associated with off-flavors, but diets containing lower level of glucosinolates (1 to 2 mg g^{-1} diet) produced no detrimental effects when compared with those containing soybean or cottonseed meals.[121]

Fishy taints[250] in eggs are known to follow the feeding of rapeseed to certain strains of laying hens, primarily (but not exclusively) those laying brown-shelled eggs. The taint is due to trimethylamine (at typical levels of 1 to 5 μg g^{-1} egg), derived mainly from the choline ester, sinapine. Glucosinolates are also intimately involved in this problem since 5-vinyloxazolidine-2-thione, presumably formed *in vivo*, depresses trimethylamine oxidation (the normal route for excretion), especially in birds possessing a single, autosomal semidominant "tainting" gene*. 5-Vinyloxazolidine-2-thione and 2-hydroxybut-3-enyl glucosinolate (200 μg g^{-1} diet) were found to produce taint when added back to control, rapeseed

---

* The similarity of this effect to that believed to be responsible for trimethylaminuria in humans has prompted speculation that glucosinolate-containing brassicas may be an aggravating factor in this distressing condition.[251]

meal-free rations, and removal of glucosinolates from rapeseed meal markedly reduced the inhibitory effects and the incidence of taint. The application of heat and anhydrous ammonia to rapeseed reduced levels of 2-hydroxybut-3-enyl glucosinolate to <120 μg g^{-1} meal.[252,253] When this was fed to laying hens, trimethylamine levels of 0.8 μg g^{-1} were recorded in eggs, this being the lower limit of taint.[254] Recently, a significant correlation has been demonstrated between taint and levels of free 5-vinyloxazolidine-2-thione in the diet, rather than that of the precursor glucosinolate.[255]

There appear to be no published data indicating any deleterious effects of glucosinolates on carcass flavor in cattle or pig meat. In the latter case, the majority of experiments have involved low glucosinolate rapeseed, and in all cases no significant increase occurred in off-flavor or fishy taint.[110,256] In view of the clear involvement of 5-vinyloxazolidine-2-thione in the development of egg taint, rapeseed has been suspected of causing taint in poultry meat. While there are several circumstantial claims to this effect, none is supported by unambiguous chemical and sensory data. Salmon et al.[257] have reported that chicken meat flavor was decreased and the incidence of off-flavors increased when broilers were fed starter and finisher rations containing 28.1% and 12.1% low-glucosinolate (canola) rapeseed meal, respectively. No such effects were noted when the same diets contained 21.0% and 9.0% canola meal, respectively. Larmond et al.[258] observed a slight increase in off-flavors for turkey meat when 7.3% canola meal was included in the ration; however, no such effects were noted at higher (14.4% and 21.1%) inclusion levels.

XI. SIGNIFICANCE OF GLUCOSINOLATES IN THE HUMAN DIET

In the human diet, glucosinolates are important as precursors of the desirable pungency of mustard, horseradish, radish, and the characteristic flavor of brassica vegetables. While concerns remain over the passage of goitrogenic products via milk and dairy products into the human body, the risk, if any, resulting from this has yet to be ascertained. While very much higher intakes of these compounds are potentially possible via direct consumption of brassicas, recent research suggests that, at least for cooked brassicas, this risk may have been overestimated.

There is much interest in identifying naturally occurring compounds which inhibit or otherwise prevent the toxic effects of other dietary species.[5,259] In this context the anticarcinogenic and antimutagenic properties of certain isothiocyanates and indoles derived from glucosinolates have been the focus of much interest.[185,260] It is clear that these effects are also produced by the intact glucosinolates,[261] although at this stage there is little information available to indicate whether this is the effect of the glucosinolate per se or of active metabolic products. Present research is directed toward the determination of dose-response behavior and in assessing whether the presence of these compounds in a meal would offer protection against the effects of other dietary components of the same meal. If the results of these studies prove encouraging, then the breeding of new cultivars of brassica for their therapeutic properties becomes a realistic possibility.

One factor which does have to be considered is the finding by Japanese workers that glucosinolate-derived indoles in Chinese cabbage may react with nitrite to produce nonvolatile compounds showing direct-acting mutagenicity.[262,264] Since the potential levels of such indole compounds in other brassicas may be much higher, given current concerns in Western Europe and elsewhere about the levels of nitrate in vegetables and the fact that physiological reduction of nitrate to nitrite is known to occur, this finding warrants further study. A nonvolatile nitrosamine is also formed by reaction of nitrite and 5-vinyloxazolidine-2-thione;[265,266] this may also have toxicological implications. Integrated chemical, physiological, and toxicological studies are needed to identify the significance in man of these *in vitro* findings. Ultimately, it is likely that it will be the balance of overall effects, beneficial and deleterious, which will be of importance.

In passing, mention should be made of the research of Kammüller and co-worker[267] regarding the etiology of Spanish toxic oil syndrome which, since 1981, has affected 20,000 people and been associated with over 350 deaths. Epidemiological and other data clearly support the association between the epidemic and the consumption of denatured rapeseed oil. These workers have suggested that isothiocyanates in rapeseed oil may cyclize with aniline, added as a denaturant, to yield a series of pharmacologically active products possessing the capability of inducing allergic or autoimmune-like diseases.

XII. OVERVIEW

In recent years, the requirements for future work on glucosinolates have been discussed,[1,7] and, in general, these remain valid. There is an urgent need, in common with studies on other natural toxicants, to determine the effects of longer-term exposure to realistic levels of glucosinolates and their products, to more fully understand the metabolism of these compounds in animals and man, and to determine the toxic potential of such metabolites. Current dietary recommendations and public trends in eating habits toward the consumption of greater quantities of leafy green vegetables have given added impetus to the examination of the protective effects of indole and certain other glucosinolates and their products. Rather more information will be needed on the factors affecting glucosinolate breakdown during processing and domestic cooking since even minor products may show highly significant biological activity either alone or after suitable activation. In this context the finding that certain glucosinolates may function as procarcinogens via interaction with nitrite should be examined further. The mutagenic effect of brassicas containing high levels of nitrate (thus taking into account the biological reduction of nitrate to nitrite) should be determined in suitable animal models.

The use of cruciferous oilseeds as animal feedstuffs will become more important as glucosinolate levels are further decreased. In many parts of the world there is resistance to the introduction of rapeseed meal in animal rations, and feed manufacturers and farmers will need to be convinced of the lack of any adverse effects by realistic, long-term (and therefore, costly) feeding trials. In view of the great differences in glucosinolate content between rapeseed samples, even between different batches of the same cultivar, the formulation of diets on the basis of "% rapeseed (meal)" may be inappropriate. Rather, diets should be formulated with an appreciation of likely intakes, e.g., in terms of glucosinolate content (both individual and total). Since different animal species and different strains within a species are likely to respond in different ways, this will require a great deal of additional animal data.

A combination of processing and plant breeding may be seen as the most realistic way of reducing glucosinolate levels to those necessary for no adverse effects to be seen. In developing such processes care must be taken to ensure that new products, having different but equally undesirable properties, are not being formed. In such areas the development of new analytical procedures may be deemed necessary. In general, however, the ability of the analytical chemist to determine glucosinolate levels and those of the obvious products, has far outstripped knowledge about the physiological, antinutritional, and pharmacological properties of those products.

Many of the above objectives will be realistically sought only by the initiation of coherent research programs involving food scientists and technologists, plant breeders, chemists, toxicologists, and pathologists. Such work, and the necessary transfer of technology and results to the benefit of both industry and society at large, will require appropriate funding over a realistic period of time.

Thus, as glucosinolate research progresses and diversifies, it remains a challenging and interesting area for the specialist, offers potential benefits to industry, and is relevant to the consumer's concern for safe, wholesome food.

ACKNOWLEDGMENT

The authors are grateful to Johanna Wrightson for her patience and diligence in preparing this manuscript.

REFERENCES

1. **Fenwick, G. R., Heaney, R. K., and Mullin, W. J.**, Glucosinolates and their breakdown products in food and food plants, *CRC Crit. Rev. Food Sci. Nutr.*, 18, 123, 1983.
2. **Ettlinger, M. G. and Kjaer, A.**, Sulphur compounds in plants, in *Recent Advances in Phytochemistry*, Vol. 1, Mabry, T. J., Alston, R. E., and Runeckles, V. C., Eds., Appleton-Century-Crofts, New York, 1968, 89.
3. **Sørensen, H., Ed.**, *Advances in the Production and Utilization of Cruciferous Crops*, Martinus Nijhoff/Junk, Dordrecht, Netherlands, 1985.
4. **Heaney, R. K. and Fenwick, G. R.**, Brassica vegetables — a major source of glucosinolates in the human diet, in *Advances in the Production and Utilization of Cruciferous Crops*, Sørensen, H., Ed., Martinus Nijhoff/Junk, Dordrecht, Netherlands, 1985, 40.
5. **Wattenberg, L. W.**, Naturally occurring inhibitors of chemical carcinogenesis, in *Naturally Occurring Carcinogens — Mutagens and Modulators of Carcinogenesis*, Miller, E. C., Miller, J. A., Hirono, I., Sugimura, T., and Takayama, S., Eds., University Park Press, Baltimore, 1979, 315.
6. **Tookey, H. L., VanEtten, C. H., and Daxenbichler, M. E.**, Glucosinolates, in *Toxic Constituents of Food Crops*, 2nd ed., Liener, I., Ed., Academic Press, New York, 1980, 103.
7. **VanEtten, C. H. and Tookey, H. L.**, Glucosinolates, in *Handbook of Naturally Occurring Food Toxicants*, Rechcigl, M., Jr., Ed., CRC Press, Boca Raton, FL, 1983, 15.
8. **Heaney, R. K. and Fenwick, G. R.**, Identifying toxins and their effects: glucosinolates, in *Natural Toxicants in Food, Progress and Prospects*, Watson, D. H., Ed., Ellis Horwood, Chichester, England, 1987, 76.
9. **Kjaer, A.**, The natural distribution of glucosinolates: a uniform class of sulphur-containing glucosides, in *Chemistry in Botanical Classification*, Bendz, G. and Santesson, J., Eds., Academic Press, London, 1974, 229.
10. **Crisp, P.**, Trends in the breeding and cultivation of cruciferous crops, in *The Biology and Chemistry of the Cruciferae*, Vaughan, J. G., MacLeod, A. J., and Jones, B. M. G., Eds., Academic Press, London, 1976, 69.
11. **Dorsch, W., Adam, O., Weber, J., and Ziegeltrum, M.**, Antiasthmatic compounds of onion extracts — detection of benzyl — and other isothiocyanates (mustard oils) as antiasthmatic compounds of plant origin, *Eur. J. Pharmacol.*, 107, 17, 1985.
12. **Gil, V., MacLeod, A. J., and Morreau, M.**, Volatile components of cocoas with particular reference to glucosinolate products, *Phytochemistry*, 23, 1937, 1984.
13. **MacLeod, A. J. and Panchasara, S. D.**, Volatile aroma components particularly glucosinolate products of cooked edible mushroom (*Agaricus bisporus*) and cooked dried mushroom, *Phytochemistry*, 22, 705, 1983.
14. **Hanley, A. B., Heaney, R. K., Lewis, J., Sones, K., Spinks, E. A., Wilkinson, A. P., and Fenwick, G. R.**, Benzyl isothiocyanate in onion (*Allium cepa* L) and mushroom (*Agaricus bisporus*) - a re-examination, *Food Chem.*, 30, 157, 1988.
15. **Bjerg, B., Fenwick, G. R., Spinks, E. A., and Sørensen, H.**, Failure to detect gluosinolates in cocoa, *Phytochemistry*, 26, 567, 1987.
16. **Ettlinger, M. G. and Lundeen, A. J.**, The structure of sinigrin and sinalbin: an enzymatic rearrangement, *J. Am. Chem. Soc.*, 78, 4172, 1956.
17. **Marsh, R. E. and Waser, J.**, Refinement of the crystal structure of sinigrin, *Acta Crystallogr. Sect. B*, 26, 1030, 1970.
18. **Ettlinger, M. G. and Lundeen, A. J.**, First synthesis of a mustard oil glucoside: the enzymatic Lossen rearrangement, *J. Am. Chem. Soc.*, 79, 1764, 1957.
19. **Linscheid, M., Wendisch, D., and Strack, D.**, The structures of sinapic acid esters and their metabolism in cotyledons of *Raphanus sativus*, *Z. Naturforsch.*, 35C, 907, 1980.
20. **Sørensen, H.**, Private communication.
21. **Hanley, A. B., Heaney, R. K., and Fenwick, G. R.**, Improved isolation of glucobrassicin and other glucosinolates, *J. Sci. Food Agric.*, 34, 869, 1983.

22. Peterka, S. and Fenwick, G. R., The use of flash chromatography for the isolation and purification of glucosinolates (mustard oil glycosides), *Fette Wiss. Technol.*, 90(2), 61, 1988.

23. Sørensen, H., Private communication.

24. McGregor, D. I., Mullin, W. J., and Fenwick, G. R., Analytical methodology for determining glucosinolate composition and content, *J. Assoc. Off. Anal. Chem.*, 66, 825, 1983.

25. Cox, I. J., Hanley, A. B., Belton, P. S., and Fenwick, G. R., NMR spectra (^{1}H, ^{13}C) of glucosinolates, *Carbohydr. Res.*, 137, 323, 1984.

26. MacLeod, A. J. and Rossiter, J. T., Non-enzymatic degradation of 2-hydroxybut-3-enylglucosinolate (progoitrin), *Phytochemistry*, 25, 855, 1986.

27. MacLeod, A. J., Panesar, S. S., and Gil, V., Thermal degradation of glucosinolates, *Phytochemistry*, 20, 977, 1981.

28. Youngs, C. G. and Perlin, A. S., Fe(II)-catalysed decomposition of sinigrin and related thioglycosides, *Can. J. Chem.*, 45, 1801, 1967.

29. Austin, F. L., Gent, C. A., and Wolff, I. A., Degradation of natural thioglucosides with ferrous salts, *J. Agric. Food Chem.*, 16, 752, 1968.

30. Gronowitz, S., Svensson, L., and Ohlsson, R., Studies of some nonenzymatic reactions of progoitrin, *J. Agric. Food Chem.*, 26, 887, 1978.

31. Maheshwari, P. N., Stanley, D. W., van de Voort, F. R., and Gray, J. I., The heat stability of allyl glucosinolate (sinigrin) in aqueous and model systems, *J. Inst. Can. Sci. Technol. Aliment.*, 13, 28, 1980.

32. MacLeod, A. J., Volatile flavour compounds of the Cruciferae, in *The Biology and Chemistry of the Cruciferae*, Vaughan, J. G., MacLeod, A. J., and Jones, B. M. G., Eds., Academic Press, London, 1976.

33. Phelan, J. R., Allen, A., and Vaughan, J. G., Myrosinase in *Raphanus sativus* L., *J. Exp. Bot.*, 35, 1558, 1984.

34. Pocock, K., Heaney, R. K., Wilkinson, A. P., Beaumont, J., Vaughan, J. G., and Fenwick, G. R., Changes in the myrosinase activity and isoenzyme pattern, glucosinolate content and the cytology of myrosin cells in the leaves of heads of three cultivars of English white cabbage, *J. Sci. Food Agric.*, 41, 245, 1987.

35. MacLeod, A. J. and Rossiter, J. T., Isolation and examination of thioglucoside glucohydrolase from seeds of *Brassica napus*, *Phytochemistry*, 25, 1047, 1986.

36. Uda, Y., Kurata, T., and Arakawa, N., Effects of pH and ferrous ion on the degradation of glucosinolates by myrosinase, *Agric. Biol. Chem.*, 50, 2735, 1986.

37. Uda, Y., Kurata, T., and Arakawa, N., Effects of thiol compounds on the formation of nitriles from glucosinolates, *Agric. Biol. Chem.*, 50, 2741, 1986.

38. Petroski, R. J. and Kwolek, W. F., Interactions of a fungal thioglucoside glucohydrolase and cruciferous plant epithiospecifier protein to form 1-cyanoepithioalkanes: implications of an allosteric mechanism, *Phytochemistry*, 24, 213, 1985.

39. MacLeod, A. J. and Rossiter, J. T., The occurrence and activity of epithiospecifier protein in some Cruciferae seeds, *Phytochemistry*, 24, 1895, 1985.

40. Lüthy, J. and Benn, M. H., The conversion of potassium allyl glucosinolate to 3,4-epithiobutane nitrile by *Crambe* seed flour, *Phytochemistry*, 18, 2028, 1979.

41. Brocker, E. R. and Benn, M. H., The intramolecular formation of epithioalkanenitriles from alkenylglucosinolates by *Crambe abyssinica* seed flour, *Phytochemistry*, 22, 770, 1983.

42. Underhill, E. W. and Wetter, L. R., Biosynthesis of glucosinolates, *Biochem. Soc. Symp.*, 38, 303, 1973.

43. Kjaer, A., Glucosinolates in the Cruciferae, in *The Biology and Chemistry of the Cruciferae*, Vaughan, J. G., MacLeod, A. J., and Jones, B. M. G., Eds., Academic Press, London, 1976, 207.

44. Underhill, E. W., Glucosinolates, in *Encyclopaedia of Plant Physiology*, Vol. 8, Bell, E. A. and Charlwood, B. V., Eds., Springer-Verlag, Heidelberg, 1980, 493.

45. Greer, M. A., The natural occurrence of goitrogenic agents, *Recent Prog. Horm. Res.*, 18, 187, 1962.

46. Lewis, J., Wyatt, G. M., Horn, N., and Fenwick, G. R., Unpublished work.

47. VanEtten, C. H., Gagne, W. E., Robbins, D. J., Booth, A. N., Daxenbichler, M. E., and Wolff, I. A., Biological evaluation of *Crambe* seed meals and derived products in rat feeding, *Cereal Chem.*, 46, 145, 1969.

48. Pearson, A. W., Greenwood, N. M., Butler, E. J., Fenwick, G. R., and Curl, C. L., Rapeseed meal and egg taint: effects of *B. campestris* meals, progoitrin and potassium thiocyanate on trimethylamine oxidation, *J. Sci. Food Agric.*, 34, 965, 1983.

49. Smith, T. K. and Campbell, L. D., Rapeseed meal glucosinolates. Metabolism and effect on performance of laying hens, *Poult. Sci.*, 55, 861, 1976.

50. Macholz, R., Ackermann, H., Diedrich, M., Henschel, K. P., Kujawa, M., Lewerenz, H.-J., Przybilski, H., Schnaak, W., Schulze, J., and Woggon, H., Studies on the degradation of glucotropaeolin and progoitrin — toxicity and reactivity of splitting products, in *Proc. Eurofoodtox II*, Schlatter, C. ed., Swiss Federal Institute of Technology and University of Zurich, Zurich, 1986, 40.

51. **Nugon-Baudon, L. and Szylit, O.,** Private communication.

52. **VanEtten, C. H. and Daxenbichler, M. E.,** Glucosinolates and derived products in cruciferous vegetables: total glucosinolates by retention on anion exchange resin and enzymatic hydrolysis to measure released glucose, *J. Assoc. Off. Anal. Chem.*, 60, 964, 1977.

53. **Heaney, R. K. and Fenwick, G. R.,** A microcolumn method for the rapid determination of total glucosinolate content of cruciferous material, *Z. Pflanzenzuecht.*, 87, 89, 1981.

54. **Møller, P., Plöger, A., and Sørensen, H.,** Quantitative analysis of total glucosinolate content in concentrated extracts from double-low rapeseed by the Pd-glucosinolate complex method, in *Advances in the Production and the Utilization of Cruciferous Crops*, Sørensen, H., Ed., Martinus Nijhoff/Junk, Dordrecht, Netherlands, 1985, 97.

55. **Thies, W.,** Determination of glucosinolate content in commercial rapeseed loads with a pocket reflectometer, *Fette Seifen Anstrichm.*, 87, 347, 1985.

56. **Smith, C. A. and Dacombe, C.,** Rapid method for determining total glucosinolates in rapeseed by measurement of enzymatically released glucose, *J. Sci. Food Agric.*, 38, 141, 1987.

57. **Starr, C., Suttle, J., Morgan, A. G., and Smith, D. B.,** A comparison of sample preparation and calibration techniques for the estimation of nitrogen, oil and glucosinolate content of rapeseed by near infrared spectroscopy, *J. Agric. Sci.*, 104, 317, 1985.

58. **Schnug, E. and Haneklaus, S.,** Rapid and precise determination of total glucosinolate content in rapeseed. Comparison of X-ray Fluorescence spectroscopy with gas chromatography and high performance liquid chromatography, *Fette Wiss. Technol.*, 89, 32, 1987.

59. Oilseeds and Oilseed Residues — Determination of Isothiocyanates and Vinyl Thiooxazolidone, ISO 5504, International Standards Organisation, Geneva, 1983.

60. **Daun, J. K.,** Glucosinolate analysis in rapeseed and Canola — an update, *Yukagaku*, 35, 426, 1986.

61. **Bachmann, M., Theus, R., Lüthy, J. and Schlatter, C. L.,** Vorkommen von Goitrogen Stoffen in Milch, *Z. Lebensm. Unters. Forsch.*, 181, 375, 1985.

62. **de Brabander, H. F. and Verbeke, R.,** Determination of oxazolidine-2-thiones in biological fluids in the ppb range, *J. Chromatogr.*, 252, 225, 1982.

62a. **Yui, S.,** Year-round cultivation in cruciferous crops in Japan, *Jpn. Agric. Res. Q.*, 20, 185, 1987.

63. **VanEtten, C. H., Daxenbichler, M. E., Tookey, H. L., Kwolek, W. F., Williams, P. H., and Yoder, O. C.,** Glucosinolates: potential toxicants in cabbage cultivars, *J. Am. Soc. Hortic. Sci.*, 105, 710, 1980.

64. **Sones, K., Heaney, R. K., and Fenwick, G. R.,** The glucosinolate content of UK vegetables — cabbage (*Brassica oleracea*), swede (*B. napus*) and turnip (*B. campestris*), *Food Additives Contam.*, 1, 289, 1984.

65. **Daxenbichler, M. E., VanEtten, C. H., and Williams, P. H.,** Glucosinolates and derived products in cruciferous vegetables: analysis of fourteen varieties of Chinese cabbage, *J. Agric. Food Chem.*, 27, 34, 1979.

66. **Heaney, R. K. and Fenwick, G. R.,** Glucosinolates in *Brassica* vegetables. Analysis of twenty two varieties of Brussels sprout (*Brassica oleracea L.* var *gemmifera*), *J. Sci. Food Agric.*, 31, 785, 1980.

67. **Gramberg, L. G., Rijk, M. A. H., Schouten, A., and deVos, R. H.,** Glucosinolates in Dutch cole crops, in *Proc. Eurofoodtox II*, Schlatter, C., ed., Swiss Federal Institute of Technology and University of Zurich, Zurich, 1986, 279.

68. **Sones, K., Heaney, R. K., and Fenwick, G. R.,** Glucosinolates in *Brassica* vegetables. Analysis of twenty seven cauliflower cultivars (*Brassica oleracea. L.* var. *botrytis*. subvar. *cauliflora.DC*), *J. Sci. Food Agric.*, 35, 762, 1984.

69. **Lewis, J. and Fenwick, G. R.,** Glucosinolate content of *Brassica* vegetables. Analysis of twenty four cultivars of calabrese (green sprouting broccoli, *Brassica oleracea L.* var. *botrytis*. subvar. *cymosa. Lam.*), *Food Chem.*, in press.

70. **Mullin, W. J., Proudfoot, K. G., and Collins, M. J.,** Glucosinolate content and clubroot of rutabaga and turnip, *Can. J. Plant Sci.*, 60, 65, 1980.

71. **Carlson, D. G., Daxenbichler, M. E., VanEtten, C. H., Tookey, H. L., and Williams, P. H.,** Glucosinolates in crucifer crops: turnips and rutabagas, *J. Agric. Food Chem.*, 29, 1235, 1981.

72. **Carlson, D. G., Daxenbichler, M. E., VanEtten, C. H., Hill, C. B., and Williams, P. H.,** Glucosinolates in radish cultivars, *J. Am. Hortic. Sci.*, 110, 634, 1985.

73. **Kojima, M., Hamada, H., and Toshimitsu, N.,** Simultaneous quantitative determination of allyl isothiocyanate and β-phenethyl isothiocyanate by gas chromatography equipped with FPD, *Nippon Shokuhin Kogyo GakkaiShi*, 33, 155, 1986.

74. **Hälvä, S., Hirvi, T., Mäkinen, S., and Honkanen, E.,** Yield and glucosinolates in mustard seeds and volatile oils in caraway seeds and coriander fruit. I. Yield and glucosinolate contents of mustard (*Sinapis* sp; *Brassica* sp.) seeds, *J. Agric. Sci. Finl.*, 58, 157, 1986.

75. **Röbbelen, G. and Thies, W.,** Variations in rapeseed glucosinolates and breeding for improved meal quality, in *Brassica Crops and Wild Allies,Biology and Breeding*, Tsunoda, S., Hinata, K., and Gomez-Campo, G., Eds., Japanese Scientific Societies Press, Tokyo, 1980, 285.

76. **Carlson, D. G., Daxenbichler, M. E., VanEtten, C. H., Kwolek, W. F., and Williams, P. H.,** Glucosinolates in crucifer vegetables — broccoli, Brussels sprouts, cauliflower, collards, kale, mustard greens and kohlrabi, *J. Am. Hortic. Sci.*, 112, 173, 1987.

77. **Carlson, D. G., Daxenbichler, M. E., Tookey, H. L., Kwolek, W. F., Hill, C. B., and Williams, P. H.,** Glucosinolates in turnip tops and roots — cultivars grown for greens or roots, *J. Am. Hortic. Sci.*, 112, 179, 1987.

78. **Bradshaw, J. E., Heaney, R. K., Fenwick, G. R., and McNaughton, I. A.,** The glucosinolate content of the leaf and stem of fodder kale (*Brassica oleracea* L.), rape (*B. napus*) and radicole (*Raphanobrassica*) *J. Sci. Food Agric.*, 34, 571, 1983.

79. **Bradshaw, J. E., Heaney, R. K., Macfarlane-Smith, W. H., Gowers, S., Gemmell, D. J., and Fenwick, G. R.,** The glucosinolate content of some fodder brasssicas, *J. Sci. Food Agric.*, 35, 977, 1984.

80. **Elfving, S.,** Studies on the naturally occurring goitrogen, 5-vinyl-2-thiooxazolidone, *Ann. Clin. Res.*, 28, 1, 1980.

81. **VanEtten, C. H., Daxenbichler, M. E., Schroeder, W., Princen, L. H., and Perry, T. W.,** Test for epi-progoitrin, derived nitriles and goitrin in body tissues of farm cattle fed crambe seed meals, *Can. J. Anim. Sci.*, 57, 75, 1977.

82. **Papas, A., Ingalls, J. R., and Campbell, L. D.,** Studies on the effects of rapeseed meal on thyroid status of cattle. Glucosinolate and iodine content of milk and other parameters, *J. Nutr.*, 109, 1129, 1979.

83. **Sones, K., Heaney, R. K., and Fenwick, G. R.,** An estimate of the mean daily intake of glucosinolates from cruciferous vegetables in the UK, *J. Sci. Food Agric.*, 35, 712, 1984.

84. **Mullin, W. J. and Sahasrabudhe, M. R.,** An estimate of the average daily intake of glucosinolates via cruciferous vegetables, *Nutr. Rep. Int.*, 18, 273, 1978.

85. Present Status of National Nutrition in Japan, Nutrition Division, Bureau of Public Health, Ministry of Health and Welfare, Daiichi Shuppan, Tokyo, 1975.

86. **Gail-Eller, R. and Gierschner, K.,** Zum Gehalt und Verhalten der Glucosinolate in Weisskohl und Sauerkraut, *Dtsch. Lebensm. Rundsch.*, 80, 341, 1984.

87. **Marquard, R. and Schlesinger, V.,** Methodische Untersuchungen zur Glucosinolatbestimmung bei Raps, *Fette Seifen Anstrichm.*, 87, 471, 1985.

88. **Heaney, R. K., Spinks, E. A., and Fenwick, G. R.,** The glucosinolate content of Brussels sprouts: factors affecting their relative abundance, *Z. Pflanzenzuecht.*, 91, 219, 1983.

89. **VanEtten, C. H., Daxenbichler, M. E., Kwolek, W. F., and Williams, P. H.,** Distribution of glucosinolates in the pith, cambial-cortex and leaves of the head in cabbage, *Brassica oleracea* L., *J. Agric. Food Chem.*, 27, 648, 1979.

90. **Kondo, H., Nozaki, H., Kawaguchi, T., Naoshima, Y., and Hayashi, S.,** The effect of germination upon the volatile components of rapeseeds (*Brassica napus* L.), *Agric. Biol. Chem.*, 48, 1891, 1984.

91. **Rahman, M. H., Stølen, O., and Sørensen, H.,** Glucosinolate content in seeds from silique at different position of *Brassica campestris*, *Acta Agric. Scand.*, 36, 318, 1986.

92. **Sang, J. P., Bluett, C. A., Elliott, B. R., and Truscott, R. J. W.,** Effect of time of sowing on oil content, erucic acid and glucosinolate contents in rapeseed (*Brassica napus* L. cv. Marnoo), *Aust. J. Exp. Agric. Anim. Husb.*, 26, 607, 1986.

93. **Parnell, A., Craig, E. A., and Draper, S. R.,** Changes in the glucosinolate content of seed of winter oilseed rape varities in successive generations, *J. Natl. Inst. Agric. Bot. G. B.*, 16, 207, 1983.

94. **Daxenbichler, M. E., VanEtten, C. H., and Williams, P. H.,** Glucosinolate products in commercial sauerkraut, *J. Agric. Food Chem.*, 28, 809, 1980.

95. **Kojima, M., Hamada, H., and Toshimitsu, N.,** Changes in isothiocyanates of *Wasabia japonica* roots by drying, *Nippon Shokuhin Kogyo GakkaiShi*, 32, 886, 1985.

96. **Maheshwari, P. N., Stanley, D. W., and Gray, J. I.,** Detoxification of rapeseed products, *J. Food Prot.*, 6, 459, 1981.

97. **Fenwick, G. R. and Curtis, R. F.,** Rapeseed meal and its use in poultry diets, a review, *Anim. Food Sci. Technol.*, 5, 255, 1980.

98. **Rundgren, M.,** Low glucosinolate rapeseed products for pigs — a review, *Anim. Food Sci. Technol.*, 9, 239, 1983.

99. **Bell, J. M.,** Nutrients and toxicants in rapeseed meal: a review, *J. Anim. Sci.*, 58, 996, 1984.

100. **Fiems, L. O. and Buysse, F. X.,** A review of the use of rapeseed meal as protein source in diets for ruminants, *Rev. Agric.*, 38, 261, 1985.

101. **McIntosh, M. K. and Aherne, F. X.,** Canola meal for starter pigs (4—8 weeks of age), in 61st Annu. Feeders Day Rep., Univ. of Alberta, Canada, 1982, 74.

102. **McIntosh, M. K. and Aherne, F. X.,** Taste perferences of piglets fed soybean meal and canola meal supplemented diets, in 61st Annu. Feeders Day Rep., Univ. of Alberta, Canada, 1982, 76.

103. **Ochetim, S.,** The feeding value of Tower rapeseed for early weaned pigs. I. Effects of methods of processing and of dietary levels, *Can. J. Anim. Sci.*, 60, 407, 1980.

104. **McKinnon, P. J. and Bowland, J. P.,** Comparison of low glucosinolate, low erucic acid rapeseed meal (*cv.* Tower), commerical rapeseed meal and soybean meal as sources of protein for starting, growing and finishing pigs and young rats, *Can. J. Anim. Sci.,* 57, 663, 1977.

105. **Bowland, J. P.,** Evaluation of low glucosinolate — low erucic acid rapeseed meals as protein supplements for young growing pigs including effects on blood constituents, *Can. J. Anim. Sci.,* 55, 409, 1975.

106. **Goold, A. T., Taverner, M. R., and Hodge, R. W.,** The effect of rapeseed meal in the diet of young pigs on digestibility of diets and performance, *Aust. J. Exp. Agric. Anim. Husb.,* 16, 372, 1976.

107. **Baidoo, S. and Aherne, F. X.,** Canola meal for starter pigs, *Agric. For. Spec. Issue Univ. Alberta,* 120, 1983.

108. **McIntosh, M. K. and Aherne, F. X.,** An evaluation of Canola meal (var. Candle) as a protein supplement for starter pigs, *Agriculture and Forestry Special Issue Univ. Alberta,* 16, 1981.

109. **Castell, A. G.,** Effects of cultivar on the utilisation of ground rapeseed in diets for growing-finishing pigs, *Can. J. Anim. Sci.,* 57, 111, 1977.

110. **Thomke, S., Elwinger, K., Rundgren, M., and Ahlstrom, B.,** Rapeseed meal of Swedish low glucosinolate type fed to broiler chickens, laying hens and growing finishing pigs, *Acta Agric. Scand.,* 33, 75, 1983.

111. **Eggum, B. O., Just, A., and Sørensen, H.,** Double low rapeseed meal in diets to growing-finshing pigs, in *Advances in the Production and Utilisation of Cruciferous Crops,* Sørensen, H., Ed., Martinus Nijhoff/ Junk, Dordrecht, Netherlands, 1985, 167.

112. **Thomke, S.,** Further experiments with RSM of a Swedish low glucosinolate type fed to growing-finishing pigs, *Swed. J. Agric. Res.,* 14, 151, 1984.

113. **Nasi, M., Alaviuhkola, T., and Suomi, K.,** Rapeseed meal of low and high glucosinolate type fed to growing-finishing pigs, *J. Agric. Sci. Fin.,* 57, 263, 1985.

114. **Thomke, S.,** Review of rapeseed meal in animal nutrition. II. Ruminant animals, *J. Am. Oil Chem. Soc.,* 58, 805, 1981.

115. **Bush, R. S., Nicholson, J. W. G., MacIntyre, T. M., and McQueen, R. E.,** A comparison of Candle and Tower rapeseed meals in lamb, sheep and beef steer rations, *Can. J. Anim. Sci.,* 58, 369, 1978.

116. **Iwarsson, K., Ekman, L., Everitt, B. R., Figueiras, H., and Nilsson, P. O.,** The effect of feeding rapeseed meal on thyroid function and morphology in growing bulls, *Acta Vet. Scand.,* 14, 620, 1973.

117. **Fiems, L. O., Boucque, C. V., Cottyn, B. G., and Buysse, F. X.,** Evaluation of rapeseed meal with low and high glucosinolate as a protein source in calf starters, *Livestock Prod. Sci.,* 12, 131, 1985.

118. **Ingalls, J. R. and Sharma, H. R.,** Feeding of Bronowski, Span and commercial rapeseed meals with or without addition of molasses or flavour in rations of lactating cows, *Can. J. Anim. Sci.,* 55, 721, 1975.

119. **Papas, A., Ingalls, J. R., and Cansfield, P.,** Effect of Tower and 1821 rapeseed meals and Tower gums on milk yield, milk composition and blood parameters of lactating dairy cows, *Can. J. Anim. Sci.,* 58, 671, 1978.

120. **Martillotti, F., Malossini, F., Carreta, A., and Lanzani, A.,** Chemical composition of the seeds and oil meals of rapeseed belonging to two varieties and their influence on milk yields and milk composition in dairy cows, *Zootech. Nutr. Anim.,* 5, 239, 1979.

121. **Sanchez, J.M. and Clayool, D. W.,** Canola meal as a protein supplement in dairy rations, *J. Diary Sci.,* 66, 80, 1983.

122. **DePeters, E. J. and Bath, D. L.,** Canola meal versus cottonseed meal as the protein supplement in dairy diets, *J. Dairy Sci.,* 69, 148, 1986.

123. **Fisher, L. J. and Walsh, D. S.,** Substitution of rapeseed meal for soybean meal as a source of protein for lactating cows, *Can. J. Anim. Sci.,* 56, 233, 1976.

124. **Stedman, J. A., Pittam, S., and Hill, R.,** The voluntary feed intake and rate of weight gain of calves given a high proportion of rapeseed meal, *Anim. Prod.,* 36, 513, 1983.

125. **Yilala, K. and Bryant, M. J.,** The effects upon the intake and performance of store lambs of supplementing grass silage with barley, fish meal and rapeseed meal, *Anim. Prod.,* 40, 111, 1985.

126. **Thomas, V. M., Katz, R. J., Auld, D. L., and Peterson, C. L.,** Value of mechanically extracted rape and safflower oilseed meals as protein supplements for growing lambs, *Anim. Food Sci. Technol.,* 11, 269, 1984.

127. **Dogan, K. and Yucelen, Y.,** Feeding value and use of rapeseed oilmeal for fattening lambs, *Doga Bilim Dergisi D1,* 8, 24, 1984.

128. **Fitzgerald, J. J. and Black, W. S. M.,** Finishing store lambs on green forage crops. I. A comparison of rape, kale, and fodder radish as sources of feed for finishing store lambs in autumn, *Ir. J. Agric. Res.,* 23, 127, 1984.

129. **Fitzgerald, J. J.,** Finishing store lambs on green forage crops. II. Effects of supplementary rape with barley on lamb performance, *Ir. J. Agric. Res.,* 23, 137, 1984.

130. **Clandinin, D. R. and Robblee, A. R.,** Rapeseed meal in animal nutrition. I. Non-ruminant animals, *J. Am. Oil Chem. Soc.,* 58, 682, 1981.

131. Stoewsand, G. S., Anderson, J. L., and Lisk, D. J., Changes in liver glutathione-S-transferase activities in Coturnix quail fed municipal sludge-grown cabbages with reduced levels of glucosinolates, *Proc. Soc. Exp. Biol. Med.*, 182, 95, 1986.

132. Marangos, A., Hill, R., Laws, B. M., and Muschamp, D., The influence of three rapeseed meals and a mustard seed meal on egg and broiler production, *Br. Poult. Sci.*, 15, 405, 1974.

133. Throckmorton, J. C., Cheeke, J. P. R., Patton, N. M., Ascott, G. H., and Jolliff, G. D., Evaluation of meadowfoam (*Limnanthes alba*) meal as a foodstuff for broiler chicks and weanling rabbits, *Can. J. Anim. Sci.*, 61, 735, 1981.

134. Blair, R., Nutritional evaluation of ammoniated mustard meal for chicks, *Poult. Sci.*, 63, 754, 1984.

135. Bock, H. D., Mieth, G., Ohff, R., Kesting, S., Heinz, T., and Kreienbring, F., Nutritive value of heat-treated extraction meal from winter rape varieties poor in erucic acid and glucosinolate. Experiments on growing albino rats and broilers, *Nahrung*, 30, 177, 1986.

136. Lodhi, G. N., Cilly, V. K., and Ichhponani, J. S., Effect of iodine supplementation in mustard cake on productive performance of White Leghorn laying pullets, *Z. Tierphysiol. Tierernaehr. Futtermittelkd.*, 44, 64, 1980.

137. Nasser, A. R. and Arscott, G. H., Canola meal for broilers and the effects of a dietary supplement of iodinated casein on performance and thyroid status, *Nutr. Rep. Int.*, 34, 791, 1986.

138. Pfirter, H. P., Halter, H. M., and Lüthy, J., Researches of rapeseed processing as feedstuffs for growing pigs and poultry. I. Growth performance and organ development, *Int. J. Vitam. Nutr. Res.*, 52, 217, 1982.

139. Lee, P. A., Pittam, S., and Hill, R., The voluntary food intake by growing pigs of diets containing treated rapeseed meals or extracts of rapeseed meals, *Br. J. Nutr.*, 52, 159, 1984.

140. Brak, B. and Henkel, H., Experiments on the use of seed and meal of rapeseed low in glucosinolate content in feeding rations for monogastric animals, *Fette Seifen Anstrichm.*, 80, 104, 1978.

141. Proudfoot, F. G., Hulan, F. W., and McRae, K. B., The effect of diets supplemented with Tower and/or Candle rapeseed meals on performance of meat chicken breeders, *Can. J. Anim. Sci.*, 62, 239, 1982.

142. Hulan, F. W. and Proudfoot, F. G., The nutritional value of rapeseed meal for caged layers, *Can. J. Anim. Sci.*, 60, 139, 1980.

143. Vermorel, M. and Baudet, J. J., Valorization of rapeseeed meal. II. Nutritive value of high- or low-glucosinolate varieties and effect of dehulling, *Reprod. Nutr. Dev.*, 27, 45, 1987.

144. Vermorel, M., Heaney, R. K., and Fenwick, G. R., Antinutritional effects of the rapeseed meals, Darmor and Jet Neuf and the glucosinolate progoitrin, together with myrosinase in the growing rat, *J. Sci. Food Agric.*, 44, 321, 1988.

145. Shires, A., Bell, J. M., Keith, M. O., and McGregor, D. I., Rapeseed dockage, effects of feeding raw and processed wild mustard and stinkweed seed on growth and feed utilization of mice, *Can. J. Anim. Sci.*, 62, 275, 1982.

146. Vermorel, M., Fayet, J. C., and Baudet, J. J., Valorisation of rapeseed meal. I. Nutritive value in growing rat after glucosinolate elimination, dehulling and galactoside extraction, *Ann. Biol. Anim. Biochim. Biophys.*, 18, 1393, 1978.

147. Vermorel, M., Davicco, M. J., and Evrard, J., Valorization of rapeseed meal. III. Effects of glucosinolate content on food intake, weight gain, liver weight and plasma thyroid hormone levels in growing rats, *Reprod. Nutr. Dev.*, 27, 57, 1987.

148. Josefsson, E. and Uppström, B., Influence of sinapine and *p*-hydroxybenzyl glucosinolate on nutritional value of rapeseed meal and white mustard meals, *J. Sci. Food Agric.*, 27, 438, 1976.

149. Vermorel, M., Heaney, R. K., and Fenwick, G. R., Nutritive value of rapeseed meal: effects of individual glucosinolates, *J. Sci. Food Agric.*, 37, 1197, 1986.

150. Bille, N., Eggum, B. O., Jacobsen, I., and Sørensen, H., Antinutritional and toxic effects in rats of individual glucosinolates (±myrosinase) added to a standard diet, *Z. Tierphysiol. Tierernaehr. Futtermittelkd.*, 49, 195, 1983.

151. Srivastava, V. K., Philbrick, D. J., and Hill, D. C., Response of rats and chicks to rapeseed meal subjected to different enzymatic treatments, *Can. J. Anim. Sci.*, 55, 331, 1975.

152. Loew, F. W., Doige, C. E., Manns, D. G., Searcy, G. P., Bell, J. M., and Jones, J. D., Evaluation of dietary rapeseed protein concentrate flours in rats and dogs, *Toxicol. Appl. Pharmacol.*, 35, 257, 1976.

153. Brown, R. G., Hoag, G. N., and Bracken, E., Utilisation of rapeseed meal in dog rations, *J. Anim. Sci.*, 43, 1225, 1976.

154. Belzile, R. J., Poliquin, L. S., and Jones, J. D., Nutritive value of rapeseed flour for mink: effects on live performance, nutrient utilisation, thyroid function and pelt quality, *Can. J. Anim. Sci.*, 54, 639, 1974.

155. Hardy, R. W. and Sullivan, C. V., Canola meal in rainbow trout (*Salmo gairdneri*) production diets, *Can. J. Fish Sci.*, 40, 281, 1983.

156. Higgs, D. A., McBride, J. R., Markert, J. R., Dosanjh, B. S., Plotnikoff, M. D., and Clark, W. C., Evaluation of Tower and Candle rapeseed (Canola) meal and Bronowski rapeseed protein concentrate as protein supplements in practical dry diets for juvenile chinook salmon (*Oncorhynchus tshawytscha*), *Aquaculture*, 29, 1, 1982.

157. **Hill, R.**, The toxic effect of rapeseed meals with observations on meal from improved varieties, *Br. Vet. J.*, 135, 3, 1979.
158. **Manns, J. G. and Bowland, J. P.**, Solvent extracted rapeseed oilmeal as a protein source for pigs and rats. I. Growth, carcass characteristics and reproduction, *Can. J. Anim. Sci.*, 43, 252, 1963.
159. **Devilat, J. and Skoknic, A.**, Feeding high levels of rapeseed meal to pregnant gilts, *Can. J. Anim. Sci.*, 51, 715, 1971.
160. **Schuld, F. W. and Bowland, J. P.**, Dietary rapeseed meal for swine reproduction, *Can. J. Anim. Sci.*, 48, 57, 1968.
161. **Marangos, A. G. and Hill, R.**, The influence of rapeseed and mustard meals on reproductive efficiency in gilts, *Br. Vet. J.*, 133, 46, 1973.
162. **Lee, P. A., Hill, R., and Ross, E. J.**, Studies on rapeseed meal from different varieties of rape in the diets of gilts. II. Effects on farrowing performance of gilts, performance of their piglets to weaning and subsequent conception of the gilt, *Br. Vet. J.*, 141, 592, 1985.
163. **Lucan, J. J., Brunstad, G. E., and Fowler, S. H.**, The relationship of altered thyroid activity to various reproductive phenomena in gilts, *J. Endocrinol.*, 17, 54, 1958.
164. **Laws, B., Stedman, J. A., and Hill, R.**, Rapeseed meal in animal feeds, *Agritrade*, February 27, 1982.
165. **Vincent, I. C., Hill, R., Williams, H. L., and Noakes, D. E.**, The absence of effect on fertility of high levels of inclusion on British rapeseed meal in the diet of yearling heifers, *Anim. Prod.*, 40, 561, 1985.
166. **Ahlin, K. A., Emanuelson, M., Edquist, L. E., Larsson, K., and Wiktorsson, H.**, Rapeseed products as feeds for dairy cows. Preliminary results from a long term study, in *Advances in the Production and Utilisation of Cruciferous Crops*, H. Sørensen, Ed., Martinus Nijhoff/Junk, Dordrecht, Netherlands, 1985, 222.
167. **Vincent, I. C., Hill, R., and Williams, H. L.**, British rapeseed meal in the diets of pregnant and lactating mature Suffolk-Mule ewes, *Anim. Prod.*, 42, 453, 1986.
168. **Hansen, C. and Sanne, S.**, Protein trials at Smedsmora, *Farskotel*, 65, 7, 1985.
169. **Griffiths, S. and Evans, P. R.**, Breeding ewes — effect of feeding on rape during the mating season on lambing performance, *Anim. Prod.*, 40, 565, 1985.
170. **Chesney, A.M., Clawson, T. A., and Webster, B.**, Endemic goitre in rabbits. I. Incidence and characteristics, *Bull. Johns Hopkins Hosp.*, 43, 261, 1928.
171. **Webster, B. and Chesney, A. M.**, Endemic goitre in rabbits. II. Effects of administration of iodine, *Bull. Johns Hopkins Hosp.*, 43, 291, 1928.
172. **Gmelin, R. and Virtanen, A. I.**, Glucobrassicin, the precursor of SCN^-, 3-indolyl acetonitrile and ascorbigen in *Brassica oleracea* species, *Ann. Acad. Sci. Fenn. Ser. AII*, 107, 3, 1961.
173. **Astwood, E. B.**, The chemical nature of compounds which inhibit function of the thyroid gland, *Pharmacol. Exp. Ther.*, 78, 79, 1943.
174. **Hercus, C. E. and Purves, H. B.**, Studies on endemic goitre, *J. Hyg.*, 36, 182, 1936.
175. **Astwood, E. B., Greer, M. A., and Ettlinger, M. G.**, L-5-Vinyl-2-thiooxazolidine, an antithyroid compound derived from yellow turnip and from *Brassica* seeds, *J. Biol. Chem.*, 181, 121, 1949.
176. **Langer, P. and Greer, M. A.**, *Antithyroid Substances and Naturally Occurring Goitrogens*, S. Karger, Basel, 1977.
177. **Smithard, R. R. and Eyre, M. D.**, The effects of dry extrusion of rapeseed with other feedstuffs upon its nutritional value and antithyroid activity, *J. Sci. Food Agric.*, 37, 136, 1986.
178. **Mukherjee, K. D., Mangold, H. K., and El Nokrashy, A. S.**, Nutritional evaluation of low glucosinolate rapeseed meals obtained by various processes, *Nutr. Metab.*, 23, 1, 1979.
179. **Bell, J. M., Benjamin, B. R., and Giovannetti, P. M.**, Histopathology of thyroids and livers of rats and mice fed diets containing *Brassica* glucosinolates, *Can. J. Anim. Sci.*, 52, 395, 1972.
180. **Wight, P. A. L. and Shannon, D. W. F.**, The morphology of the thyroid glands of quails and fowls maintained on diets containing rapeseed, *Avian Pathol.*, 14, 383, 1985.
181. **Laarveld, B., Brockman, R. P., and Christensen, D. A.**, The goitrogenic potential of Tower and Midas rapeseed meals in dairy cows determined by thyrotrophin releasing hormone test, *Can. J. Anim. Sci.*, 61, 141, 1981.
182. **Dabrowski, K., Evans, R., Czarnocki, J., and Kozlowska, H.**, Rapeseed meal in the diet of common carp reared in treated waters. IV. Iodine (^{121}I) accumulation and thyroid histology, *Z. Tierphysiol. Tierernaehr. Futtermittelkd.*, 48, 1, 1982.
183. **Marks, H. L.**, Growth responses of selected quail lines to goitrogenic stress, *Poult. Sci.*, 53, 1762, 1974.
184. **Langer, J. P. and Michajlovskij, N.**, Studies on the antithyroid activity of naturally occurring l-5-vinyl-2-thiooxazolidine and its urinary metabolite in rats, *Acta Endocrinol. (Copenhagen)*, 62, 21, 1969.
185. **McDannell, R., McLean, A. E. M., Hanley, A. B., Heaney, R. K., and Fenwick, G. R.**, Chemical and biological properties of indole glucosinolates (glucobrassicins) — a review, *Food Chem. Toxicol.*, 26(1), 59, 1988.

186. **Laarveld, B., Brockman, R. P., and Christensen, D. A.**, The effects of Tower and Midas rapeseed meals on milk production and concentrations of goitrogens and iodine in milk, *Can. J. Anim. Sci.*, 61, 131, 1981.

187. **Schone, F. and Padtzelt, H.**, Excretion of thiocyanate ion in urine in growing pigs after rapeseed meal feeding, *Nahrung*, 29, 541, 1985.

188. **Jahreis, G., Hesse, V., Schöne, F., Lüdke, W., Hennig, A., and Mehnert, E.**, Einfluss von Nitrat und pflanzlichen Goitrogenen auf die Schilddrüsenhormone, den Somatomedinstatus und das Wachtum beim Schwein, *Monatsh. Veterinaermed.*, 41, 528, 1986.

189. **Paik, I. K., Robblee, A. R., and Clandinin, D. R.**, The effect of sodium thiosulphate and hydroxy-cobalamin on rats fed nitrile-rich or goitrin-rich meals, *Can. J. Anim. Sci.*, 60, 1003, 1980.

190. **Paxman, P. J. and Hill, R.**, The goitrogenicity of kale and its relation to thiocyanate content, *J. Sci. Food Agric.*, 25, 329, 1974.

191. **March, B. E., Biely, J., and Soong, R.**, Rapeseed meal in chicken breeder diet. Effects on production, mortality, hatchability and progeny, *Poult. Sci.*, 51, 1589, 1972.

192. **Goh, Y. K. and Clandinin, D. R.**, Transfer of ^{125}I to eggs in hens fed on diets containing high and low-glucosinolate rapeseed meals, *Br. Poult. Sci.*, 18, 705, 1977.

193. **Papas, A., Campbell, L. D., Cansfield, P. E., and Ingalls, J. R.**, The effect of glucosinolates on egg iodine and thyroid status of poultry, *Can. J. Anim. Sci.*, 59, 119, 1979.

194. **Iwarsson, K. L. and Nilsson, P. O.**, Rapeseed meal as a protein supplement for dairy cows. II. Investigations in rats on the goitrogenic properties of milk from cows fed rapeseed meal, *Acta Vet. Scand.*, 14, 595, 1973.

195. **Fritz, A., Kinal, S., and Fuchs, B.**, Complete feed with large proportion of rapeseed oilmeal from Start 00 and Quinta cultivars for fattening pigs, *Biul. Inf. Przem. Paszowego*, 22, 15, 1983.

196. **Bourdon, D., Perez, J.-M., and Baudet, J. J.**, Utilization of new types of rapeseed meals in growing-finishing pigs. Influence of glucosinolates and dehulling process, *J. Rech. Porc. France*, 12, 245, 1980.

197. **Witowski, A., Radomski, L., Pytel, S., and Pilarski, W.**, An analysis of the weight of certain internal organs of calves receiving a diet containing solvent-extracted rapeseed oilmeal and an addition of exogenous thyroxine, *Pr. Mater. Zootechh.*, 25, 61, 1981.

198. **Barry, T. N., Reid, T. C., Millar, K. R., and Sadler, W. A.**, Nutritional evaluation of kale (*Brassica oleracea*) diets. II. Copper deficiency, thyroid function and selenium status in young cattle and sheep fed kale for prolonged periods, *J. Agric. Sci.*, 96, 269, 1981.

199. **Blafeld, S. J.**, An experiment on grazing lambs on marrow stem kale and rape, *J. Agric. Res. Icel.*, 8, 66, 1976.

200. **Barry, T. N., Duncan, S. J., Sadler, W. A., Millar, K. R., and Sheppard, A. D.**, Iodine metabolism and thyroid hormone relationships in growing sheep fed on kale (*Brassica oleracea*) and ryegrass (*Lolium perenne*) — clover (*Trifolium repens*) fresh forage diets, *Br. J. Nutr.*, 49, 241, 1983.

201. **Bougon, M. and Guyen, N.**, Effects of rapeseed oilmeal on the performance of chickens, *Bull. Inf. Stn. Exp. Avic. Ploufragan*, 25, 170, 1985.

202. **Elwinger, K.**, Continued experiments with rapeseed meal of a Swedish low glucosinolate type fed to poultry. II. An experiment with laying hens, *Swed. J. Agric. Res.*, 16, 35, 1986.

203. **Pearson, A. W., Greenwood, N. M., Butler, E. J., and Fenwick, G. R.**, Biochemical changes in layer and broiler chickens when fed on a high glucosinolate rapeseed meal, *Br. Poult. Sci.*, 24, 417, 1983.

204. **Miller, K. W. and Stoewsand, G. S.**, Comparison of the effects of Brussels sprouts, glucosinolates and glucosinolate metabolite consumption on rat hepatic polysubstrate monoxygenases, *Dev. Toxicol. Environ. Sci.*, 11, 341, 1983.

205. **Bille, N., Eggum, B. O., Jacobsen, I., Olsen, O., and Sørensen, H.**, The effects of processing on antinutritional constituents and nutritive value of double low rapeseed meal, *Z. Tierphysiol. Tierernaehr. Futtermittelkd.*, 49, 148, 1983.

206. **Gould, D. H., Gumbmann, M. R., and Daxenbichler, M. E.**, Pathological changes in rats fed the crambe meal glucosinolate hydrolytic products 2S-1-cyano-2-hydroxy-3,4-epithiobutanes (*erythro* and *threo*) for 90 days, *Food Cosmet. Toxicol.*, 18, 619, 1980.

207. **Nishie, K. and Daxenbichler, M. E.**, Toxicology of glucosinolates, related compounds (nitriles, R-goitrin, isothiocyanates) and vitamin U found in cruciferae, *Food Cosmet. Toxicol.*, 18, 159, 1980.

208. **Brocker, E. R., Benn, M. H., Lüthy, J., and von Däniken, A.**, Metabolism and distribution of 3,4-epithiobutanenitrile in the rat, *Food Cosmet. Toxicol.*, 22, 227, 1984.

209. **Ibrahim, I. K., Hodges, R. J. C., and Hill, R.**, Haemorrhagic liver syndrome in laying fowls fed diets containing rapeseed meal, *Res. Vet. Sci.*, 29, 68, 1980.

210. **Campbell, L. D.**, Incidence of liver haemorrhage amongst white leghorn strains fed on diets containing different types of rapeseed meals, *Br. Poult. Sci.*, 20, 239, 1979.

211. **Yamashiro, S., Bhatnagar, M. K., Scott, J. R., and Slinger, S. J.**, Fatty haemorrhagic syndrome in laying hens on diets supplemented with rapeseed products, *Res. Vet. Sci.*, 19, 312, 1975.

212. **Bromidge, E. S., Wells, J. W., and Wight, P. A. L.,** Elevated bile acids in the plasma of laying birds fed rapeseed meal, *Res.Vet. Sci.,* 39, 378, 1985.

213. **Martland, M. F., Butler, E. J., and Fenwick, G. R.,** Rapeseed induced liver haemorrhage, reticulolysis and biochemical changes in laying hens, the effects of feeding high and low glucosinolate meals, *Res. Vet. Sci.,* 36, 298, 1984.

214. **Campbell, L. D. and Smith, T. K.,** Response of growing chickens to high dietary content of rapeseed meal, *Br. Poult. Sci.,* 20, 231, 1979.

215. **Ibrahim, I. K. and Hill, R.,** The effects of rapeseed meals from *Brassica napus* varieties and the variety Tower on the production and health of laying fowl, *Br. Poult. Sci.,* 21, 423, 1980.

216. **Papas, A., Campbell, L. D., and Cansfield, P. E.,** A study of the association of glucosinolates to rapeseed meal-induced haemorrhagic liver in poultry and the influence of supplemental vitamin K, *Can. J. Anim. Sci.,* 59, 133, 1979.

217. **Wight, P. A. L., Shannon, D. W. F., Wells, J. W., and Mawson, R.,** The role of glucosinolates in the causation of liver haemorrhage in laying hens fed water-extracted or heat-treated rapeseed cakes, *Res. Vet. Sci.,* in press.

218. **Timms, L. M.,** Forms of leg abnormality observed in male broilers fed on a diet containing 12.5 per cent rapeseed meal, *Res. Vet. Sci.,* 35, 182, 1983.

219. **Moody, D. L., Slinger, S. J., Leeson, S., and Summers, J. D.,** Utilisation of dietary Tower rapeseed products by growing turkeys, *Can. J. Anim. Sci.,* 58, 585, 1978.

220. **Holmes, W. B. and Roberts, R.,** A perotic syndrome in chicks fed extracted rapeseed meal, *Poult. Sci.,* 42, 803, 1963.

221. **Mawson, R.,** Unpublished observation.

222. **Timms, L. M.,** Influence of a 12.5 per cent rapeseed diet and an avian reovirus on the production of leg abnormalities in male broiler chickens, *Res. Vet. Sci.,* 38, 69, 1985.

223. **MacMillan, M., Spinks, E. A., and Fenwick, G. R.,** Preliminary observations on the effect of dietary Brussels sprouts on thyroid function, *Hum. Toxicol.,* 5, 15, 1986.

224. **Dahlberg, P. A., Bergmark, A., Björck, L., Bruce, A., Hambraeus, L., and Claesson, O.,** Intake of thiocyanate by way of milk and its possible effect on thyroid function, *Am. J. Clin. Nutr.,* 39, 416, 1984.

225. **Bell, J. M., Anderson, D. M., and Shires, A.,** Evaluation of Candle rapeseed meal as a protein supplement for swine, *Can. J. Anim. Sci.,* 61, 453, 1981.

226. **Narendran, R., Bowman, G. H., Leeson, S., and Pfeiffer, W.,** Effect of different levels of Tower rapeseed meal in corn-soybean meal based diets on growing finishing pig performance, *Can. J. Anim. Sci.,* 61, 213, 1981.

227. **Hill, R.,** The toxic substances of rapeseed meal and their effect on animals, *J. Sci. Food Agric.,* 29, 413, 1978.

228. **Lindell, L. and Knutsson, P. G.,** Rapeseed meal in rations for dairy cows. I. Comparison of three levels of rapeseed meal, *Swed. J. Agric. Res.,* 6, 55, 1976.

229. **Sharma, H. R., Ingalls, J. R., and McKirdy, J. A.,** Effect of feeding a high level of Tower rapeseed meal in dairy rations on feed intake and milk production, *Can. J. Anim. Sci.,* 57, 653, 1977.

230. **Moss, B. R.,** Mustard meal in dairy rations, *J. Dairy Sci.,* 58, 1682, 1975.

231. **Sharma, H. R., Ingalls, J. R., and McKirdy, J. A.,** Nutritive value of formaldehyde-treated rapeseed meal for dairy calves, *Can. J. Anim. Sci.,* 52, 363, 1972.

232. **Sharma, H. R. and Ingalls, J. R.,** Comparative value of soybean-rapeseed- and formaldehyde-treated rapeseed meals in urea containing calf rations, *Can. J. Anim. Sci.,* 53, 273, 1973.

233. **Claypool, D. W., Hoffman, C. H., Oldfield, J. E., and Adams, H. P.,** Canola meal, cottonseed and soybean meals as protein supplements for calves, *J. Dairy Sci.,* 68, 67, 1985.

234. **Gorrill, A. D. L., Jones, J. D., and Nicholson, J. W. G.,** Low and high glucosinolate rapeseed flours and rapeseed oil in milk replacers for calves; their effect on growth, nutrient digestion and nitrogen retention, *Can. J. Anim. Sci.,* 56, 409, 1976.

235. **Gorrill, A. D. L., Seoane, J. R., Jones, J. D., and Nicholson, J. W. G.,** Nutrient digestion and nitrogen retention by lambs fed milk replacers containing solvent extracted or full-fat products from different rapeseed cultivars, *Can. J. Anim. Sci.,* 56, 401, 1976.

236. **Young, N. E., Austin, A. R., Orr, R. J., Newton, J. E., and Taylor, R. J.,** A comparison of a hybrid stubble turnip (*cv.* Appin) with other cruciferous catch crops for lamb fattening. II. Animal performance and toxicological evaluation, *Grass Forage Sci.,* 37, 39, 1982.

237. **Srisangnam, C., Salunke, D. K., Reddy, N. R., and Dull, G. G.,** Quality of cabbage. II. Physical, chemical and biochemical modification in processing treatments to improve flavour of blanched cabbage (*Brassica oleracea* L.), *J. Food Qual.,* 3, 233, 1980.

238. **Fenwick, G. R., Griffiths, N. M., and Heaney, R. K.,** Bitterness in Brussels sprouts (*Brassica oleracea* L. var. *gemmifera*). The role of glucosinolates and their breakdown products, *J. Sci. Food Agric.,* 34, 73, 1983.

239. **Uda, Y. and Maeda, Y.,** Volatile constituents occurring in autolysed leaves of three cruciferous vegetables, *Agric. Biol. Chem.,* 50, 205, 1986.
240. **Kjaer, A., Øgaard Madsen, J., Maeda, Y., Ozawa, Y., and Uda, Y.,** Volatiles in distillates of processed radish of Japanese origin, *Agric. Biol. Chem.,* 42, 1989, 1978.
241. **Hardman, J. A. and Ellis, P. R.,** Host plant factors influencing the susceptibility of cruciferous crops to cabbage root fly attack, *Entomol. Exp. Appl.,* 24, 193, 1978.
242. **Åhman, I.,** Toxicities of host secondary compounds to eggs of the *Brassica* specialist *Dasineura brassicae, J. Chem. Ecol.,* 12, 1481, 1986.
243. **Mithen, R. F., Lewis, B. G., and Fenwick, G. R.,** The *in vitro* activity of glucosinolates and their products against *Leptosphaeria maculans, Trans. Br. Mycol. Soc.,* 87, 433, 1986.
244. **Mithen, R. F., Lewis, B. G., Heaney, R. K., and Fenwick, G. R.,** Resistance of leaves of *Brassica* species to *Leptosphaeria maculans, Trans. Br. Mycol. Soc.,* 88, 525, 1987.
245. **Hanley, A. B. and Fenwick, G. R.,** The genus Allium. III, *CRC Crit. Rev. Food Sci. Nutr.,* 23, 1, 1986.
246. **Butcher, D. N., Chamberlain, K., Rausch, T., and Searle, L. M.,** Changes in indole metabolism during the development of clubroot symptoms in brassicas, British Plant Growth Regulator Group, Monogr. 11, 1984, 91.
247. **Chong, C., Chiang, S. M., and Crete, R.,** Studies on glucosinolates in clubroot resistant selections and susceptible commercial cultivars of cabbages, *Euphytica,* 34, 65, 1985.
248. **Walker, N. J. and Gray, I. K.,** The glucosinolates of land cress (*Coronopus didymus*) and its enzymatic degradation products as precursors of off-flavour in milk — a review, *J. Agric. Food Chem.,* 18, 346, 1970.
249. **Frediksen, H. J., Andersen, P. E., Mortensen, B. K., and Jensen, F.,** *Statens Husdyrbrugstors. Copenhagen Medd.,* 280, 1979.
250. **Butler, E. J. and Fenwick, G. R.,** Trimethylamine and fishy taint in eggs, *World's Poult. Sci. J.,* 40, 38, 1984.
251. **Fenwick, G. R., Butler, E. J., and Brewster, M. A.,** Are brassica vegetables aggravating factors in trimethylaminuria (fish odour syndrome)?, *Lancet,* p. 916, 1983.
252. **Goh, Y. K., Shires, A., Robblee, A. R., and Clandinin, D. R.,** Effect of ammoniation of rapeseed meal on the sinapine content of the meal, *Br. Poult. Sci.,* 23, 121, 1982.
253. **Goh, Y. K., Robblee, A. R., and Clandinin, D. R.,** Effect of ammoniation on the fishy odour and trimethylamine contents of eggs produced by brown egg layers, *Poult. Sci.,* 63, 706, 1984.
254. **Goh, Y. K., Robblee, A. R., and Clandinin, D. R.,** Influence of glucosinolates and free oxazolidinethiones in a laying diet containing a constant amount of sinapine on the trimethylamine content and fish odour from brown shelled egg layers, *Can. J. Anim. Sci.,* 63, 671, 1983.
255. **Goh, Y. K., Robblee, A. R., and Clandinin, D. R.,** Influence of glucosinolates and free oxazolidinethiones in a laying diet containing a constant amount of sinapine on the thyroid size and hepatic trimethylamine oxidase activity of brown egg layers, *Can. J. Anim. Sci.,* 65, 921, 1985.
256. **Dransfield, E., Nute, G. R., Mottram, D. S., Rowan, T. G., and Lawrence, T. L. J.,** Pork quality from pigs fed on low glucosinolate rapeseed meal: influence of level in the diet, sex and ultimate pH, *J. Sci. Food Agric.,* 36, 546, 1985.
257. **Salmon, R. E., Gardiner, E. E., Klein, K. K., and Larmond, E.,** Effect of canola (low glucosinolate rapeseed) meal, protein and nutrient density on performance, carcass grade and meat yield, and of canola meal on sensory quality of broilers, *Poult. Sci.,* 60, 2519, 1980.
258. **Larmond, E., Salmon, R. E., and Klein, K. K.,** Effects of canola meal on the sensory quality of turkey meat, *Poult. Sci.,* 62, 397, 1983.
259. **Wattenberg, L. W.,** Inhibition of carcinogenic effects of polycyclic hydrocarbons by benzyl isothiocyanate and related compounds, *J. Natl. Cancer Inst.,* 58, 395, 1977.
260. **Pantuck, E., Pantuck, C. B., Garland, W. A., Mins, B., Wattenberg, L. W., Anderson, K. E., Kappas, A., and Conney, A. H.,** Stimulatory effect of Brussels sprouts and cabbage on human drug metabolism, *Clin. Pharmacol. Ther.,* 25, 88, 1979.
261. **Wattenberg, L. W., Hanley, A. B., Barany, G., Sparnins, V. L., Lam, L. K. T., and Fenwick, G. R.,** Inhibition of carcinogenesis by some minor dietary constituents, in *Diet, Nutrition and Cancer,* Hayashi, Y., Ed., Japanese Scientific Societies Press, Tokyo/VNU Science Press, Utrecht, 1986, 193.
262. **Wakabayashi, K., Nagao, M., Tahira, T., Saito, H., Katayama, M., Marumo, S., and Sugimura, T.,** 1-Nitrosoindole-3-acetonitrile, a mutagen produced by nitrite treatment of indole-3-acetonitrile, *Proc. J. Jpn.,* 61B, 190, 1985.
263. **Wakabayashi, K., Nagao, M., Ochiai, M., Tahira, T., Zamaizumi, Z., and Sugimura, T.,** A mutagen precursor in Chinese cabbage, indole-3-acetonitrile, which becomes mutagenic on nitrite treatment, *Mutat. Res.,* 143, 17, 1985.
264. **Wakabayashi, K., Nagao, M., Tahira, T., Yamaizumi, Z., Katayama, M., Marumo, S., and Sugimura, T.,** 4-Methoxyindole derivatives as nitrosatable precursors of mutagens in Chinese cabbage, *Mutagenesis,* 1, 423, 1986.

265. **Lüthy, J., Carden, B., Friedrich, U., and Bachmann, M.,** Goitrin — a nitrosatable constituent of plant foodstuffs, *Experientia,* 40, 452, 1984.

266. **Lüthy, J.,** Identifizierung und Mutagenität des Reaktionsproduktes von Goitrin und Nitrit, *Mitt. Geb. Lebensmittelunters. Hyg.,* 75, 101, 1984.

267. **Kammüller, M. E. and Seinen, W.,** On the possible etiological role of isothiocyanate-derived heterocyclic compounds in the pathogenesis of Sapinsh Toxic Oil Syndrome and other allergic and autoimmune-like diseases — hypothesis, in *Proc. Eurofoodtox II,* Schlatter, C., ed., Swiss Federal Institute of Technology and Univrsity of Zurich, Zurich, 1986, 300.

Chapter 2

CYANOGENIC GLYCOSIDES

Olumide O. Tewe and Eustace A. Iyayi

TABLE OF CONTENTS

I. INTRODUCTION

Cyanogenic glycosides are important natural toxicants in both human and animal nutrition. Chronic toxicological effects occur in humans consuming cassava in tropical countries, while the cyanide production potential is of concern in several other food crops. Livestock poisonings are associated with consumption of forage sorghums and cyanogen-containing plants such as wild chokecherry.

II. OCCURRENCE

Cyanogenic glycosides are compounds which on treatment with acid or appropriate hydrolytic enzymes produce hydrocyanic acid (HCN). These glycosides usually contain glucose as their sugar component, although they may contain instead a disaccharide. Many organisms possess the ability to evolve cyanide either during their growth or on traumatization of their tissues. The occurrence of these cyanogenic glycosides is quite widespread, ranging from the simple bacteria and fungi to the much more specialized higher plants as well as some classes of animals. In bacteria, they are largely restricted to the Pseudomonadaceae family, while in fungi, occurrence is in the genera *Clitocybe, Marasmius, Tricholoma*, and *Mucor*. Two classes of animals, Myriapoda and Insecta, are known to be cyanogenic. Cyanogenic glycoside are found in over 1000 species of higher plants.[1] The cyanophoric bacteria and fungi produce their cyanide from labile compounds which are glucosidic but of which isolation and characterization has been restricted. However, in Myriapoda, Insecta, and higher plants, the cyanogens have been isolated and characterized as stable glucosidic compounds of α-hydroxynitriles or cyanohydrins. Cyanolipids are another group of cyanide precursors which have been isolated from the lipid fraction of seeds of *Sapindaceae*. Apart from HCN, the cyanolipids also yield long-chain fatty acid moieties and an isoprenoid hydroxy or dihydroxy residue on hydrolysis.[2,3]

Aspects of cyanide production by fungi and bacteria have been summarized by Vennesland et al.[4] Microbial cyanide metabolism has also been reviewed by Knowles and Bunch.[5]

A. Cyanogenesis in Fungi and Algae

Cyanogens have been reported in about 30 fungi species from five families: Agaricaceae, Cortinariaceae, Polyporaceae, Rhodophyllaceae, and Tricholomataceae.[6] In fungal cyanogenesis, the direct precursors of hydrogen cyanide are unstable. Attempts to elucidate the structure of the cyanogenic compounds prepared from psychrophilic basidiomycetes have not been very successful. To date, the only cyanogens isolated from fungi (Figure 1) are the cyanohydrins of pyruvic acid and glyoxylic acid which were isolated from unidentified basidiomycete by Tapper and Macdonald.[7]

Glycine has been implicated in the formation of cyanide in fungi. Production proceeds via glyoxylic acid cyanohydrin, a compound which is unstable and has cyanogenic properties. Small amounts of glucosidic cyanogens have been detected[8] in psychrophilic basidiomycetes when valine and isoleucine were fed. A stable acetylenic nitrite has been isolated and characterized in *Clitocybe diatreta* without any success at elucidation of a possible link between this compound and the hydrogen cyanide of this fungus.

Chlorella vulgaris is the major alga known to exhibit cyanogenesis. This alga produces low levels of hydrogen cyanide from a variety of amino acids. The addition of D-histidine leads to the formation of higher levels of cyanide. To a large extent the nature of the precursor cyanogens formed by these organisms is unknown, and attempts to isolate and characterize them have been rather unsuccessful. The production of HCN by algae has been reviewed by Vennesland et al.[4]

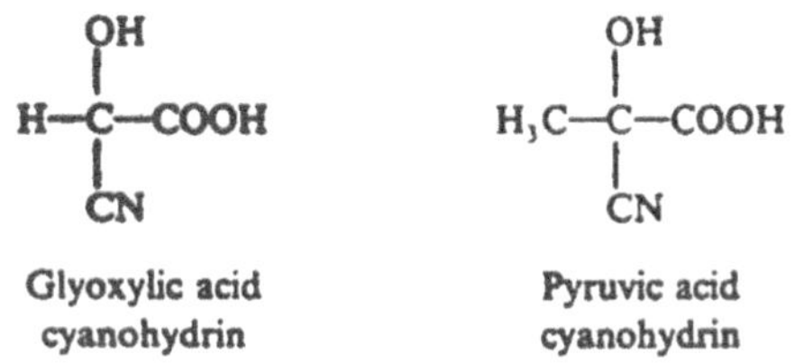

$$\begin{array}{ccc}
& \text{OH} & \\
& | & \\
\text{H--C--COOH} & & \text{H}_3\text{C--C--COOH} \\
& | & \\
& \text{CN} & \text{CN}
\end{array}$$

Glyoxylic acid   Pyruvic acid
cyanohydrin   cyanohydrin

FIGURE 1.   Cyanohydrin of basidiomycetes.

B. Cyanogenesis in Bacteria

Cyanogenesis in bacteria was first observed by Emerson et al.[9] The taxonomic distribution of cyanogenic bacteria as recorded by Clawson and Young[10] includes *Bacillus fluorescens* (*Pseudomonas fluorescens*), *B. pyocyaneus* (*P. aeruginosa*), *B. violaceus* (*Chromobacterium violaceum*), and a bacterial strain resembling *P. aurofaciens*. Thus, the phenomenon is a common feature with the pseudomonads and chromobacteria.

The culture conditions of the pseudomonads determine the optimal production of cyanide. The metabolic precursor of cyanide in bacteria as in fungi is glycine. This was first reported with *P. aeruginosa*[11] and later with *P. violaceum*.[12] When *P. violaceum* was grown on a glutamate-salts medium containing L-threonine, Collins et al.[13] noted a stimulation of HCN production. According to Castric,[14] this organism presumably possesses an enzyme capable of converting L-threonine to glycine. This was demonstrated radioactively by the same author in *P. aeruginosa*. L-phenylalanine stimulates cyanide production in *P. aeruginosa*, although the conversion of the amino acid to HCN is only slight when compared to threonine and valine.

In *C. violaceum*, the source of the carbon atom in cyanide is the methylene group of glycine, while the origin of the cyanide nitrogen is the amino nitrogen of glycine. The utilization of the carboxyl group as an HCN precursor in *P. aeruginosa* according to Castric[14] points to (1) a fundamental difference in the mode of conversion of glycine to cyanide by this organism as compared to *C. violaceum* or (2) intermediary metabolism of glycine in *P. aeruginosa* which results in a partial randomization of label in glycine prior to the conversion of glycine to cyanide. The latter possibility might occur if glycine was metabolized to a purine which then might be degraded to glyoxylate. Transamination of glyoxylate would yield glycine. The addition of glycine and methionine in culture media of bacteria produces a synergistic action which enhances cyanide production. Compounds which are methyl donors in biosynthetic pathways, like betaine, choline, and *N,N*-dimethyl glycine, can replace methionine.[15] Proposed pathways for the biosynthesis of HCN by bacteria have been reviewed by Castric.[14]

C. Occurrence and Classification of Cyanogens in Plants

The majority of the direct precursors of hydrogen cyanide in cyanogenic plants are glucosidic, and they occur in virtually every part of the plant. However, despite the widespread distribution of these cyanogenic glucosides in the plant kingdom only about 55 such compounds including the pseudocyanogenic glycosides have been isolated and characterized as direct precursors of cyanide in plant tissues. While the occurrence of cyanogens in plants is a function of genotype, phenotypic variations such as geographical distribution and location, climatic conditions, soil type, and other ecological factors also contribute to their occurrence (see Table 1). Cassava (*Manihot esculenta*) is the major food crop containing cyanogens. The two major cyanogens so far isolated in cassava are linamarin and its homologue lotaustralin. The former is identical with phaseolunatin which occurs in *Phaseolus lunatus* and *Linum usitatissimum*. Moreover, an important biochemical feature of the cassava plant is that it synthesizes and accumulates the cyanogenic compounds throughout its entire life cycle.

Table 1
DISTRIBUTION OF SOME CYANOGENIC GLYCOSIDES IN PLANTS

Glycoside	Plant (genus, species)	Common name
Amygdalin	*Prunus* spp.	Almond, cherry, peach, plum
	Malus spp.	Apple
Dhurrin	*Sorghum* spp.	Forage sorghums
Linamarin	*Linum usitatissimum*	Flax (linseed)
	Trifolium repens	White clover
	Lotus spp.	Trefoils
	Manihot esculenta	Cassava
	Phaseolus lunatus	Lima bean
Lotaustralin	*Linum usitatissimum*	Flax
	Lotus spp.	Trefoils
	Phaseolus lunatus	Lima bean
	Trifolium repens	White clover
Prunasin	*Acaia* spp.	Acacias
	Eucalyptus spp.	Eucalyptus
	Prunus spp.	Cherry, almond
	Pteridium aqualinum	Bracken
Triglochinin	*Triglochin* spp.	Arrowgrass
Vicianin	*Vicia* spp.	Vetches

The extensive occurrence of cyanogens in the plant kingdom has some taxonomic significance in the resolution of problems of classification within families, genera, and on the species level.[16] However, there is no definite pattern of distribution of cyanogens. This is largely a result of the fact that these glycosides occur in over 50 species of plants. Linamarin and lotaustralin occur together in a wide variety of higher plant species belonging to the families of Mimosaceae, Fabaceae, Linaceae, Euphorbiaceae, and Compositae. There is a common occurrence of two of the cyanogens, amygdalin and prunasin, in the Rosaceae, while sambunigrin and vicianin occur together in *Sambucus nigra L.* of the Caprifoliaceae family. Both linamarin and lotaustralin occur in *Linum* species, while in *Trifoluim repens,* lotaustralin appears to be in higher concentration than linamarin. In common crop plant species like sorghum and kaffir corns, dhurrin is the main cyanogen, while triglochinin is found in arrow grass (*Triglochinin maritimum L.*). The occurrence of cyanogenic compounds in higher plants was reviewed comprehensively by Poulton.[17]

D. Cyanogenesis in Animals

Unlike in the plant kingdom where cyanogenic glycosides are taxonomically widespread, cyanogens in animals appear to be restricted to the phyllum Arthropoda; the subject has been reviewed by Duffey.[18] The classes of this phyllum where cyanogenesis occurs include the Chilopoda (centipedes), Diplopoda (millipedes), and Insecta (insects). However, many animal species including the protistans, coelenterates, annelids, echinoderms, molluscans as well as some vertebrates can produce or sequester other products such as batrachotoxin, cardenolides, cobrataxin, murexine, tetrodotoxin, and benzoquinones.[18] These products are usually more toxic than HCN. Through a variety of physical and/or physiological devices, these organisms can tolerate their own toxins.

Studies of Cimino et al.[19] have shown that sponges in the family Vergonidae metabolize tyrosine to the nitrile aerophysinin-1 and the cyclic oxime homoaerothionin. Evidence of an oxime being an intermediate in the biosynthesis of the nitrile aerophysinin-1 stems from the fact that other derivatives of tyrosine, 4-hydroxyphenylpyruvic acid, oxime, and 3,5-dibromotyrosine, occur in these sponges. The occurrence of formamide, isonitrile, or isothiocyanate groups in sponges has led to the postulation[20] that the formamide is a precursor in the biogenesis of the isonitrile. Though oximes and nitriles have thus been detected in sponges, the present state of knowledge does not show instances of cyanogenesis in animals other than arthropods.

In approximately 3000 species of centipedes, cyanogenesis has been observed in seven species: six geophilids and a scolopendrid. However, in the 7500 species of the millipedes, cyanogenesis is much more prevalent.[18] In Insecta, of about 750,000 species, only ten species are known to be cyanogenic, including three beetles, three butterflies, and four moths.[17] Cyanogens have been reported in several zygaenid moths, some species of chrysomelid beetle larva, as well as in geophilomorphid centipedes. The work of Duffey and Towers[21] has further shown the natural occurrence of benzoyl cyanide in various polydesmoids and centipedes.

In the moth *Z. filipendulae*, linamarin and lotaustralin make up to 2% of the wet weight of the freshly emerged gravid females (400 µg HCN per adult); the glucosides are highly compartmentalized in the eggs so they are absent in the adult males.[18] In the polydesmoid *Harpaphe haydeniana*, *R*-mandelonitrile has been identified as the major cyanogen.[21] About 0.2 mol (8 µg HCN) of mandelonitrile were contained in each millipede. The pronounced activity of α-hydroxynitrile lyase confirmed that most of the HCN formation in this millipede resulted from the dissociation of stored mandelonitrile from the paranota where the cyanogenic apparatus is housed. The mechanisms of cyanogenesis in arthropods have been reviewed in detail by Duffey.[18]

III. CHEMICAL STRUCTURES OF THE CYANOGENIC GLYCOSIDES

In the plant kingdom, about 55 stable compounds of cyanogenic glycosides have been isolated and characterized. Figure 2 shows the chemical structures of some of the more common cyanogenic glucosides. Structurally, all the cyanogens are related, with all of them except lucumin containing a glucose molecule as well as the cyanide ion; hence, they are called cyanogenic glycosides. Only proteacin, amygdalin, and vicianin contain two glucose molecules, and in proteacin, they are joined together by an aromatic protein amino acid. Furthermore, apart from linamarin, lotaustralin, acacipetalin, and triglochinin, the other cyanogens contain aromatic amino acid residue(s) in the side chain. With the exception of a few of the cyanogens like proteacin, vicianin, and gynocardin, the methyl group is a prominent biochemical component of the cyanogenic glycosides. For a more complete list of cyanogens, the review of Siegler[22] should be consulted.

IV. METABOLISM OF CYANOGENIC GLYCOSIDES

A. Biosynthesis

As in the lower plants such as bacteria, fungi, and ferns, the cyanogenic glycosides in plants originate mainly from amino acids. There appears to be a specificity between particular compounds and their precursor amino acids. The similarity between the aglycone moieties of these compounds and corresponding amino acids provides good evidence that these moieties are indeed derived from the amino acids. Butler and Conn[23] found that the biosynthesis of linamarin and lotaustralin proceeds through the conversion of valine and isoleucine, respectively. The use of labeled L-valine-^{14}C and L-isoleucine-^{14}C has shown that the labeled C atoms of the amino acids are incorporated in linamarin and lotaustralin. In bitter almond and cherry laurel, amygdalin and prunasin are synthesized from phenylalanine. Tyrosine is used in the formation of dhurrin and taxiphyllin which occur in sorghum and yew, respectively. Acacipetalin is formed from L-leucine via dihydroacacipetalin. As in bacterial and fungal cyanogenesis, higher plants utilize two aromatic protein amino acids

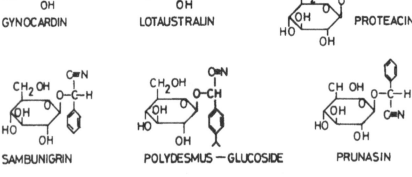

FIGURE 2.   Chemical structures of some cyanogenic glycosides.[15,22]

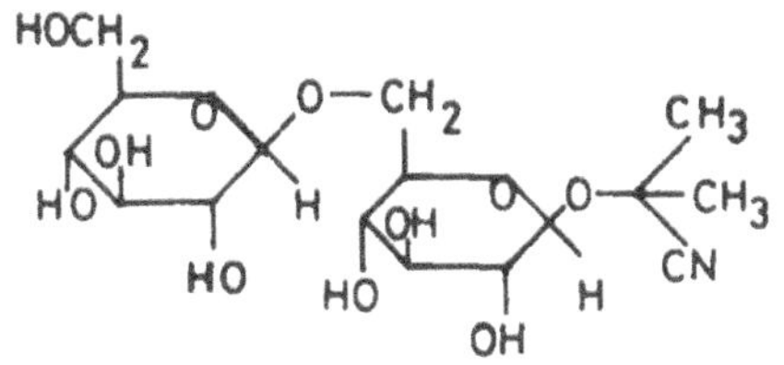

LINUSTATIN

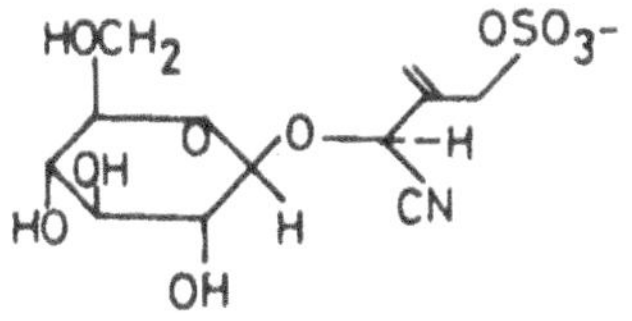

(S)— CARDIOSPERMIN
SULFATE

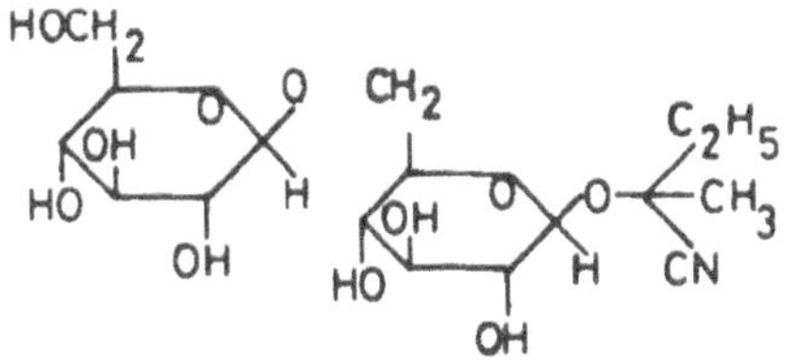

NEOLINUSTATIN

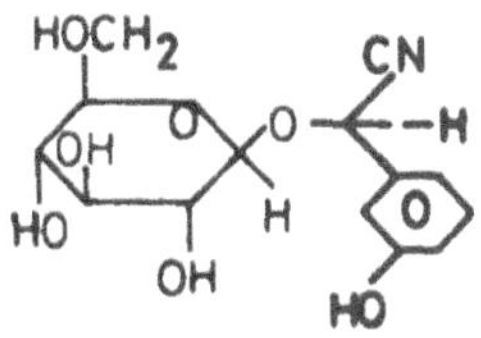

(R) HOLOCALIN

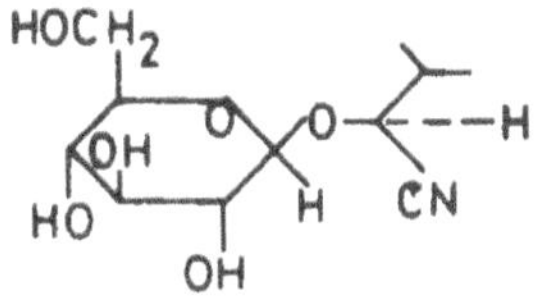

(S)—HETERODENDRIN

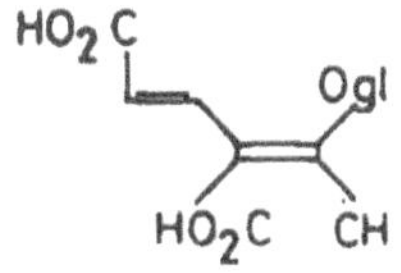

ISOTRIGLOCHININ

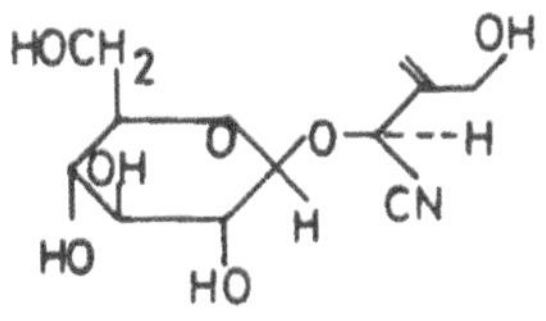

(S)— CARDIOSPERMIN

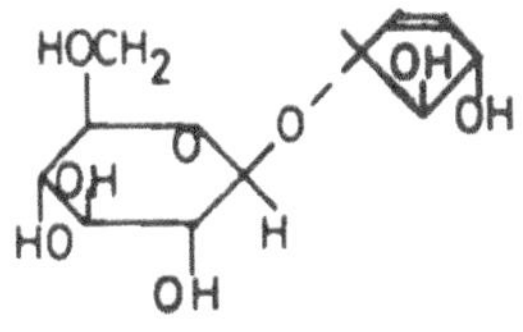

GYNOCARDIN

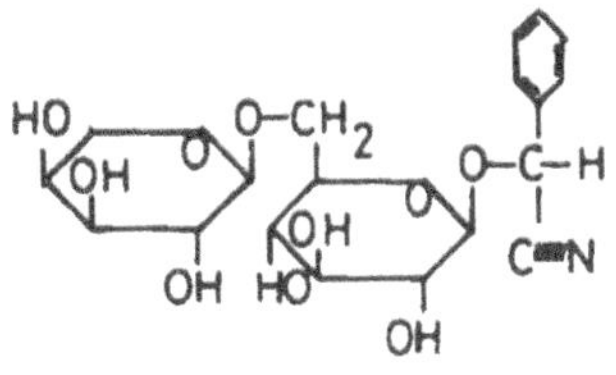

VICIANIN

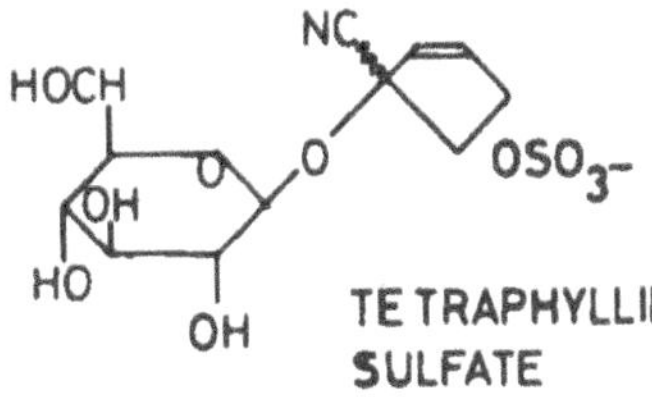

TETRAPHYLLIN β
SULFATE

FIGURE 2 (continued)

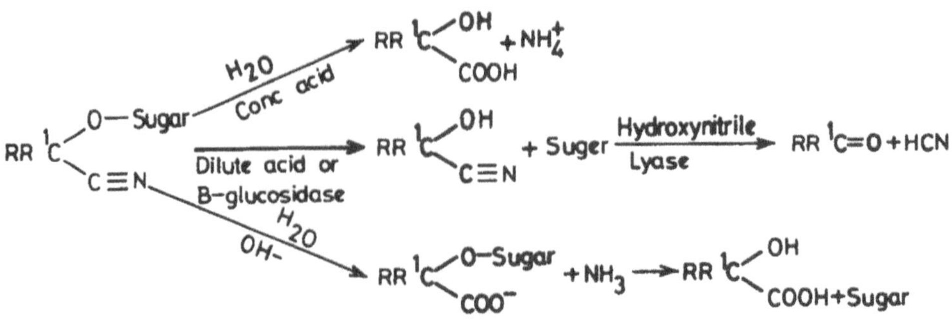

FIGURE 3.   Degradations of cyanogenic glycosides by acid, base, and enzyme catalysis. (Adapted from Nartey, F., *Abstracts on Cassava*, Vol. 4, Ser. 08EC-4, 1978. With permission.)

Table 2
HYDROLYTIC PRODUCTS OF SOME CYANOGENIC GLYCOSIDES

Glycoside	Hydrolytic products
Linamarin	D-Glucose, HCN, and acetone
Lotaustralin	D-Glucose, HCN, and 2-butanone
Dhurrin	D-Glucose, HCN, and hydroxybenzaldehyde
Amygdalin	Gentobiose, HCN, and benzaldehyde
Taxiphyllin	D-Glucose, HCN, and hydroxybenzaldehyde
Vicianin	Vicianose, HCN, and benzaldehyde
Triglochinin	D-Glucose, HCN, and triglochininc acid
Prunasin	D-Glucose, HCN, and benzaldehyde

and three aliphatic branched-chain protein amino acids (including leucine) for the synthesis of most of the cyanogenic glycosides. The hydrogen cyanide evolved is derived from the α-carbon and nitrogen atoms of these amino acids without prior cleavage of the carbon-nitrogen bond.[15]

B. Degradation

The degradation of cyanogenic glycosides takes place under enzymatic acid and base hydrolysis (Figure 3). In many of the cyanophoric plants where these glycosides occur, there are enzymes which are highly specific for the β-glycosidic linkage. Table 2 shows the products of hydrolysis of some of the cyanogenic glycosides. In most of the hydrolytic processes, HCN, a sugar molecule, and a ketone or an aldehyde are produced.

In most plants where linamarin and lotaustralin occur, there is the presence of the endogenous enzyme linamarase or linase which causes the hydrolysis of these glucosides with production of HCN. According to Nartey,[15] products of dissociation of the cyanogenic glycosides under concentrated acid and base conditions do not include HCN. Furthermore, in the dissociation of the cyanogenic glycosides by their specific β-glycosidases or dilute acid, the reaction proceeds via two steps. In the first, the cyanogenic glycoside is degraded to sugar and cyanohydrin (α-hydroxynitrile). The latter is further dissociated to the aldehyde or ketone and HCN. Thus, in the degradation of linamarin, for example, the glucoside is first hydrolyzed by linamarase to produce β-D-glucopyranose and 2-hydroxyisobutyronitrile or acetone cyanohydrin (Figure 4). The latter dissociates, catalyzed by a hydroxynitrile lyase, to produce acetone and HCN.

C. Cyanide and Cytochrome Inhibition

The lethality of cyanide has been ascribed to its ability to inhibit respiration. Cyanide

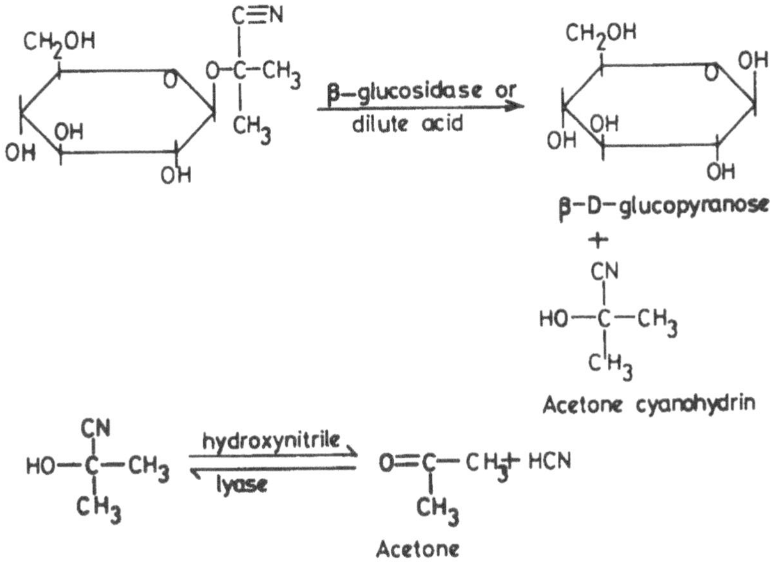

FIGURE 4.   Hydrolysis of linamarin.

toxicosis is due to the inhibition of the cytochrome oxidase of the respiratory chain. The cyanide binds with both the oxidized and reduced form of the enzyme, but the binding is tighter and slower with the oxidized form than the reduced form. Earlier opinion in the literature indicated that cyanide inhibition of cytochrome oxidase involved the copper ions. However, recent evidence from X-ray absorption-edge studies indicate that binding is with the iron edge. Although the liver cytochrome oxidase is not inhibited by cyanide, the brain cytochrome oxidase may be the site of lethal action.[24] Earlier *in vivo* studies by Olsen and Klein[25] support this view. The mechanisms of cyanide intoxication were reviewed by Way[26] and Poulton.[17]

D. Cyanide and Metallo-Enzymes

Nitrate reductase, nitrogenase, xanthine oxidase, and sulfite oxidase are among the mo-lybdenum-containing enzymes for which activities are inhibited by cyanide. The inhibition of nitrate reductase is of metabolic importance in photosynthetic organisms where such inhibition causes a metabolic control of nitrate assimilation.[27,28] Reactivation of the inhibited enzyme always results in the release of the bound cyanide from the complex. Inhibition usually results at the molybdenum centers of the enzymes. The inhibition of nitrate reductase as well as reversibility of such inhibition makes cyanide a natural regulator of the enzyme.

Other molybdo-enzymes usually differ in their sensitivity to cyanide inhibition. The re-versibility of cyanide inhibition of sulfite oxidase, xanthine dehydrogenase, aldehyde oxi-dase, and xanthine oxidase has been demonstrated by Coughlan,[29] who also showed that absorption spectra were consistent with the molybdenum centers as the binding site for cyanide. When such inhibition occurs, the molybdenum center is rendered incapable of participating in the transfer of electrons to the natural acceptors. The inactivation of the molybdenum enzymes xanthine oxidase, xanthine dehydrogenase, aldehyde oxidase, and some hydroxylases containing molybdenum is usually in the oxidized form. Thus, this inactivation, unlike that of the reduced enzymes, is irreversible. The sulfur released during the reactivation of the inhibited molybdo-enzymes originates with a persulfide (RSS) group which is associated with the molybdenum center.[24] The possibility of an active center cysteine residue as the source of the cyanolyzable sulfur has also been indicated. In such reactivation reactions a major product is thiocyanate. Thus, certain molybdo-enzymes can be inactivated in the reduced form, while others are in the oxidized form. Those inhibited in the reduced form can be reactivated when treated with an oxidant. Those inhibited in the oxidized form can be reactivated, in part, by treatment with sodium sulfide to give thiocyanate as a product.

Apart from molybdo-enzymes, enzymes containing metals like copper and zinc are susceptible to cyanide inhibition. Enzymes containing zinc in their complexes include alkaline phosphatase and carbonic anhydrase. Alkaline phosphatase of leukocytes has been shown to be inhibited by cyanide.[29] The inhibition of carbonic anhydrase, an enzyme which is important for colinergic functions, is due to the binding of the cyanide to the zinc center. Plastocyanin, which is a copper protein that functions as an electron carrier between photosystems, is also inhibited by cyanide. This inhibition proceeds by the removal of copper from the enzyme. The inactive apoplastocyanin so produced is not capable of electron transfer. Ascorbate oxidase, a blue copper protein, is inhibited by cyanide, perhaps by removal of the copper moiety.

The inactivation and bleaching of iron-sulfur proteins like clostridial ferredoxin were reported by Wallace and Rabinowitz.[30] In this inactivation thiocyanate is produced. The inhibition of seleno-enzymes by the removal of selenium when such enzymes are treated with cyanide is a probable reason why selenium poisoning is protected against by the consumption of food sources containing cyanogenic glycosides.[31]

E. Cyanide Inhibition of Other Types of Enzymes

Enzymes without metals can also be inhibited by cyanide. Such enzyme reactions which involve the production of the Schiff-base intermediate are usually susceptible. In such reactions the interaction is between the Schiff-base intermediate and the cyanide. Ribulose diphosphate carboxylase or oxygenase is an example of such an enzyme that is inhibited by cyanide. A Schiff-base intermediate such as hepatic 2-keto-4-hydroxyglutarate aldolase, an enzyme which catalyzes the terminal step in the catabolism of hydroxyproline, is inhibited by cyanide. Enzymes with pyridoxal phosphate which normally go through the Schiff-base intermediate are potentially inhibited by cyanide. Glutamate decarboxylase and α-aminobutyric acid transaminase are examples of such enzymes.

As an enzyme stimulator, cyanide stimulates the guanylate cyclase activity in tissues such as the stomach, kidney, liver, lung, etc.[32]

V. TOXICITY OF CYANOGENIC GLYCOSIDES IN HUMANS

The ability of the cyanogenic glycosides to release cyanide implicates them in a number of diseases encountered in populations which depend primarily on those food sources that contain cyanogenic glycosides.

The consumption of cassava, an important staple food in the tropics, has been reported to cause neurological and endocrinological diseases in many tropical countries. Where the cassava products are not well processed to remove most of the HCN, deaths have occurred. The minimum lethal dose of HCN for humans has been estimated at between 0.5 and 3.5 mg/kg of body weight.[33] This is equivalent to an oral dose of between 30 and 210 mg for a 60-kg adult. Cassava in the fresh form is known to contain high levels of HCN which can be reduced considerably with processing. The HCN contents of different cassava varieties and its products are shown in Table 3.

HCN is rapidly absorbed from the gastrointestinal tract and produces recognizable symptoms at both lethal and sublethal levels of ingestion. With ingestion of lethal doses, death results from the anoxic condition caused by the inhibition of cytochrome oxidase by cyanide as earlier described. The toxicity of cyanide under such conditions is described as acute toxicity. While specific minimum levels of cyanide in cassava have not been correlated to

Table 3
HYDROCYANIC ACID CONTENT OF
DIFFERENT NIGERIAN CASSAVA
VARIETIES AND THEIR PRODUCTS

Cassava/product	Hydrocyanide acid content (ppm)	Ref.
Garri	11	35
Purupuru	40—60	35
Fresh whole root	88.3—416.3	42
Fresh pulp	34.3—201.3	42
Fresh peel	364.2—814.7	42
Sun-dried whole root	23.1—41.6	42
Sun-dried pulp	17.3—26.7	42
Sun-dried peel	264.3—321.5	42
Oven-dried whole root	51.7—63.7	42
Oven-dried pulp	23.7—31.3	42
Oven-dried peel	666.8—1250.0	42
Dried cassava waste (peels and discarded small tubers)	240	52

acute toxicity in humans, levels lower than 100 gm/kg food have not been reported to be fatal. While levels of HCN in well-processed cassava products are usually lower than this, failure to adhere strictly to established traditional processes which usually involve soaking, fermentation, and drying can cause retention of high and sometimes fatal levels of HCN in the products. Occasionally, processing techniques which do not involve fermentation reduce hydrolysis and result in retention of high levels of the intact cyanogenic glycosides in the final product. This can be eventually hydrolyzed into HCN in the subject in the presence of glucosidases or dilute acids. It is therefore important in the determination of HCN of foods to determine the free and bound cyanide (total HCN) which involves the addition of extraneous enzyme to the product being analyzed. Several methods are available for determining both the presence and the amount of cyanogenic glucosides in plants.[34] A recent technique which allows the determination of total cyanide content of cassava and cassava products has been developed at the International Institute for Tropical Agriculture (IITA).[34] It is important to standardize methods of HCN determination in foods to allow a clear demarcation of levels that can be regarded as fatal in different human foods. Where nonlethal doses of cyanide are consumed, the inhibition of cellular respiration can be reversed, due to the removal of HCN by respiratory exchange or by a detoxification process. Many pathways for cyanide detoxification have been proposed; the major pathway is by reaction with thiosulfate to form thiocyanate and sulfite. Rhodanese, or thiosulfate sulfurtransferase, the enzyme that catalyzes this reaction, is widely distributed in animal tissues. The thiocyanate so produced is excreted in the urine. In spite of this mechanism of detoxification, the residual cyanide that exists in sublethal doses in cassava and other cyanogenic plants has been incriminated in the etiology of some diseases. The symptoms manifested are classified under chronic cyanide intoxication. Osuntokun[35,36] has shown considerable epidemiological and pathological evidence linking the degenerative disease known as tropical ataxic neuropathy (TAN) to chronic cyanide intoxication of dietary origin. This results from the consumption of casssava products, notably "purupuru" and to some extent "gari". Levels of HCN in gari have been reported to be below 20 mg/kg[34,35] and those in purupuru to range beween 40 to 60 mg/kg. With a daily consumption of up to 750 g, an intake of 32 to 48 mg HCN can occur. At a lower limit of 30 mg HCN earlier computed for a 60-kg adult, the HCN consumed from purupuru can be considered to be at least sublethal. The consumption of these cassava products has been reported to cause an elevation of plasma cyanide and thiocyanate concentrations. A low level of sulfur amino acids was also reported in the patients suffering from TAN. The damage done to the nervous tissues might be related to the inhibition of oxidative systems by cyanide whose potency can be increased in the absence of sufficient sulfur amino acids to detoxify it to thiocyanate. It should be noted that related neuropathies have been described in Jamaica and Malaya where cassava is consumed. Ambylopia (blindness) reported in prisoners of war consuming rice, cassava, sweet potato, and mung bean[37] might be related to chronic cyanide intoxication. The possibility that malnutrition increases susceptibility to trace amounts of cyanide in these subjects has been suggested.[34,38]

Thiocyanate has also been reported to be a goitrogenic factor. Thiocyanate inhibits the intrathyroidal transfer of iodine and causes an elevation of the thyroid stimulating hormone (TSH). Ekpechi[39] reported the occurrence of goiter among inhabitants of eastern Nigeria where cassava is a staple food. Studies in Zaire have shown that cassava has a definite antithyroid action in humans resulting in the development of endemic goiter and cretinism. The predisposition to these anomalies also depends on the iodine consumption of the subject. It has been shown that thyroid function is clearly impaired in adults, especially when the iodine/thiocyanate ratio is lower than a critical threshhold of about two. In severe endemic goiter, infants are more at risk of overt thyroid failure than adults.[40] Furthermore, cyanide may be implicated in diabetes. There is a high incidence of diabetes in an area near the Zaire/Zambian border, a predominantly cassava-eating zone. This has been confirmed by the study of McMillan and Geevarghese,[41] that the association of malnutrition diabetes in the tropics with food cyanogens is mediated by cyanide in its pathogenesis. Their evidence supports the concept that a low protein intake, combined with ingestion of cyanide or a cyanide precursor, may lead to exocrine and endocrine pancreatic calcification with a loss of β-cell function. The implication of protein malnutrition in the neurological and endocrinological anomalies associated with cassava consumption has also been suggested.[42] A review of cassava utilization in Nigeria by Iyayi and Tewe[43] also suggests that palm oil has a beneficial role in reducing cassava toxicity in humans. In this regard, regions of Nigeria with consumption of cassava products in which palm oil is incorporated (Bendel State) do not manifest various anomalies associated with cassava consumption as opposed to reports from areas where cassava is a staple without incorporation of palm oil.

VI. TOXICITY OF CYANOGENIC GLYCOSIDES IN ANIMALS

As in humans, the ingestion of high concentrations of cyanogenic glycosides from cyanophoric plants has resulted in mortality in numerous species of animals. Death has been reported more in grazing and browsing animal species such as cattle, sheep, and goats than in those usually kept under intensive systems of management like pigs and poultry. This is because it is possible to consume larger doses of the glycosides from fresh, cyanophoric plant material than in processed products usually given to pigs and poultry. Maceration by the animal of the fresh plant tissue as it is ingested initiates the enzymatic breakdown of the glycoside by the glycosidases. Hence, the animal needs merely to eat enough of a plant that is sufficiently rich in cyanogen and enzymes to be poisoned. Leaves of cherry, cyanogenic acacia species, and young sorghum leaves have caused death frequently in grazing animals. Similarly, cassava leaves have been reported to cause death in goats in Nigeria.[44] For nonruminant animals like pigs and poultry, mortality occurs only when the cyanophoric plants are offered in the fresh, unprocessed forms. Thus, death has occurred in pigs offered fresh, uncooked cassava tubers and leaves. As the HCN level varies widely in cassava tubers, death has been more common with the bitter varieties containing HCN levels higher than 500 ppm.

As shown in Table 3, the HCN contents of cassava products vary widely depending on processing techniques. Oven-dried samples retain higher HCN than sun-dried samples because of the rapid loss of moisture and destruction of enzyme, thus resulting in retention of high levels of unhydrolyzed cyanogenic glycosides. This can be degraded to free HCN in the presence of glycosidases and/or dilute acids.

Acute cyanide poisoning in livestock, therefore, depends on the cyanophoric plants, the pH of the stomach, presence of microbial organisms capable of degrading glycosides in the gut as in ruminants, and the concentration of total HCN present in feeds.

The capacity to hydrolyze cyanogenic glycosides is widely distributed among rumen bacteria, but rumen microbes also have the ability to detoxify cyanide via thiocyanate.[45] Recently, the hydrolysis of prunasin, linamarin, and amygdalin by rumen microorganisms was studied.[46] Prunasin had the fastest rate of hydrolysis, suggesting that prunasin-containing forages are more toxic than those which accumulate the others.

The microbial hydrolysis of β-D-glucosides such as miserotoxin or prunasin occurs very slowly in the digestive tract of nonruminants, so the intact glycoside is usually absorbed.[47] This accounts in large part for the lower toxicity of these glycosides in nonruminants than in ruminants. Inhibition of glycoside hydrolysis may offer an additional route for detoxification and elimination of cyanogenic glycosides in ruminants.[47]

Chronic cyanide intoxication due to consumption of sublethal doses of HCN has been reported, mainly in rats and pigs. Oyenuga and Amazigo[48] observed paralysis of the hind limbs and subsequent loss of weight when pigs were fed uncooked cassava diets. Pond et al.[49] have also reported parakeratosis in pigs fed cassava-flour-based rations. This symptom was attributed to zinc deficiency which can be aggravated by the cyanide in cassava diets. In breeding pigs a level of 500 ppm HCN in cassava-based diets for gestating pigs caused an elevation in fetal thiocyanate concentration. However, there was no reduction in performance of the offspring during the postnatal phase of life.[42] In the same study, proliferation of the glomerulus cell of the kidney, reduced thyroid activity and increased milk thiocyanate were observed in the sows consuming the 500-ppm-HCN cassava diets during the gestation phase. Ingestion of sublethal doses of cyanide by growing pigs depresses growth rate because of the utilization of the sulfur amino acids to detoxify cyanide to thiocyanate through the rhodanese pathway.[50] Depression in growth is aggravated on cassava diets when these are also deficient in protein.[51] Serum thyroxin levels are significantly reduced in growing pigs with cyanide levels above 500 ppm in cassava diets.[52] Iyayi[53] in a series of studies with pigs has indicated that serum testosterone production is depressed with cassava diets containing 400 ppm of cyanide. Serum estradiol was not affected in female pigs on these diets.

Toxicity in rats has also been demonstrated by Hill,[54] who reported that with doses of linamarin of 50 mg/100 g body weight, mortality was produced. There were also alterations in several biochemical parameters measured in the blood and in heart tissues and in electrocardiograms. It is interesting to note that with cyanide levels as high as 1500 to 2400 ppm cyanide (as KCN) in cassava diets, no mortality was recorded in rats, with the main effects being reduction in feed consumption and body weight gain. Thus, species difference appears to play a role as the rat has an exceptional ability to tolerate a high intake of cyanide as compared to other animal species including humans. Feeding rations containing up to 1000 ppm cyanide to gestating rats did not reduce their performance during gestation, neither was there a carryover effect of gestation diets on the pups produced.[51] However, there was a reduction in feed intake, protein efficiency ratio, and increase in serum thiocyanate in the lactating rats.

Thiocyanate has an inhibitory effect on the development of the mammary glands in rats in the growing phase from 3 weeks to 3 months of age and also during gestation and lactation.[55] Large doses of thiocyanate cause a reduction of both the DNA and RNA in the mammary glands with a consequent decrease in metabolic activity. Osuntokun[36] carried out an 18-month feeding trial with rats and reported chronic cyanide intoxication due to consumption of a cassava product, "purupuru". This caused segmental demyelination in single nerve-fiber preparations from sciatic nerves. He also reported alopecia in the rats, probably due to nutritional imbalance of the 100% cassava diet, particularly due to a lack of the sulfur amino acids. A study with cassava diets in rats by Delange et al.[56] showed that in the neonatal period, there was an occurrence of severe hypothyroidism with alterations in the maturation of the central nervous system.

In poultry, very few reports have focused on the toxicity of cyanide. However, depression in growth rate has been reported in poultry with inclusion of cassava in the diets.[57] This is aggravated when cassava is used to replace maize in rations without supplementation with protein due to a lower protein content in cassava. In a series of studies by Tewe,[58] performance of poultry on cassava diets was satisfactory as long as the HCN content in the final ration did not exceed 100 ppm. Such rations must, however, be nutritionally balanced, especially in terms of calories and the essential amino acids.

A number of *Eucalyptus* spp. of Australia contain cyanogenic glycosides,[59] including *E. cladocalyx* (sugar gum) and *E. viminalis* (manna gum). These plants may cause livestock poisoning when the trees are felled in a pasture or when suckers sprout from stumps. Webber et al.[60] reported a recent incident of poisoning of goats which consumed *E. cladocalyx* leaves containing very high levels of cyanogenic glycosides. Koalas are sometimes poisoned in the wild when they consume the young regrowth, very high in cyanogens, of eucalyptus trees sprouting after bush fires.[59] Cyanide poisoning of koalas in zoos is sometimes encountered, when young foliage with a high cyanogen content is fed.[61]

Bracken fern (*Pteridium aqualinum*) contains the cyanogenic glycoside prunasin, which appears to function as a deterrent of herbivory. Cooper-Driver and Swain[62] observed that bracken plants lacking either prunasin or its hydrolytic enzyme were heavily grazed by deer and sheep, whereas those plants with the capability of cyanide production were untouched.

In North America, the principal livestock poisonings from cyanogens involve sudan grass and other forage sorghums, chokecherries (*Prunus* spp.), arrowgrass (*Triglochin* spp.),[63] and the Saskatoon serviceberry (*Amelanchier alnifolia*).[64,65] Wildlife (deer) may also be poisoned by these plants.[66]

VII. PHARMACOLOGICAL PROPERTIES OF CYANOGENIC GLYCOSIDES

Since the employment of cyanide in preparations for homicides and suicides in the Nineteenth Century, pharmacological application of cyanide has been widespread. It was a potent military chemical weapon during World War I and II. Medicinally, small doses were given over long periods without harm and even with apparent benefit. Pharmaceutical preparations (cherry laurel) of HCN were used in the treatment of pulmonary diseases. It was used in the early Twentieth Century for treatment of chest complaints, asthma, catarrh, coughs, and as a sedative as well as a palliative in tuberculosis. Pharmaceutical preparations of cyanide from peach stones and almonds have also been used as flavoring agents. The anticarcinogenic property of the cyanogenic glucosides has been suggested. The general hypothesis is that the cyanide released from such glycosides at the neoplastic site selectively attacks the cancer cell which is presumed to be low in rhodanese. On the other hand, normal cells which possess enough rhodanese are capable of detoxifying the cyanide. The efficacy of this treatment (with amygdalin or laetrile) has not been widely accepted by the medical profession.

Cyanide can produce activation of the chemoreceptors in the carotid body causing bradycardia in dogs. The cyanide stimulates the peripheral chemoreceptors to stimulate ventilation. Cyanide can also produce electroencephalographic[67] and electrocardiographic[68] conditions. Neuroelectrical changes with an abrupt loss of electrical activity[69] have been reported. Cyanide causes extensive alterations in cardiac physiology with an effect on the contractile components of the myocardium.

VIII. MODIFICATION OF CYANOGENIC GLYCOSIDE CONTENT BY PLANT BREEDING

The cyanogenic glycoside content of plants is influenced by plant maturity and agronomic conditions. The ability to modify the content of cyanogens of plants genetically has been reported in literature. Increased light intensity can cause an increase in the amount of cyanoglycoside of plants while variation in the production of HCN due to differences in the genotype of plants also occurs. Thus, under the same environmental conditions, seedling shoots of different varieties of a plant can contain significantly different amounts of HCN.[70] This variation occurs in many polymorphic plants. In nonpolymorphic plants like sorghum species, the number of genes and dominance relationships of these allele genes in the production of cyanogenic glycosides are variously reported. A low-HCN-producing gene has been shown to be dominant in sudangrass,[71] whereas in hybrids of sorghum with sudangrass, Barnett and Caviness[72] showed a multigenicity with the high-HCN-producing gene being dominant. In grain sorghum, older hybrid plants with significant changes in their HCN content can be made by selecting their parents on the basis of the HCN content of their seedlings.[73] In white clover (*Trifolium repens*), with the two glucosides linamarin and lotaustralin and where inheritance of cyanogenesis is diploid, the presence of one functional allele of a gene results in the production of both glycosides. In varieties where two nonfunctional alleles are present, there is no production of measurable amounts of glycosides. This same trend is true for the linamarase enzyme that degrades the cyanogens to produce HCN. Since the segregation of the genes for the cyanogens and enzymes is separate, it will be desirable to breed varieties with two nonfunctional alleles of the gene. In polymorphic plants such as the lotus species, the recognition of two phenotypes, cyanogenic and acyanogenic, has been reported although there is a possibility of four phenotypes. In this regard segregation of alleles of a single or more genes plays a significant role in the exhibition of phenotypes, cyanogenicity inclusive. Apart from the type of gene that is present in a plant, the number of alleles which produce the phenotype of cyanogenicity is important. As far as the nature of these genetic controls are concerned, where an allele of a gene is not an enzyme inhibitor, two possible alternatives for the nature of such a gene are postulated. According to Hughes,[74] such a gene (1) may be the site which specifies the structure of the enzyme or (2) may represent a site which controls the synthesis of the enzyme. Where a mutation of an allele of a gene occurs, such a mutation can be reflected in the alteration of the enzyme site without a change in its immunological properties. Feeding of ^{14}C-valine and ^{14}C from labeled isobutyraldoxine to produce linamarin in white clover plants[74] shows that at least two steps in the biosynthetic pathway of the cyanoglycoside are missing. This leads to a possible hypothesis that where cyanoglycoside biosynthesis involves a membrane-bound or particulate system, the loss or alteration of a single polypeptide would lead to lack of more than one step in the glycoside biosynthetic pathway. A relationship between glycoside content of hybrid progenes bears a direct relationship with that of the parents. Thus, cyanoglycoside content of plants with functional alleles in their genes can be inherited.

This heritability has been shown by Magoon[75] to be low in cassava. The presence of cyanogenic variability in cassava has led to intensified research aimed at lowering the glycosides of this important foodstuff. In Indonesia, varieties with as low as 6 mg/kg cyanide have been reported. Efforts to find zero-cyanide clones from the germ-plasm bank at the Centro International de Agricultura Tropical (CIAT)[76] and in Brazil have not proved successful. At the IITA,[77] two cultivars with low cyanide content of about 16 ppm have been successfully selected. There seems to be a positive relationship between cyanide level and root yield. Bruijn[78] reported a weak positive correlation (r = 0.20, p <0.1) between root yield and HCN content of peeled roots on a dry-matter basis. Cooke et al.[79] reported similar findings. However, the work at CIAT[76] and IITA[77] failed to confirm this observation. The work at IITA showed that a number of low-cyanide clones gave relatively high root yield. If this is the case, a combination between high root yield and low cyanide content is possible and would be desirable for breeders. Apart from the cyanogenic glycosides, genetic variations in the distribution of the enzyme activity which degrade the glycosides are also desirable.

There are other factors apart from the genetic ones which affect the glycoside content of cassava. Nitrogen fertilization increases the cyanoglycoside content of cassava.[80,81] Severe drought has been reported to increase glycoside levels,[78] while in Ivory Coast, glycoside contents are reported to be high at the beginning of the rainy season. Pruning of the aerial part before harvesting,[82] agricultural chemical application,[83] and high plant population density have all been found to decrease HCN content.

REFERENCES

1. **Hegnauer, R.**, *Chemotaxonomie der Pflanzen*, Band II, Berkhauer, Basel, 1963.
2. **Eyjolfsson, R.**, Recent advances in the chemistry of cyanogenic glucosides, *Fortschr. Chem. Org. Naturst.*, 28, 74, 1970.
3. **Seigler, D. S.**, Determination of cyanolipids in seed oils of the Sapindaceae by means of their NMR, *Spectra Phytochem.*, 13, 841, 1974.
4. **Vennesland, B., Pistorius, E. K., and Gewitz, H.**, HCN production by microalgae, in *Cyanide in Biology*, Vennesland, B., Conn, E. E., Knowles, C. J., Westley, J., and Wissing, F., Eds., Academic Press, London, 1981, 349.
5. **Knowles, C. J. and Bunch, A. W.**, Microbial cyanide metabolism, *Adv. Microb. Physiol.*, 27, 73, 1986.
6. **Singer, R.**, *The Agricales in Modern Taxonomy*, 3rd ed., Cramer, J., Ed., Ganter Verlag, Verdun, France, 1975.
7. **Tapper, B. A. and MacDonald, M. A.**, Cyanogenic compounds in cultures of a psychrophilic basidiomycete (snow mould), *Can. J. Microbiol.*, 20, 563, 1974.
8. **Stevens, D. L. and Strobel, G. L.**, Origin of cyanide in cultures of a psychrophilic Basidiomycete, *J. Bacteriol.*, 95, 1094, 1968.
9. **Emerson, H. W., Cady, H. P., and Bailey, E. H.**, On the formation of hydrocyanic acid from proteins, *J. Biol. Chem.*, 15, 415, 1913.
10. **Clawson, B. J. and Young, C. C.**, Preliminary report on the production of hydrocyanic acid by bacteria, *J. Biol. Chem.*, 15, 419, 1913.
11. **Lorck, H.**, Production of hydrocyanic acid by bacteria, *Physiol. Plant.*, 1, 142, 1948.
12. **Michaels, R. and Corpe, W. A.**, Cyanide formation by *Chromobacterium violaceum*, *J. Bacteriol.*, 89, 106, 1965.
13. **Collins, P. A., Rodgers, P. B., and Knowles, C. J.**, in *Cyanide in Biology*, Vennesland, B., Conn, E. E., Knowles, C. J., Westley, J., and Wissing, F., Eds., Academic Press, London, 1981.
14. **Castric, A. P.**, Metabolism of HCN by bacteria, in *Cyanide in Biology*, Vennesland, B., Conn, E. E., Knowles, C. J., Westley, J., and Wissing, F., Eds., Academic Press, London, 1981.
15. **Nartey, F.**, *Manihot esculenta* (Cassava): cyanogenesis, ultrastructure and seed germination, in *Abstracts on Cassava*, Vol. 4, Ser. 08EC-4, C.I.A.T. Publication, Columbia, 1978.
16. **Gibbs, R. D.**, History of chemical taxonomy, in *Chemical Plant Taxonomy*, Swain, T., Ed., Academic Press, London, 1963, 48.
17. **Poulton, J. E.**, Cyanogenic compounds in plants and their toxic effects, in *Handbook of Natural Toxins*, Vol. 1, Keeler, R. F., and Tu, A. T.,Eds., Marcel Dekker, New York, 1983.
18. **Duffey, S. S.**, Cyanide and arthropods, in *Cyanide in Biology*, Vennesland, B., Conn, E. E., Knowles, C. J., Westley, J., and Wissing, F., Eds., Academic Press, London, 1981, 385.
19. **Cimino, G., De Stafano, S., Minale, L., and Sodano, G.**, Metabolism in Porifera. III. Chemical patterns and the classification of the Desmospongiae, *Comp. Biochem. Physiol.*, 50B, 279, 1975.
20. **Lengo, A., Santacroce, C., and Sodano, G.**, Metabolism in Porifera. X. On the intermediary of a formamide moiety in the biosynthesis of isonitrile terpenoids in sponges, *Experientia*, 35, 10, 1979.
21. **Duffey, S. S. and Towers, G. H. N.**, On the biochemical basis of HCN production in the millipede Harpaphe haydeniana. (Xystodesmidae: Polydesmida), *Can. J. Zool.*, 56, 7, 1978.
22. **Siegler, D. S.**, Cyanogenic glycosides and lipids: structural types and distribution, in *Cyanide in Biology*, Vennesland, B., Conn, E. E., Knowles, C. J., Westley, J., and Wissing, F., Eds., Academc Press, London, 1981, 133.

23. **Butler, G. W. and Conn, E. E.,** Biosynthesis of the cyanogenic glucosides Linamarin and Lotaustralin. I. Labeling studies *in vivo* with *Linum usitatissimum, J. Biol. Chem.,* 239, 1674, 1964.

24. **Solomonson, L. P.** Metabolism of sulphur compounds, in *Metabolic Pathways,* Vol. 7, 3rd ed., Greenburg, C. M., Ed., Academic Press, New York, 1978, 433.

25. **Olsen, E. M. and Klein, R. J.,** Effects of cyanide on the concentration of lactate and phosphate in brain, *J. Biol. Chem.,* 167, 739, 1947.

26. **Way, J. L.,** Cyanide intoxication and its mechanism of antagonism, *Annu. Rev. Pharmacol. Toxicol.,* 24, 451, 1984.

27. **Solomonson, L. P. and Spehar, A. M.,** Stimulation of cyanide formation by ADP and its possible role in the regulation of nitrate reductase, *Nature (London),* 265, 373, 1977.

28. **Solomonson, L. P. and Spehar, A. M.,** Model for the regulation of nitrate assimilation, *J. Biol. Chem.,* 254, 2176, 1979.

29. **Coughlan, R. H.,** cited by **Solomonson, L. P.,** in *Cyanide in Biology,* Vennesland, B., Conn, E. E., Knowles, C. J., Westley, J., and Wissing, F., Eds., Academic Press, London, 1981, 133.

30. **Wallace, E. F. and Rabinowitz, J. C.,** Reaction of Clostridial ferredioxin with cyanide, *Arch. Biochem. Biophys.,* 146, 400, 1971.

31. **Palmer, I. S., Olson, O. E., Halverson, A. W., Miller, R., and Smith, C.,** Isolation of factors in linseed oil meal protective against chronic selenosis in rats, *J. Nutr.,* 110, 145, 1980.

32. **Veseley, D.L., Benson, W. R., Sheinin, E. B., and Levey, G. S.,** Two components of laetrile (mandelonitrile and cyanide) stimulate guanylate cyclase activity, *Proc. Soc. Exp. Biol. Med.,* 161, 319, 1979.

33. **Montgomery, R. D.,** Cyanogens, in *Toxic Constituents of Plant Foodstuffs,* Liener, I. E., Ed., Academic Press, New York, 1969.

34. **Rao, P. V. and Hahn, S. K.,** An automated enzymic assay for determining the cyanide content of cassava *Manihot esculenta* (Krantz) and cassava products, *J. Sci. Food Agric.,* 35, 426, 1984.

35. **Osuntokun, B. O.,** An ataxic neuropathy in Nigeria; a clinical biochemical and electrophysiological study, *Brain,* 91, 215, 1968.

36. **Osuntokun, B. O.,** Cassava diet and cyanide metabolism in Wistar rats, *Br. J. Nutr.,* 24, 377, 1970.

37. **Montgomery, R. D.,** The medical significance of cyanogen in plant foodstuffs, *Am. J. Clin. Nutr.,* 17, 103, 1965.

38. **Conn, E. E.,** Cyanogenetic glycosides, in *Toxicants Occurring Naturally in Foods,* National Academy of Sciences, Washington, D.C., 1973.

39. **Ekpechi, O. E.,** Pathogenesis of endemic goitre in Eastern Nigeria, *Br. J. Nutr.,* 21, 537, 1966.

40. **Delange, F.,** Endemic goitre and thyroid function in Central Africa, in *Monographs in Paediatrics,* Vol. 2, S. Karger, Basel, 1974.

41. **McMillan, D. S. and Geevarghese, P. J.,** Dietary cyanide and tropical malnutrition diabetes, *Diabetes Care,* 2(2), 1979.

42. **Tewe, O. O.,** Thyroid cassava toxicity in animals, in *Cassava Toxicity and Thyroid. Research and Public Health Issues,* Delange, F. and Ahiuwatia, R., Eds., IDRC, Ottawa, 1982, IDRC-207e, 114.

43. **Iyayi, E. A. and Tewe, O. O.,** Preparation of cassava foods: removal of its toxic factor, in Proc. Natl. Sem. Nigerian Food Culture, Institute of African Studies, University of Ibadan, Nigera, Nov. 26 to 28, 1986.

44. **Obioha, F. S.,** Utilization of cassava as human food. A literature review and research recommendations on cassava, AID Contract No. CSD/2497, 131, 1972.

45. **Majak, W. and Cheng, K.-J.,** Cyanogenesis in bovine rumen fluid and pure cultures of rumen bacteria, *J. Anim. Sci.,* 59, 784, 1984.

46. **Majak, M. and Cheng, K.-J.,** Hydrolysis of the cyanogenic glycosides amygdalin, prunasin, and linamarin by ruminal microorganisms, *Can. J. Anim. Sci.,* 67, 1133, 1987.

47. **Majak, W., Cheng, K.-J., and Muir, A. D.,** Analysis and metabolism of nitrotoxins in cattle and sheep, in *Plant Toxicology,* Seawright, A. A., Hegarty, M. P., James, L. F., and Keeler, R. F., Eds., Animal Research Institute, Veerongpilly, Australia, 1985.

48. **Oyenuga, V. A. and Amazigo, E. O.,** A note on the hydrocyanic acid content of cassava (*Manihot utilisima* Pohl.), *West Afr. J. Biol. Chem.,* 1(2), 39, 1957.

49. **Pond, W. G., Maust, L. E., Warner, E. G., and McDowell, R. E.,** Rice bran — cassava meal as a carbohydrate feed for growing pigs, *J. Anim. Sci.,* 29 (Abstr.), 140, 1969.

50. **Maner, J. H. and Gomez, G.,** Implications of cyanide toxicity in animal feeding studies using high cassava rations, in *Chronic Cassava Toxicity: Proc. of the Interdisciplinary Workshop,* Nestle, B. and MacIntyre, R., Eds., IDRC, Ottawa, 1973, IDRC-010e, 113.

51. **Tewe, O. O. and Maner, G. H.,** Cyanide, protein and iodine interaction in the performance, metabolism and pathology of pigs, *Res. Vet. Sci.,* 29, 271, 1980.

52. **Tewe, O. O., Afolabi, A. O., Grisson, F. E., Littleton, G. K., and Oke, O. L.,** Effect of varying dietary cyanide levels on serum thyroxine and protein metabolites in pigs, *Nutr. Rep. Int.,* 30, 1245, 1984.

53. Iyayi, E. A., Effects of varying dietary cyanide and protein levels on the performance of growing pigs, Ph.D. thesis, University of Ibadan, Nigeria, 1986.
54. Hill, D. C., Physiological and biochemical responses of rats given potassium cyanide or linamarin, in *Cassava as Animal Feed*, 1977, 33.
55. Pyska, H., Effect of thiocyanate on mammary gland growth in rats, *J. Dairy Res.*, 44, 427, 1977.
56. Delange, F., Bourdoux, P. E., Collnet, E., Courtoise, P., and Ermans, A. M., Nutritional factors involved in the goitrogenic action of cassava, in Proc. Cassava Toxicity and Thyroid. Research in Public Health Issues, Ottawa, Canada, 1983, 18.
57. Job, T. A., Oluyemi, J. A., Awopeju, A. F., and Odeyemi, T. O., Optimal level of cassava flour in the diet of growing chick, *Zentralbl. Veterinaermed.*, 27, 669, 1980.
58. Tewe, O. O., Energy and protein sources in poultry feeds, in Poultry Seminar on Soyabean, Poultry Association of Nigeria, Ibadan, 1984, 52.
59. Everist, S. L., *Poisonous Plants of Australia*, Angus and Robertson Publishers, Sydney, 1981.
60. Webber, J. J., Roycroft, C. R., and Callinan, J. D., Cyanide poisoning of goats from sugar gums (Eucalyptus cladocalyx), *Aust. Vet. J.*, 62, 28, 1985.
61. Collins, L. and Roberts, M., Arboreal folivores in captivity — maintenance of a delicate minority, in *The Ecology of Arboreal Folivores*, Montgomery, G. G., Ed., Smithsonian Institution Press, Washington, D.C., 1978, 5.
62. Cooper-Driver, G. A. and Swain, T., Cyanogenic polymorphism in bracken in relation to herbivore predation, *Nature (London)*, 260, 604, 1976.
63. Majak, W., McDiarmid, R. E., Hall, J. W., and Van Ryswyk, A. L., Seasonal variation in the cyanide potential of arrowgrass (*Triglochin maritima*), *Can. J. Plant Sci.*, 60, 1235, 1980.
64. Majak, W., Ubenberg, T., Clark, L. J., and McLean, A., Toxicity of Saskatoon serviceberry to cattle, *Can. Vet. J.*, 21, 74, 1980.
65. Majak, W., McDiarmid, R. E., and Hall, J. W., The cyanide potential of saskatoon serviceberry (*Amelanchier alnifolia*) and chokecherry (*Prunus virginiana*), *Can. J. Anim. Sci.*, 61, 681, 1981.
66. Quinton, D. A., Saskatoon serviceberry toxic to deer, *J. Wildl. Manage.*, 49, 326, 1985.
67. Ivanov, J. P., The effect of elevated oxygen pressure on animals poisoned with potassium cyanide, *Pharmacol Toxicol. (U.S.S.R.)*, 22, 476, 1959.
68. Cope, C., The importance of oxygen in the treatment of cyanide poisoning, *J. Am. Med. Assoc.*, 175, 1061, 1961.
69. Burrows, G. E., Liu, D. H. W., and Way, J. L., Effect of oxygen on cyanide intoxication. V. Physiologic effects, *J. Pharmacol. Exp. Ther.*, 184, 739, 1973.
70. Trione, E., cited by Hughes, A. M., The genetic control of plant cyanogenesis, in *Cyanide in Biology*, Vennesland, B., Conn, E. E., Knowles, C. J., Westley, J., and Wissing, F., Eds., Academic Press, London, 1981, 495.
71. Snyder, F. B., Inheritance and association of hydrocyanic acid potential, disease reactions and other characteristics in Sudangrass, Ph.D. thesis, University of Wisconsin, Madison, 1950.
72. Barnett, R. D. and Caviness, C. E., Inheritance of hydrocyanic acid production in two sorghum x sudangrass crosses, *Crop Sci.*, 8, 89, 1968.
73. Eck, H. V., Gilmore, E. C., Fergusson, D. B., and Wilson, G. C., Heritability of nitrate reductase and cyanide levels in seedlings of grain sorghum cultivars, *Crop Sci.*, 15, 421, 1975.
74. Hughes, A. M., Genetic control of plant cyanogenesis, in *Cyanide in Biology*, Vennesland, B., Conn, E. E., Knowles, C. J., Westley, J., and Wissing, F., Eds., Academic Press, London, 1981, 495.
75. Magoon, M. L., Problems and prospects in the genetic improvement of cassava in India, in *Tropical Root Tuber Crop Tomorrow, Vol. 1, Proc. 2nd Int. Symp. Tropical Root and Tuber Crops*, Univ. of Hawaii, Honolulu, 1970, 58.
76. Centro Internacional de Agricultura Tropical (CIAT), Annual Report, Cali, Colombia, 1975.
77. International Institute of Tropical Agriculture (IITA), Annual Report, Ibadan, Nigeria, 1981.
78. Bruijn, G. H., Towards lower levels of cyanogenesis in cassava, in *Cassava Toxicity and Thyroid. Research and Public Health Issues*, Delange, F. and Ahiuwatia, R., Eds., IDRC, Ottawa, 1982, 103.
79. Cooke, R. D., Howland, A. K., and Han, S. K., Screening cassava for low cyanide using an enzymatic assay, *Exp. Agric.*, 14, 367, 1978.
80. Obigbesan, G. O., The influence of potassium nutrition on yield and chemical composition of some tropical root and tuber crops, in 10th Int. Potash Inst. Colloq., Abidjan, Ivory Coast, 1973, 439.
81. Nugroho, J. H. and Dharmaputra, T. S., The effect of nitrogen fertilizer and organic matter on the yield of Mukibat cassava, *Agrivita*, 2(6), 21, 1979.
82. Lorenzi, J. O., Gutierrez, L. E., Normanha, E. S., and Clone, J., Variacao de carboidratose acido cianidrico em raizes de mandioca, apos a poda de parte aerea (Variation in carbohydrate and HCN content in cassava roots after pruning the aerial part), *Bragantia*, 37(16), 139, 1978.
83. Shanmugan, A. and Shanmughavelu, K. G., Influence of ethrel on growth and yield of tapioca, *Indian J. Plant Physiol.*, 17, 44, 1974.

Chapter 3

CARDIAC GLYCOSIDES

J. P. J. Joubert

TABLE OF CONTENTS

I. INTRODUCTION

Cardiac glycosides have been given this name owing to their specific and powerful effect on the cardiac musculature. This ability to increase the force of myocardial contraction or positive inotropic action places cardiac glycosides in an unrivaled position for the treatment of congestive heart failure in human medicine.[1-3]

In veterinary medicine, cardiac-glycoside-containing plants are known to cause numerous deaths among domestic herbivorous livestock each year.[2,4-13] Thus, cardiac glycosides are looked upon mainly as toxins in the animal industry as they are recognized as dangerous but effective medicinal remedies for humans.[1-13]

From ancient times, humans have used cardiac-glycoside-containing plants and their crude extracts as arrow, ordeal, homicidal, suicidal, and rat poisons, heart tonics, diuretics and emetics.[1-4] The Romans and Egyptians, for instance, used squill, an extract of *Scilla* or *Urginea maritima*, as a rat poison and, to a lesser extent, medicinally.[1] In Africa, the lethal effects of cardiac-glycoside-containing plants on animals probably prompted the use of them as arrow and ordeal poisons.[4]

Welsh physicians were writing about digitalis (as cardiac glycosides are often called in association with *Digitalis* spp. which are well known for their cardiac glycosides) or foxglove, in 1250.[1] The foxglove was described botanically in 1542 by Fuchsius who named it *Digitalis purpurea*.[1]

Withering described the medical uses of foxglove and symptoms of poisoning in 1785.[14] He recognized the salutary effects on the heart, although he saw its main use in the treatment of dropsy. In 1799 John Ferriar put the effect of digitalis on the heart as primary and the diuretic effect as secondary to that. Throughout the Nineteenth Century little attention was paid to sound advice of Withering and Ferriar, and digitalis was used indiscriminately for many disorders, often at toxic doses. During the early Twentieth Century the drug gradually became looked upon as specific for the treatment of congestive heart failure.[1]

Cardiac-glycoside-containing poisonous plants grow all over the world, but the widest variety of them grow in Africa south of the Sahara desert.[2,4] In South Africa, for instance, cardiac-glycoside-plant poisoning is regarded as one of the six most important syndromes of poisoning by plants. There are many poisonous members of the families Liliaceae, Iridaceae, Crassulaceae, Apocynaceae, Scrophulariaceae, Santalaceae, Asclepiadaceae, and Melianthaceae causing livestock losses due to cardiac glycoside poisoning.[2,4-6,8,9,12,15,16] *Nerium* spp. and *Thevetia* spp., natives of India, are seen worldwide as they are planted as ornamental shrubs in gardens.[4,17] *Digitalis* spp. grow throughout Europe, the British Isles, and North America and mainly invade disturbed soil along roads and overgrazed pastures. *Asclepias* spp. also have a worldwide distribution and can be seen growing along roads and in bare patches in natural grasslands.[4,10,13,17] In Australia, some members of the families Apocynaceae, Liliaceae, and Iridaceae grow.[2,11] Two of the southern African Iridaceae have become established there and are becoming threats to livestock, while the others seldom cause animal deaths.[2,11] In most countries, cardiac-glycoside poisoning by plants results in only occasional deaths as these plants are not normally eaten by livestock. With droughts and overgrazing, the *Asclepias* spp. have become established in natural grasslands of the western part of North America and some South American countries where they constitute a problem to grazing stock.[4,10,13,17]

Animals mostly affected are cattle, sheep, and goats, while horses and game animals such as the deer can be poisoned also.[2,4,7-10,12,13] Most wild herbivores instinctively avoid these poisonous plants, and cases of poisoning are rare among them.[2,8,9] Dogs and humans were poisoned by eating raw or undercooked meat and intestines of animals which died of cardiac-glycoside-plant poisoning.[2,4,8,9] Dogs can also be poisoned when they bite frogs of the family Bufonidae.[2] Fowls are rarely poisoned by these plants, and birds are poisoned when they catch insects which feed on cardiac-glycoside-containing plants.[2,4,10,13]

Arrow poisons in Africa are made with crude extracts and the tissues of seeds, leaves, fruits, and wood of *Acokanthera* spp., *Strophanthus* spp., *Adenium* spp., and many other cardiac-glycoside-containing plants. The poison is applied between the back of the barbs of the arrow and the shaft. It is protected from weathering by leather or bark binding.[4]

II. CHEMISTRY

A. Nomenclature and General Characteristics

Glycosides which have the property of increasing the force of contraction of heart muscle are called cardiac glycosides. They consist of a steroidal aglycone or genin and a sugar portion. The aglycone is the pharmacological (or toxicological) active part of the molecule, while the sugars (usually there are 1 to 4 sugar molecules bonded to the aglycone) only influence the action of the aglycone by affecting the lipid solubility and cell penetrability of the entire molecule.[1,2]

The aglycone can be released from the molecule by enzymatic, alkaline, or acid hydrolysis. They are chemically related to bile acids, sterols, and steroidal hormones. An aglycone is essentially a cyclopentanoperhydrophenanthrene nucleus to which is added an unsaturated lactone ring at C-17. To this aglycone, methyl, hydroxyl and aldehyde groups are attached at different positions. These groups also influence the lipid or water solubility and activity of the aglycone as well as its protein binding. Naturally occurring aglycones have a hydroxyl group at C-14 and often at C-3 also, where the sugars are usually attached. The hydroxyl group at C-3 is highly reactive, and this is made use of to produce semisynthetic aglycones.[1-3]

Cardiac glycosides are also divided into cardenolides and bufadienolides. Cardenolides have a composite name, with the first part "card" referring to its cardiac effect and the second part "enolide" to the singly unsaturated five-membered lactone ring at C-17 of the aglycone.[1-3] *Digitalis* and *Strophanthus* spp. contain typical examples of cardenolides. In most, but not all, cardenolides, one or more rare sugars are attached to the aglycone and to that, one or more glucose molecules. Enzymatic hydrolysis (if not curbed) can remove glucose molecules at an early stage of extraction. Bufadienolides, on the other hand, are named after the toad of the genus *Bufo* from which these aglycones can be extracted. The second part of the name describes the doubly unsaturated six-membered lactone ring at C-17 of the molecule. *Urginea, Homeria,* and *Tylecodon* spp. yield bufadienolides.[1-4,6,10,12]

The structural formulas of digitoxin (cardenolide) and proscillaridin (bufadienolide) can be seen in Figure 1. There are structural differences among the cardenolides, such as those existing between cardenolides of *Digitalis* and *Asclepias* spp. Cardenolides of *Digitalis* spp. have 1 to 3 sugars attached to the aglycone through a hydroxyl group at C-3, while those of *Asclepias* spp. have one sugar only which is attached through hydroxyl groups at both C-3 and C-2 of the aglycone. As a result of the attachment to C-2 and C-3, these molecules are markedly resistant to acid hydrolysis. Another difference is in the stereochemistry of the A/B ring fusion of the aglycone. It has a *cis* formation for cardenolides of *Digitalis* spp. and a *trans* formation in the other group.

An example of each of these cardenolides is shown in Figure 2.[4,10,13,18] Among the bufadienolides there are some interesting differences. The best known bufadienolide is proscillaridin (isolated from *Urginea maritima* and still in commercial use) which has one sugar attached through a hydroxyl group to C-3 of the aglycone.[1-4] The sugars of bufadienolides of *Tylecodon grandiflorus* are attached to their aglycones at C-2 as well as C-3 with acetal and hemiacetal bands. In the orbicusides of *Cotyledon orbiculata*, the sugars are linked with three ether bridges to C-2 and C-3 of the aglycone. In addition to this, lanceotoxin A, isolated from *Kalanchoe lanceolata*, is a naturally occurring bufadienolide glyconate. In Figure 3 these interesting structures are shown.[2,5,6,12,19-27]

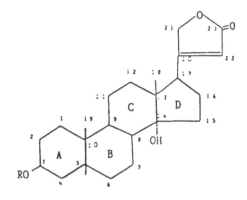

PROSCILLARIDIN

DIGITOXIN

GENERAL STRUCTURE OF A CARDENOLIDE

FIGURE 1. Structures of digitoxin (cardenolide), proscillaridin (bufadienolide), and the general structure of a cardenolide showing the numerical
distribution of the carbon atoms and rings.

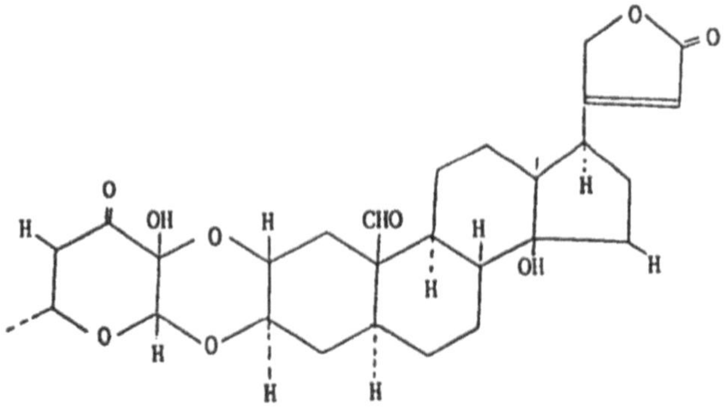

USCHARIDIN

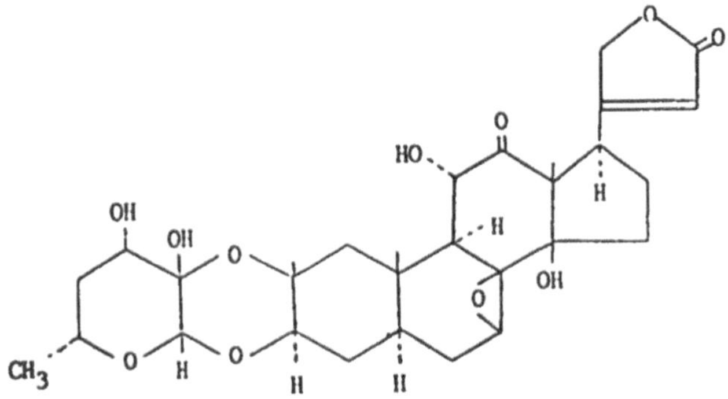

ERIOCARPIN

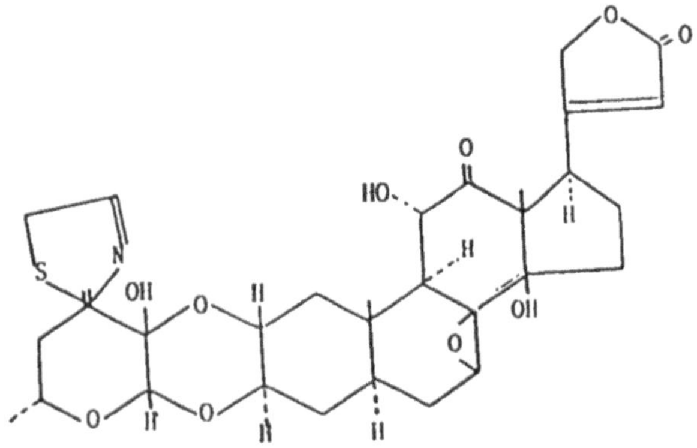

LABRIFORMIN

FIGURE 2.   Structural formulas of cardenolides uscharidin, eriocarpin, and labriformin.

PROSCILLARIDIN

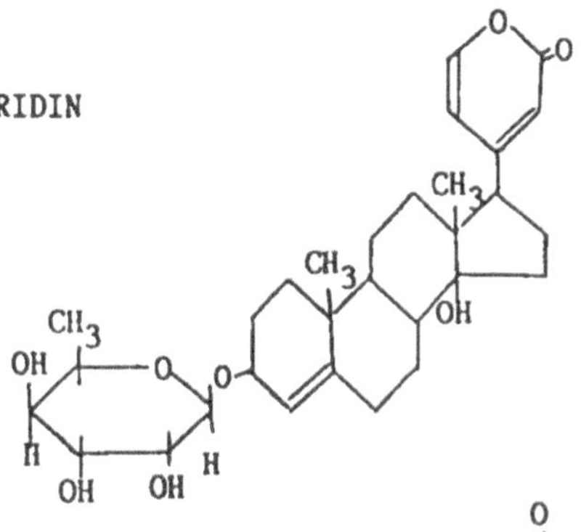

LANCEOTOXINA

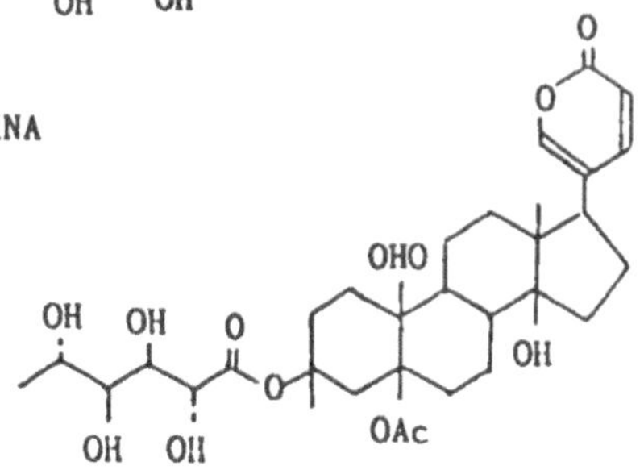

TYLEDOSIDE A

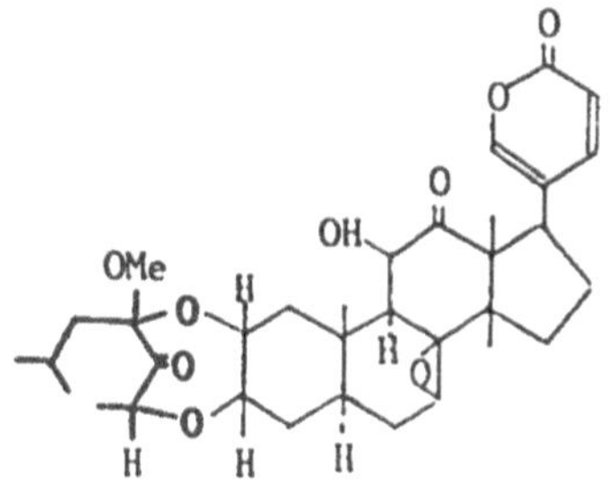

ORBICUSIDE A

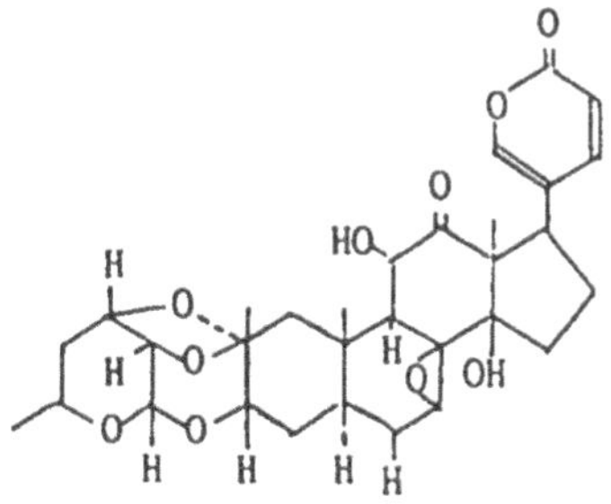

FIGURE 3. Structural formulas of bufadienolides proscillaridin, lanceo-toxin A, tyledoside A, and orbicuside A.

Saturation of the lactone ring at C-17 of any cardiac glycoside reduces the pharmacological activity tenfold or more and increases the onset of cardiac effects. Opening of the lactone ring completely abolishes activity. This indicates that the pharmacological and toxicological properties of the aglycone are vested in the unsaturated lactone ring at C-17.[1-3] Another factor to keep in mind at this stage is that two nontoxic bufadienolides isolated from *Urginea physodes* lack the hydroxyl group at C-14.[6] This fact emphasizes the importance of hydroxyl and other groups and sugars to the aglycone.

A rather confusing characteristic of cardiac glycosides is that their names refer to their botanical or zoological origin and not to chemical structures.[1] A list of most, but not all, known cardiac glycosides and their botanical origins is given in Table 1.[1-6,10,12,13,17-36]

B. Isolation

Chemical and biochemical methods are used to isolate the cardiac glycosides from plant material. Some cardiac glycosides are more and others less heat labile, and this influences the choice of an extraction method. Furthermore, most workers have their own trusted procedures and chemicals which they prefer to use. In view of these factors, only a few methods are set out in this chapter to illustrate how extractions can be done.

A method described for the isolation of bufadienolides of *Tylecodon grandiflorus* is useful for *Kalanchoe lanceolata* and *Cotyledon orbiculata* as well. It is done as follows. Fresh plant material is minced and then extracted in a Waring blender with ethyl acetate (three times). The solvent is evaporated under reduced pressure, resulting in a syrup which is partitioned between 95% methanol (5 l) and petroleum ether. Both extracts are evaporated to dryness; the methanol extract then contains the poisonous substance. Repeated chromatography of the toxic residue on silica gel with chloroform-acetone (7:3 v/v), chloroform-acetone-methanol (60:40:1.5 v/v/v), and benzene-methanol (78:22 v/v) usually yields crystalline toxic compounds.[6,19-23]

For the isolation of 1α-,2α-epoxyscillirosidine out of *Homeria glauca*, Naudé advocates the use of mild biochemical methods. Finely milled air-dried plant material is soaked in 0.2 M acetic acid; extraction with chloroform (three times) is the next step, achieved by shaking at room temperature. The chloroform extract is filtered through cotton wool into saturated sodium bicarbonate. Sodium sulfate is used at this stage to dry the extract to which is added 50% ethanol:citrate buffer (pH 3.25) in equal parts. This extraction is repeated three times, then filtered, and again extracted three times with chloroform. After drying with sodium sulfate and evaporation, the toxic residue is cleaned with repeated thin-layer chromatography (TLC) using 8% of absolute ethanol in chloroform on silica gel.[12,33,35]

A method used for the extraction of cardenolides is as follows: Plant material is dried, finely milled, and then soaked in cold distilled water. After 24 h the extract is pressed out and treated with an excess of saturated basic lead acetate, the voluminous precipitate is filtered off, and excess lead is removed from the filtrate with sodium phosphate. The filtrate is then fanned down and shaken with chloroform to produce a fairly purified cardiac glycoside.[28,31,34]

These methods, with a few changes for individual plants, can be used to isolate cardiac glycosides from plant material.

C. Identification

The cardiac glycoside extracts from plants are exhaustively cleaned after isolation from plant material with the aid of TLC, high-pressure liquid chromatography (HPLC), and gas chromatography-mass spectrometric methods. After this, identification is done with the aid of melting points (MPs), ultraviolet (UV) and infrared (IR) absorption spectra, nuclear magnetic resonance (NMR), mass spectrometry (MS), as well as Lierbermann- and anthrone-color reactions.[2,4-6,10,13,18-27,29-31,33-35,37]

Analysis of high-field, radioactive-marked hydrogen- and carbon-NMR spectra can be done to elucidate the structure of a cardiac glycoside. Anderson et al.[19] and Steyn et al.[20,21] advocate the use of extensive homonuclear decoupling experiments to assign the radioactive-hydrogen-NMR spectra and radioactive-carbon-NMR spectra with broad-band, proton-decoupled and single-frequency, off-resonance, proton-decoupled experiments as well as reported chemical shifts of related compounds. They find that further residual splittings in a series of off-resonance, proton-decoupled, radioactive-carbon-NMR experiments aid in cor-

Table 1
CARDIAC GLYCOSIDES AND THEIR BOTANICAL
SOURCES

Plant family	Genus and species	Cardiac glycoside
Apocynaceae	*Acokanthera*	
	schimperi	Ouabain (strophantoside G)
	venenata	Venenatin, ouabain, acovenoside A—C
	deflersii	Acovenoside A, acolongifloroside
	friesiorum	Acovenoside
	longiflora	Acolongifloroside G, H, and K, acovenoside A—C
	spectabilis	Spectabilin, ovabain acokantherin
	Adenium	
	boehmianum	Echujin, somalin, abobioside
	somalense	Somalin (= hongheloside G)
	honghel	Hongheloside A, G, and C
	multiflorum	Hongheloside C
	Apocynum camrabinum	Cymarin
	Cerbera odollam	Cerberin (= acetylneriifolin)
	Nerium	
	oleander	Oleandrin (= oleandroside)
	odorum	Odoroside A—G
	indicum	Neriin (= nerioside)
	Pachypodium leali	Pachypodiin
	Strophantus	
	gratus	Ouabain
	kombé	Strophanthi-K, cymarin
	sarmentosus	Sarmentoside A, C, D, and E
	amboensis	Sarveroside
	courmontii	Courmonticide, sarveroside, sarmentocymarin
	eminii	K-Strophantin, periplocymarin, cymarin
	gerrardi	Sarveroside, sarmentocymarin
	grandiflorus	Sarmentocymarin
	hispidus	Strophanthin, cymarin
	hypoleuceus	Periplocymarin
	intermedius	Intermedioside, pantstroside
	mirabilis	Cymarin, periplocymarin
	nicholsonii	Cymarin, periplocymarin
	petersianus	Sarmentocymarin
	schuchardtii	Sarveroside
	speciosus	Desgluco-ditalinum-verum
	verrucosus	Strophanthin
	welwitshii	Intermedioside, pantstroside
	Thanginia venenifera	Tanghinin, desacetyltanghinin
	Thevetia	
	peruviana	Thevetin, neriifolin
	yccotli	Thevetin
	neriifolia	Thevetin, neriifolin
	Urechitis suberecta	Urechitin, urechitoxin
Asclepiadaceae	*Asclepias*	
	eriocarpa	Eriocarpin, desglucosyrioside, labriformin
	fruticosa	Gomphoside, afroside
	labriformis	Labriformin, eriocarpin, desglucosyrioside
	syriaca	Syrioside, syrioboside, desglucosyrioside

Table 1 (continued)
CARDIAC GLYCOSIDES AND THEIR BOTANICAL SOURCES

Plant family	Genus and species	Cardiac glycoside
	erosa	Eriocarpin, labriformin
	speciosa	Desglucosyrioside, syrioside
	curassavica	Calotropin, uscharin, calactin
	Calotropis	
	gigantea	Calotoxin
	procera	Calotropin, calactin, uzarin
	Cryptostegia grandiflora	Cryptograndoside A, B, digital-inum-verum
	Periploca graeca	Periplocin
	Xysmalobium undulatum	Xysmalobin, uzarin
Crassulaceae	*Cotyledon orbiculata*	Orbicuside A—C
	Kalanchoe lanceolata	Lanceotoxin A, B, and 3-*O*-Ace-tylhellebrigenin
	Tylecodon	
	wallichii	Cotyledoside
	grandiflorus	Tyledosides A—D, F, and G
Celastraceae	*Euonymus europea*	Euonoside
Cruciferae	*Cheiranthus cheiri*	Cheirotoxin, cheiroside A
	Erysimum canescens	Erysimin
Iridaceae	*Homeria glauca* (= *Homeria pallida*)	1α,2αepoxyscillirosidine
Leguminoseae	*Coronilla glauca*	Corotoxigenin, coroglaucigenin
Liliaceae	*Bowiea volubilis*	Bovoside A—C
	Convallaria majalis	Convallatoxin, convalloside
	Urginea	
	burkei	Transvaalin
	depressa	Hellebrigenin - β - D - glucoside - ⟨1,5⟩
	rubella	Rubellin
	maritima (varralla)	Scillaren A, B, proscillaridin
	maritima (varrosea)	Scilliroside, scillirosidine
Melianthaceae	*Melianthus comosus*	Hellebrigenin 3-acetate
Moroideae	*Artiaris toxicaria*	α and β-Antiarin
Ranunculaceae	*Adonis*	
	amurensis	Cymarin
	vernalis	Adoniside, adonitoxin, cymarin
	Helleborus niger	Hellebrin
Rosaceae	*Crataegus oxyacantha*	Flavonoid glycosides
Santalaceae	*Thesium lineatum*	Thesiuside
Scrophulariaceae	*Digitalis*	
	purpurea	Digitoxin, gitoxin, gitalin
	lanata	Digoxin, digitoxin, gitoxin
Tiliaceae	*Corchorus*	
	capsularis	Corchortoxin
	olitorius	Corchorin (= strophanthidin)

relating signals of the protonbearing carbon atoms with specific carbon resonances.[19-21] Accurate identification of chemical structures of cardiac glycosides can thus be done. Comparisons with data of previously isolated cardiac glycosides aid in determining whether a new or well-known cardiac glycoside is isolated.

III. INTERRELATIONSHIPS

A. Plant-Livestock Interactions

Most of the interactions of cardiac-glycoside-containing plants and livestock are the poisoning of animals eating such plants. The plants in the flowering stage are the most attractive to grazing animals; probably that is why they are, in general, more poisonous during this stage. Most of these plants have a bitter taste, and the taste plus low-grade poisoning effects may be the factors which deter animals from eating these plants once they learn to "recognize" them as poisonous plants. No definite work has yet been done to elucidate how animals learn to avoid poisonous plants.[2,4,7-9,12,38-42] Some of the plants are not nromally eaten by livestock, such as the *Asclepias* and *Melianthus* spp.[2,4,5,9,10,13,18] The reason why they are not normally attractive to livestock may be bitter taste, unattractiveness, and leathery or sinewy leaves. Whether any odors play a role is not known.

Many of these plants retain their poisonous nature even when dry. At this stage they are unattractive and would not be eaten under natural grazing conditions, but poisoning occurs when the dry plants are included in hay, especially milled hay (e.g., *Asclepias, Moraea,* and *Homeria* spp.)[2,4,9,10,12,13,17,18]

Some animals, such as the spring hare (*Pedetes capensis*) and Cape porcupine (*Hystrix africae-australis*)[41] eat cardiac-glycoside-containing plants with impunity. In southern Africa, they eat the bulbs of *Urginea, Ornithoglossum, Homeria,* and *Moraea* spp. The eradication of these animals on farms is one of the factors why these plants seem to invade and increase under natural grazing conditions.[41] Other factors which lead to their increase are overgrazing, droughts, and plowing of grasslands where Moraea and Homeria are plentiful.[2,4,9,10,13,18]

B. Plant-Plant Interactions

An interesting interaction is recorded with *Loranthus* spp., which are parasitic plants that grow on stems and branches of shrubs and trees. These *Loranthus* spp. are not poisonous when growing on shrubs or trees which are regarded as good nutritious grazing. When they parasitize the *Melianthus* spp. they are still eaten readily by stock, but contain the cardiac glycosides of the *Melianthus* spp. In this way poisoning occurs through the interaction of this parasitic plant and poisonous one.[2,6,41]

With the Santalaceae, *Thesium* spp. are known as parasitic plants growing on the roots of many shrubs and bushes. The shrubs on which they grow are known to be edible non-poisonous plants (e.g., *Lycium* spp.), but the *Thesium* spp. growing on them contain cardiac glycosides which can cause death of livestock.[43] In many instances it appears as if the *Thesium*-spp. bushes grow entirely on their own, but the rootstock of a feeder plant can be found if it is carefully looked for.[41] *Thesium* spp. in southern Africa are normally grazed down well, and on some farms there is a defintne tendency for them to be poisonous to stock during the winter months (from autumn through winter into early spring) and where they grow on the southern slopes of hills. Unfortunately, on many other farms there are no seasonal or site preferences, and stock losses are caused by *Thesium* spp. throughout the year, even on northern slopes.[43]

With some of the other cardiac-glycoside-containing plants the toxicity also varies among geographical areas and soil type, while the climate and stage of growth may also cause variations in their toxicity.[2,4,8-10,17,41]

C. Plant-Insect Interactions

Caterpillars of the monarch butterfly (*Danaus plexippus L*) feed on *Asclepias* spp., and cardenolide cardiac glycosides accumulate in their bodies. As a result, the caterpillar, pupa, butterfly, and its eggs contain detectable quantities of cardiac glycosides. If birds eat these caterpillars or butterfiles, they will vomit due to a central nervous system stimulatory effect of the cardiac glycosides.[2,10,13,18] The vomiting may prevent the death of the bird and will teach it to avoid these insects. The bitter lingering taste of these cardiac glycosides may also induce birds to avoid these insects.

Some of the cardiac glycosides are stored unchanged in the bodies of caterpillars of the monarch butterfly, but others are modified in the gut. Calactin and calotropin are stored intact while others such as uscharidin are metabolized to a mixture of calactin and calotropin. Calactin and calotropin are probably more stable in the insect gut, and it may be a detoxication action for the caterpillar.[10] In the butterfly the cardenolides are distributed throughout the body, as compared to the grasshopper which stores the toxins in poison glands from which they can be actively ejected.

The wings of the butterfly have the highest content of cardenolides, but they are the least emetic. The higher concentration in the wings may function to deter birds from eating the insect when caught by its wings. Evidence in support of this is that some monarch butterfiles are found with beak marks in their wings.[2,4,10,13,18] One butterfly can contain about 0.2 mg of a cardiac glycoside. A number of other insects feeding on *Asclepias* spp. sequester the cardiac glycosides in their bodies for defense. Most of them are brightly colored and unpalatable to some vertebrates.

The insect *Aphis nerii* feeds selectively on the phloem of *Nerium oleander* and *Asclepias curassavica* and stores the cardenolides obtained there. These cardenolides may differ from the ones in the tissues and latex of these plants.[10a]

The grasshopper (*Poekiloceros bufonius*) from Africa is toxic to its predators when feeding on *Asclepias* spp., as proved experimentally with the European jay and white mice. These grasshoppers have poison glands which contain secretions of histamines and cardiac glycosides. This secretion can be ejected as a spray during the hopper stage or as a foam in the adult stage.[2,10] The European hedgehog, which is insensitive to cardiac glycosides, can eat these poisonous grasshoppers with impunity.[10,13,18]

IV. PHYSIOLOGICAL AND PHARMACOLOGICAL PROPERTIES

Cardiac glycosides have been studied intensively in human medicine as they are widely used in the treatment of congestive heart failure. Digoxin, digitoxin, and ouabain (G-strophantoside) are the ones used most often, and most of the information available concerns these cardiac glycosides.[1,3]

A. Mode of Action

The most probable explanation for the positive inotropic effect of cardiac glycosides on the heart is their ability to bind at the extracellular side to the membrane-bound Na^+, K^+-adenosine triphosphatase (Na^+,K^+-ATPase).[1,3,37,44-47] The hydrolysis of adenosine triphosphate (ATP) by these enzyme provides energy for the sodium pump, which actively transports sodium out of and potassium into cells, and also in myocardial fibers. In this way the pump maintains a high concentration of potassium ions intracellularly and a high concentration of sodium ions extracellularly, which is required for normal impulse-conductivity of the cell membrane. The inhibition of the ATPase by cardiac glycosides results in an accumulation of sodium ions intracellularly. This in turn enhances the influx of extracellular calcium ions and the release of bound, intracellular calcium ions from sites such as the sarcoplasmic reticulum. The way in which cardiac glycosides influence the movement of calcium ions is not yet fully understood, but their effects are explained by the observation that the force of cardiac contraction is roughly proportional to the extracellular ratio of calcium ion concentration to sodium ion concentration. It was shown further that the magnitude of the positive inotropic effect of digitalis is proportional to the degree of inhibition of Na^+,K^+-ATPase.[1,3,7,44-48] The increased concentration of free clacium ions in the sarcoplasm causes a more forceful contraction of the myocardium, the positive inotropic effect.

In therapeutic doses digitalis increases the contractility of atrial and ventricular muscle of normal and failing hearts. This results in improved cardiac output with a more complete emptying of the ventricle at systole, a reduction of elevated end-diastolic ventricular pressure, and a reduction in the size of the dilated heart. At the same time the heart is capable of pumping more venous blood owing to the increased force of systolic contraction and the shortened systolic phase. The cardiac glycoside-treated heart then gets longer rest periods between contractions and more time to fill the ventricles with venous blood, which means a decreased venous pressure will follow and with that, improved peripheral circulation. Secondary to the improved function of the heart, kidney function improves with resultant diuresis and relief of edema (where it arose as a result of congestive heart failure).[1,3,4,37,44-48] Cardiac-glycoside-induced improvement of the failing heart is illustrated in Figure 4. The cardiac glycosides depress conduction in the heart by a direct effect on the atrioventricular node and bundle of His and indirectly by increasing vagal action. These combined effects slow the ventricular rate.[1,3]

B. Absorption and Distribution

When given as tablets, digoxin is absorbed to about 70% in the gastrointestinal tract. The rate and extent of absorption is influenced by the dissolution rate of the tablet, particle size, formulation, and other biological factors. In healthy, fasting humans, digoxin taken p.o. reaches peak plasma concentrations within 30 to 60 min, while if taken after a meal, it reaches the same concentrations after 2 to 8 h.[3] The absorbed digoxin is 20 to 30%-bound to plasma proteins (while digitoxin is 90%-bound) and has a large volume of distribution which ranges from 3 to 10 l/kg body weight. The apparent volume of distribution of digoxin can be defined as the volume of body water which would be required to contain the amount of drug in the body if it were uniformly present in the same concentrations in which it is in the blood. The large volume of distribution indicates the extent of its binding to tissues. Digoxin has been detected in pleural fluid, cerebrospinal fluid, breast milk, and in the fetal fluids. Therapeutic plasma levels of digoxin range between 0.5 to 2.5 ng/ml. The calculated distribution of digoxin in body tissues is about 65% in skeletal muscle, 13% in liver, 4% in heart, 3% in brain, and 1.5% in kidneys. Reported ratios between myocardial tissues and plasma digoxin concentrations vary between 39:1 and 155:1.[1,3,37]

C. Metabolism and Elimination

Some metabolism of digoxin takes place in the liver, and limited enterohepatic recirculation occurs, but it is excreted mainly unchanged through the kidneys. Digoxin is both filtered at the glomerulus and secreted by the tubules, and some reabsorption from the tubular lumen occurs. In addition to this route, about 15% of a total dose of digoxin is excreted unchanged in the feces.

Some patients differ from the norm in that digoxin undergoes extensive metabolism in their livers. Metabolites which are thus formed can be detected in the urine. They include digoxigenin, dihydrodigoxigenin, mono- and bisdigitoxosides of digoxigenin, and dihydrodigoxin. They cross-react to an extent with the antidigoxin antibody used in the radioimmunoassay (RIA) for digoxin. Digoxigenin (mono- and bisdigitoxosides) are cardioactive while dihydrodigoxin is less cardioactive than digoxin. Digoxin has an elimination half-life of 1 to 2 d, compared to 7 d for digitoxin. Daily loss of digoxin in a "digitalized" patient is about 30%, and for digitoxin it is 10%. Digoxin has limited enterohepatic recycling, while digitoxin undergoes fairly extensive recycling and hepatic metabolism.[1,3,37]

There is no detailed information available on the absorption, volume of distribution, metabolism, half-life, and excretion of the multitude of cardiac glycosides found in plants toxic to animals. Some of them may have a very short half-life like digoxin or ouabain (e.g., labriformin and eriocarpin), while those of some Crassulaceae (which cause a chronic, lingering paresis called krimpsiekte) may have half-lives equal to that of digitoxin or even longer.[13,43]

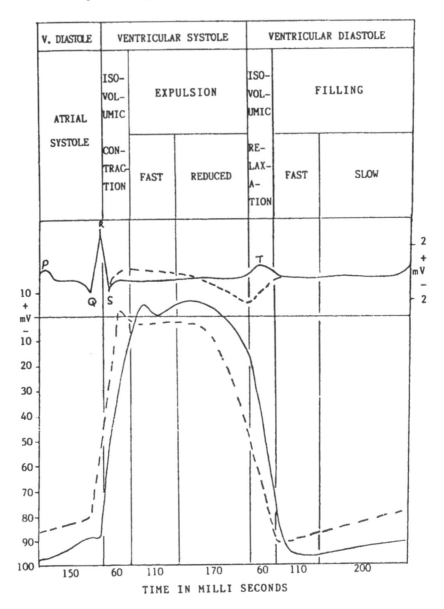

FIGURE 4.   Normal electrocardiograph (ECG) of a unipolar lead and action potential of
left ventricular myocard in relation to one full cardiac cycle and the effect of digitalis on the
ECG and myocardial action potential (shown with dashed lines). Under the influence of
digitalis the ECG shows a change in the S-T segment and T wave and a decrease in the R-
T segment. It results in an action potential which develops faster, is of shorter duration,
decreased maximal potential, and with increased depolarization. The effect of these changes
is a shortened ventricular systole.

V. DISTRIBUTION IN PLANTS

In Table 1 a number of plants are listed along with some of the cardiac glycosides extracted
from them. Not all of the known and isolated cardiac glycosides are mentioned in the list,
and many more genera and species of these plants are suspected to contain cardiac glycosides,
as they cause typical syndromes of cardiac-glycoside poisoning. The list does, however,
give an indication of the wide variety of plant families which contain cardiac glycosides.

FIGURE 5. Sketches of cardenolide containing plants: *Acokanthera oblongifolia, Asclepias fruticosa, Strophanthus sarmentosus,* and a *Digitalis* sp.

A description of the more important plant genera is given here. Sketches of some cardenolide- and bufadienolide-containing plants are shown in Figures 5 and 6, respectively.

A. *Acokanthera* Species[4,28,30,32,41]

Distribution — They grow throughout Africa in the higher rainfall areas.

Description — *Acokanthera* spp. can be shrubs or trees up to 6 m high. The leaves are thick, leathery, dark green, and elliptical, while the flowers are white, small, and bunched together. Fruits are green to red and eventually purple in color. Latex oozes profusely out of stems when they are cut off.

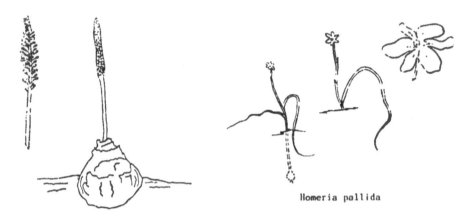

Urginea sanguinea

Homeria pallida

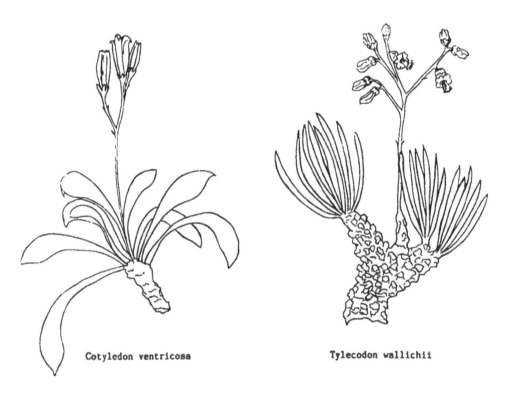

Cotyledon ventricosa

Tylecodon wallichii

FIGURE 6.   Sketches of bufadienolide containing plants: *Cotyledon ventricosa*, *Tylecodon wallichii*, *Homeria pallida*, and *Urginea sanguinea*.

Importance — They cause incidental stock losses when they are most poisonous, which is during the dry summer months when they are more likely to be eaten. Some species are used for the extraction of ouabain for commercial use. Initially, they were used as an arrow poison.

Distribution of cardiac glycosides — Cardiac glycosides are found in highest concentrations in fresh young leaves, stems, flowers, seeds, latex, and root bark. Older leaves and branches are less poisonous, and the fruit is bitter but edible. Drying does not destroy the toxicity of the plant material.

B. *Adenium* Species[2,4,41]

Distribution — These plants grow in lower- and high-rainfall areas of Africa south of the Sahara Desert.

Description — Some of them are shrubs while others can be trees up to 3 m tall. The leaves are large, about 20 cm in length, elliptical, dark green, and leathery. Three or four flowers, with diameters about 3 to 5 cm, white to pink in color, are displayed at the end of stems or twigs.

Importance — Extracts were used as arrow poisons and fish poison. Stock losses are incidental and infrequent.

Distribution of cardiac glycosides in plants — They can be extracted from the roots, stems, milky latex, and seeds.

C. *Nerium* Species[2-4,9,11]

Distribution — These plants are indigenous to India and Asia and now grow all over the world as ornamental shrubs and hedges and trees.

Description — *Nerium* species vary in size from shrubs to trees, 2 to 3 m high. *N. indicum* has sweetly scented single and double pink flowers. Leaves are thin, elongated, dark green, leathery, about 1 cm wide, and 8 to 10 cm long.

Importance — Incidental stock losses occur throughout the world, and humans have died where children gathered flowers and chewed them or where the wood was used as skewers. A few leaves, 60 g (2 oz), can kill a sheep or an ox in half an hour.

Distribution of cardiac glycosides in plant — Flowers, leaves, twigs, latex, roots, and wood contain cardiac glycosides.

D. *Strophanthus* Species[1,3,4,31]

Distribution — These plants grow throughout Africa.

Description — *Strophanthus* spp. can be shrubs or trees up to 3 m tall. The leaves are dark green, leathery, oval, and 6 to 8 cm in length. Members of this genus have a variety of flowers, all having five petioles, some thin and elongated, others small with long, thin dropping ends, and some with fairly big, rounded petioles. These flowers can be whitish, yellow to pink in color. Seeds are carried in two elongated, thin or short thick pods.

Importance — These are used for the extraction of ouabain for commercial use. Incidental stock losses do occur, and it was used as an arrow poison.

Distribution of cardiac glycoside in plants — Seeds and fruits have been used for extractions.

E. *Asclepias* Species[4,9,10,13,17,18]

Distribution — *Asclepias* spp. have spread throughout the world and grow as weeds along roads, railway lines, and on denuded patches in grasslands.

Description — They grow as shrubs up to 1.5 m tall, have white latex (therefore, called milkweed) and greyish-green stems covered with velvety hair. Leaves are opposite, thin and elongated, or broad. Flowers are small, white to yellowish-green, bunched together at the end of stems. Seeds are carried in yellowish-green pods, blown up, with hairy appendages on the outside surface.

Importance — *Asclepias* spp. cause stock losses during droughts and on overgrazed rangeland, and are an important cause of stock losses in the western states of the U.S.

Distribution of cardiac glycosides in plants — They are found mainly in latex but also in stems, leaves, and roots.

F. *Cotyledon, Kalanchoe,* and *Tylecodon* Species[2,4,9,15,16,19-24]

Distribution — They are found throughout the southern half of Africa.

Description — These species are succulents of the family Crassulaceae. *Cotyledon* spp. have thick, flat, rounded greyish-green leaves with a red edge. They are about 60 cm high, forming a single stem which carries the bell-shaped, drooping pink flowers. *Kalanchoe* spp. form erect single stems up to 1 m high with leaves opposite in pairs and flowers at the top which vary from yellowish-green to pink and red. *Tylecodon* spp. have round needle-like to thick leaves, grow about 60 cm high, and form single flower heads with drooping bell-shaped, yellowish-green to pink flowers. Most of them bloom in spring and summer.

Importance — Large stock losses occur annually, particularly in the mountainous southern Karoo of southern Africa where Angora goats are kept. Young stock are more susceptible and will eat them more readily than mature animals.

Distribution of cardiac glycosides in plants — Flowers, leaves, and stems contain cardiac glycosides; the concentration varies being less when the leaves are dropping and at its highest during flowering.

G. *Homeria* and *Moraea* Species[2,4,9,11,12,33,35,38,41]

Distribution — *Homeria* spp. (37 in total) are endemic to South Africa, while the 90 species of *Moraea* occur throughout Africa and Australia. Two of the *Homeria* spp. (*H. breyniana* and *H. miniata*) have been introduced into Australia and New Zealand where they are now well established.

Description — These plants have corms, 2 to 4 cm in diameter and about 30 cm deep in the soil. One or more long filamentous leaves are formed, and in spring or in autumn a single or widely branched flower head appears. *M. polystachya* has blue flowers, while *H. pallida* has yellow star-shaped flowers.

Importance — Losses especially among cattle occur annually, and these plants constitute a severe problem to stock owners throughout Africa and particularly in South Africa.

Distribution of cardiac glycosides in plants — Corms, leaves, stems, and flowers contain cardiac glycosides, and they retain their toxicity when dry.

H. *Urginea* Species[2-4,9,11,25,27,39-41]

Distribution — They have a worldwide distribution, but the most important poisonous ones are those growing in Africa as well as the well-known *U. maritima* from the Mediterranean countries.

Description — *Urginea* spp. are bulbous plants, the bulbs being red, brown, or yellowish-green in color, and up to 15 cm in diameter. The bulbs are just below the surface of the soil. In spring a single green flowerhead appears first, and the name ''slangkop'' (snake's head) is derived from this. Flowers are white, small, numerous, and carried on the top half of the flowerhead. The leaves appear later, and they are greyish-green, about 2.5 cm wide, and 30 cm long.

Importance — Cattle, sheep, and goats often die in spring when it is dry, and the ''snake's heads'' are the only green vegetation available. The plant is an important cause of stock losses, mainly in the summer rainfall areas. *U. maritima* is commecially used for the extraction of proscillaridin and scillirosidine. Crude extracts of its bulbs are also used as a rat poison.

Distribution of cardiac glycosides in plants — The highest concentration is in the flower head, but the bulbs and leaves contain cardiac glycosides also.

I. *Melianthus comosus*[2,5,9,41]

Distribution — It grows throughout Africa along dry riverbeds and slopes of hills and mountains. Sometimes it is cultivated as ornamental shrub.

Description — *M. comosus* is a shrub which can grow up to 3 m high. Young stems and branches are yellowish-grey in color; the leaves are greyish-green, bunched together with pale brown to red flowers tucked in under the leaves. Flowers appear in autumn, winter, and spring.

Importance — It is very unpalatable, and, as a result, it is only rarely eaten by stock and seldom causes losses. It is more of academic interest.

Distribution of cardiac glycosides in plants — Flowers, leaves, seedpods, and stems contain cardiac glycosides.

J. *Thesium* Species[4,6,43]

Distribution — These plants grow in the southern half of Africa.

Description — These plants can be small shrubs of about 25 cm high or bushes up to 2 m tall. They grow on the roots of other plants, such as *Lycium* spp., *Felicia muricata, F. filifolia, Chrysocoma tenuifolia, Pteronia sordida,* and *Melianthus comosus. Thesium* spp. have long, thin, greyish-green stems and branches. Some have tiny triangular scale-like leaves, while others have dark, green, thin, elongated leaves up to 2 cm long. The flowers are yellowish white, tiny, and seen near the tips of branches, while the fruits are white when ripe. These plants grow along dry riverbeds, on open plains, and along slopes of hills.

Importance — *Thesium* spp. cause stock losses throughout the year in sheep, goats, and cattle in the Karoo area of southern Africa. More animals are lost during the winter in some areas, and some *Thesium* plants are well grazed without seeming to be poisonous.

Distribution of cardiac glycosides in plants — Stems, branches, flowers, and fruits contain cardiac glycosides.

K. *Digitalis* Species[1-4,7,11,14,17,34]

Distribution — They are indigenous to Europe and the British Isles and have spread all over the world as garden plants. They are also seen as invaders of bare patches along roads and railroads.

Description — *Digitalis* spp. are biennial plants producing a rosette of leaves in the first year, and then the stem elongates in the second year to carry the tall handsome inflorescence. These plants can grow to about 1 m in height, with a single stem and heart-shaped leaves varying between 5 and 12 cm in length. The pretty flowers are bell-shaped, situated on the top half of the stem and drooping downward.

Importance — These plants are cultivated for the commercial extraction of cardiac glycosides such as digoxin and digitoxin. Very few cases of stock poisoning occur with these plants as they are unpalatable and seldom eaten.

Distribution of cardiac glycosides in plant — Leaves, flowers, and stems contain cardiac glycosides, but leaves and seeds are used to produce digoxin and other cardenolides.

VI. OTHER SOURCES OF CARDIAC GLYCOSIDES

Toads of the family Bufonidae, for instance, *Bufo marinus,* secrete a potent toxin containing bufadienolide cardiac glycosides. These toads have well-defined skin glands dorsal and caudal of each eye, and some of the other species have skin glands over the legs. When the toad is endangered, for instance, being mouthed by a dog, the skin glands secrete the bufadienolide-containing venom. The secretion immediately induces profuse salivation in the dog, followed by cardiac arrhythmia, pulmonary edema, prostration, convulsions, and death. A dog may die within 15 min of mouthing *B. marinus.* In 1969, Otani et al.[49] estimated that about 50 dogs may be killed annually by toad venom in Hawaii. Humans have died where scratches were exposed to the venom and where it contaminated their food.[2,49] According to Chen and Kovarikova,[50] the venom contains:

1.   Bufagins, which are aglycones similar to those extracted from plants

2.   Bufotoxins, aglycones with suberyl arginine at C-14
3.   Apart from the bufadienolides, up to 5% of adrenaline and noradrenaline as well as indole alkylamines

The combination of bufadienolides with the catecholamines has a marked synergistic effect.[2,29,50]

The poisonous toads are found in the tropics, and some nonpoisonous *Bufo* spp. are seen in the more temperate zones.[2]

VII. TOXICOSES OF HUMANS

A. Uses of Cardiac Glycosides

Medicinally they are used to treat congestive heart failure involving left or right ventricle failure, or both. Furthermore, they can be applied where there is atrial fibrillation or flutter with a very fast ventricular rate.

At one time cardiac glycosides were used to treat edematous conditions and as diuretics, but these practices have been proven wrong, as their effects are directly on the heart. The other more visible effects are secondary to improved cardiac function.[1,3,30,37]

Although cardiac glycosides are dangerous drugs to use because of the small difference between a therapeutic dose and a poisonous one, they are widely used in health care, and digoxin is the fifth most commonly prescribed drug in North America.[1,3]

B. Abuses of Cardiac Glycosides

These drugs are sometimes misused for suicidal or homicidal purposes.[1,3,51-57] Extracts of *Scilla maritima* containing proscillaridin prepared as a rat poison have caused a few deaths through negligence.[1,4] Arrow poisons and ordeal poisons are rarely used in Africa anymore.[4] Most cases of cardiac-glycoside poisoning in humans today result from a slight overdose in treatment, especially where a large loading dose is used or where the patient's condition deteriorates while under therapy.[1,3,58]

C. Monitoring of Cardiac-Glycoside Levels

Digoxin and some often used cardiac glycosides can be measured accurately with RIAs which have been developed in the last 15 to 20 years.[1,37,51-69]

1. Development of a Radioimmunoassay

First, antibodies to the glycoside (e.g., digoxin) must be produced. To achieve this, digoxin, which is a small molecule or hapten, is conjugated to bovine serum albumin (BSA), and this conjugate is used as an antigen for the production of hyperimmune sera for these assays.[37,51-69] The conjugation is done with a terminal sugar of digoxin to an amino group of albumin by periodate oxidation of two hydroxyl groups of the terminal sugar, resulting in aldehydes which are coupled to the amine group of BSA with the aid of pH changes and stablized in the presence of sodium borohydrate.[65-70] Figure 7 illustrates the conjugation of digitoxin to BSA. Antibodies produced against this conjugate are directed mostly at the aglycone which is then furthest away from the protein.

A high degree of specificity of immunity is observed, as antibodies to digoxin show 32-fold less affinity to digitoxin, while the only difference between the two molecules is one hydroxyl group at the C-12 position in digoxin (and not in digitoxin).[65] Endogenous steroids of mammals, such as cholesterol, cortisol, dehydroepiandrosterone, 17 β-estradiol, progesterone, and testosterone, usually show minimal or no tendency to cross-react in hapten-displacement studies even when present in concentrations 1000- to 10,000-fold greater than the homologous hapten (digoxin).[65]

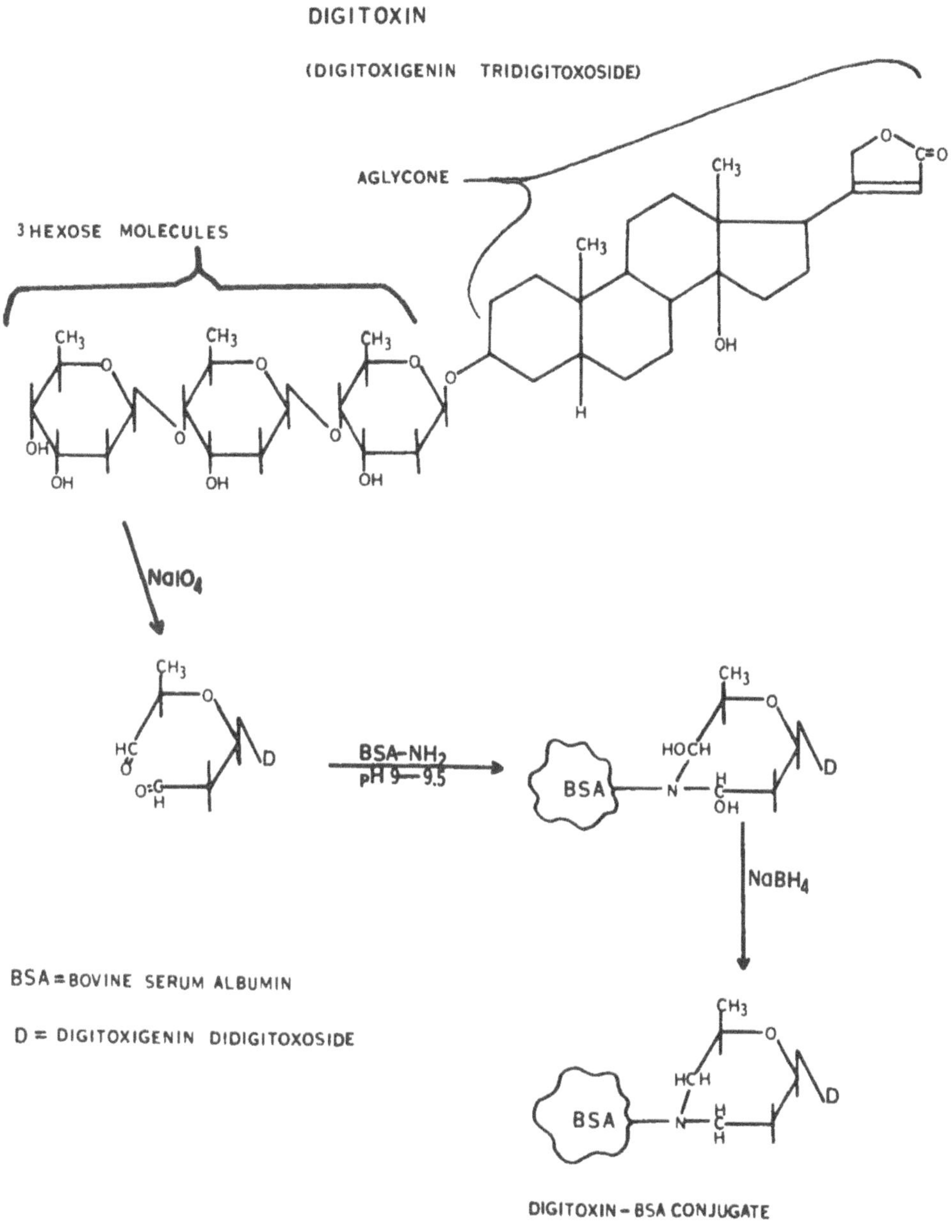

FIGURE 7. An illustration of the conjugation of digitoxin to bovine serum albumin (BSA).

2. A Radioimmunoassay Method[65-70]

The assay procedure can be explained as follows:

A serum specimen containing an unknown quantity of digoxin is mixed with a buffer solution, and to this is added a known quantity of digoxin-specific antibodies. Time is allowed for equilibration (labeled and unlabeled digoxin compete for binding sites on the antibodies). After sufficient time has elapsed, dextran-coated charcoal is added. Unbound labeled and unlabeled digoxin are adsorbed onto it, and the charcoal plus free digoxin is then removed by centrifugation. The supernatant is then added to a suitable scintillation counting medium to measure the concentration of bound labeled digoxin. The concentration of labeled digoxin which was replaced by unlabeled digoxin is then determined comparing the binding of the radiolabeled digoxin in the test specimen to that obtained in a control run without unlabeled digoxin and which is presented in a standard graph.

3. Practical Problems Encountered with Radioimmunoassays

Soldin[37] indicates that in many patients serum digoxin levels as determined with RIA do not correlate well with the effect of digoxin in the patient. Apparently, more extensive metabolism of digoxin takes place in some patients than was expected. Many of the metabolites cross-react with the antibodies of the RIA. The metabolites digoxigenin, bis- and monodigitoxoside, and digoxigenin react to the antibodies while dihydrodigoxigenin and dihydrodigoxin do not. Most commercial assays state this cross-reaction on their package insert, but the extent of metabolism of digoxin in some patients is more than expected and causes problems of interpretation of assay results.[37]

Immunoreactivity to digoxin has been found in the past few years in human plasma and urine of individuals who have never taken digoxin.[37,63,64] Some researchers postulate that there may exist natriuretic hormones or endoxin, as they have demonstrated such factors which cross-react with digoxin RIAs, inhibit Na^+, K^+-ATPase, and also cause natriuresis and diuresis. These factors or compounds, called endogenous digoxin by some researchers, might be linoleic and oleic acids and may play a role as endogenous regulators of the enzyme Na^+, K^+-ATPase.[37,63,64]

4. Other Methods to Determine Cardiac Glycosides in Serum

Soldin and co-workers developed an antigen by conjugating digoxin, not at the terminal sugar, but at the lactone ring with human serum albumin.[37] This different site of conjugation was chosen to allow the digitoxose sugars to act as antigenic determinants. Antibodies to this conjugate show minimal cross-reactivity to digoxigenin, bisdigitoxoside, monodigitoxoside, digoxigenin, and digitoxin. Unfortunately, it does cross-react with dihydrodigoxin which comprises about 12% of digoxin metabolites. They propose a double antibody approach for RIAs.[37] A competitive solid-phase enzyme immunoassay is proposed by Hinds et al.[62]

Liquid chromatographic, gas chromatographic, and mass spectrometric procedures have been described but are not used widely in clinical routine analyses for digoxin in serum and organs owing to lengthy sample preparation and lack of sensitivity. These procedures are more often used in the extraction and purifying of cardiac glycosides out of plant material.[6,19,37]

D. Symptoms of Poisoning

1. Introduction and General Characteristics

The cardiac glycosides are the fifth most prescribed drugs in North America, and they have the lowest margin of safety. Therapeutic and toxic concentrations often overlap, and individual variation in susceptibility to toxic effects increases the hazards of prescribing cardiac glycosides. In general, toxic effects can be seen when digoxin serum levels are 3 ng/ml or more, but have also been noted with therapeutic serum levels (0.5 to 2.5 ng/ml). Careful clinical judgment, therefore, is of utmost importance in the use of cardiac glycosides.[1,3]

No significant differences in the toxic:therapeutic dose ratios have been found for five cardiac glycosides tested under identical conditions.[1] Digitoxin may lead to toxic manifestations more easily than digoxin because of its relatively long half-life (5 to 7 d). This can cause overdose in time. The similarity of toxic to therapeutic dose ratios of cardiac glycosides are what can be expected if they are due to a common mechanism such as the inhibition of Na^+, K^+-ATPase.[1,3,37,45-48,58]

Most cases of cardiac glycoside intoxication in humans result from slight overdosage, and if treatment is discontinued as soon as signs of poisoning occur, the symptoms usually disappear in a few days.[1,3] The simultaneous use of diuretics may cause depletion of serum potassium and thereby enhance toxicity of cardiac glycosides. If the serum level of cardiac glycosides of each patient is closely monitored, together with clinical observations, the appearance of overdosage can be determined early and serious intoxication obviated.[1,3]

2. Symptoms in General

Withering[14] described the symptoms of toxicity with foxglove as follows: "When given in very large and quickly repeated doses it occasions sickness, vomiting, purging, giddiness, confused vision, objects appearing green or yellow, increased secretion of urine with frequent motions to part with it, slow pulse even as low as 35 in a minute, cold sweats, convulsions, syncope death."

3. Cardiovascular System

The most serious effect is on the heart, and the existing condition for which treatment is initiated, such as congestive heart failure, can aggravate the toxicity. Early signs of overdosage are atrial and/or ventricular arrhythmias and defects of conduction (e.g., atrioventricular conduction), and complete atrioventricular block may occur. The QRS complex of the electrocardiograph (ECG) is not prolonged directly by cardiac glycosides, but the P wave can be prolonged. Complete sinoauricular block can occur as well as ventricular tachycardia and ventricular fibrillation. On the other hand, a marked sinus bradycardia may develop.[1,3,58]

4. Gastrointestinal Effects

Anorexia, nausea, and vomiting are of the earliest signs of overdosage. Vomiting may develop without preliminary nausea and anorexia, where large doses of digoxin are given. Episodes of nausea and vomiting may come and go with increasing severity. Vomiting is induced owing to stimulation of a chemoreceptor trigger zone in the medulla oblongata (brain stem). In some patients, diarrhea, abdominal pain, salivation, and sweating can occur.

5. Neurological Effects

Headaches, fatigue, malaise, drowsiness, generalized muscle weakness, and neuralgic pain of the lower third of the face are early manifestations of toxicity. Paresthesias of lower extremities sometimes occur with muscular pains.

Disorientation, confusion, hallucinations, and delirium can be seen. Convulsions can occur in some cases. Vision is often blurred. Color vision is disturbed, and yellow and green or, less often, red and blue may be seen. White borders or halos may appear on dark objects.[1,3,14]

6. Symptoms of Less Importance

Rarely, other reactions to cardiac glycosides are seen, such as skin rashes, eosinophilia, and gynecomastia, which, may occur where the patient has been using a cardiac glycoside for several years. Blood coagulation was seen in experimental animals, and yet no adverse effect on coagulation time or on heparin can be demonstrated. Thrombocytopenia very rarely occurs.[3]

7. Factors Which Predispose to Toxicity

- Too large a maintenance dose is the most important factor.[1,3,37,58]
- Frequently, the simultaneous use of a diuretic, owing to its depletion of potassium, will predispose to toxicity.[1,3,58,71,72]
- A change in formulation of the tablets of digoxin, with different dissolution rates, can be a factor.[3]

- Decreased elimination rate, especially a decrease in renal and, to a lesser extent, hepatic function can aggravate conditions.[1,3]
- A decrease of plasma potassium with dialysis can also increase the changes of toxicity.[1,3,37]
- High plasma calcium and low plasma magnesium may contribute to a toxic affect.[1,3,37,44,48,58,71,73]
- Hypothyroidism increases susceptibility to toxic effects.[3]

E. Diagnosis of Cardiac-Glycoside Poisoning

As cardiac glycosides are used in patients with cardiac disease, it is often difficult to distinguish between toxicity and symptoms of disease. Another problem often occurs where a patient with cogestive heart failure or an arrhythmia is admitted to a hospital, and no history of recent use of cardiac glycosides can be obtained. Administration of a loading dose to such a patient can be lethal if the arrhythmia was caused by digitalis. RIA can be used to determine the presence and concentration of the often used cardiac glycosides such as digoxin, digitoxin, and ouabain. More hazardous methods have been used, such as the administration of an extremely short-acting cardiac glycoside (acetyl strophanthidin) to detect previous digitalization. The introduction of a single electrical stimulus to the ventricles is as hazardous. Therapeutic and toxic concentrations often overlap and due to individual variation of susceptibility, RIA results are not always sufficient. Careful, judicious clinical appraisal is still an important diagnostic tool.[1,3,37]

Potassium and calcium ion levels in saliva are used to determine toxicity, but conflicting results can be obtained.[3] Plasma potassium, magnesium, and calcium can give an indication of toxicity. Hyperkalemia occurs as a symptom in a large number of cases of acute overdosage or massive poisoning, while hypokalemia often is seen where diuretics have depleted potassium levels.[1,3,44,48,58,71-73] Sodium concentration increases and potassium decreases in red cells where digitalis toxicity occurs. This change in red-cell sodium and potassium concentrations has been used successfully to diagnose digoxin toxicity in 32 out of 34 patients compared with 30 out of 34 positive diagnoses with the determination of plasma digoxin levels.[3] Plasma digoxin of 3 ng/ml, potassium greater than 5 mmol/l, creatinine more than 150 μmol/l, age more than 60 years, and maintenance dose of more than 6 μg/kg body weight are factors to be used as guidelines of potential toxicity. However, with these criteria an overdiagnosis of toxicity of up to 41% can occur.[3] Hypomagnesemia can be indicative of toxicity or at least enhance toxic affects. Aberrant ventricular complexes on the ECG, particularly if they appear in groups, and premature ventricular beats are indicative of cardiac-glycoside overdosage. Other ECG indications of toxicity are total bradycardia, ventricular tachycardia, ventricular fibrillation, sinoatrial block, and atrioventricular block.[1,3,58]

F. Treatment of Cardiac-Glycoside Overdosage and Poisoning in Humans

1. General Discussion of Treatment

Withdrawal of the cardiac glycoside used for a day or two followed by adjustment of subsequent doses according to the needs of the patient are usually sufficient in a case of slight overdosage.[1,3,58] Serum electrolytes should be monitored as well as the ECG. The stomach should be emptied in early stages of acute poisoning. Activated charcoal or cholestyramine can be used with a lavage for the stomach, as the agent can adsorb cardiac glycosides still present.[1,3,52,53,55,56,58]

Where hypokalemia occurs, potassium chloride must be given, and if hypomagnesemia occurs, magnesium sulfate can be given. Plasma levels must be monitored continuously as overdosage of these are also detrimental. An excess of calcium can be removed with disodium EDTA. Unfortunately, disodium-EDTA treatment results in a transient improvement only, and if it is used continuously, it can damage the kidneys with calcium deposits and cause tetany owing to depletion of Ca ions.[58]

The use of diuretics is controversial, especially as electrolyte imbalances may occur. Hypokalemia is usually associated with prolonged use of diuretics.[1,3,58] Attempts to remove cardiac glycosides by means of dialysis have not been successful.[3] Hemoperfusion with dextran-coated activated charcoal as a filter has been used successfully in massive overdosage. This is a tedious operation, and it takes about 8 h to remove sufficient cardiac glycosides.

ECG monitoring must be continued while arrhythmias are treated with drugs such as phenytoin (diphenylhydantoin), propranolol, or procainamide. In patients with heart block or bradycardia, atropine can be given intravenously.[1,3,58,71,72,74]

Progressive hyperkalemia in massive overdosage indicates a very grave prognosis.[58,72]

Activated charcoal or cholestyramine can also be dosed in order to adsorb cardiac glycosides still present in the stomach and intestines. They can also bind those cardiac glycosides which enter the small intestine via enterohepatic recirculation.[52,53,55-57,71,72,76]

2. Digoxin-Specific Antibodies and FAB

Acute and massive overdosage (of digoxin) in suicidal cases has been treated successfully with the use of FAB (antigen binding fragments) of hyperimmune digoxin-specific antibodies.[1,3,37,51,54,59,61,70,75] Hyperimmune serum is produced as described for use in the RIA, and the antibodies, in this case IgG (immunoglobulin G), are further treated with papain to remove the FAB. This process is illustrated in Figure 8.

The FAB are more useful for treatment of critically poisoned patients than the antiserum or whole antibodies for the following reasons. They are less likely to cause allergic or anaphylactic reactions as they are much smaller molecules (molecular weight of 50,000 compared to 160,000 of IgG), and for the same reason they can be excreted in the urine when complexed with digoxin while IgG-digoxin complexes cannot. The resultant half-life of FAB-digoxin complexes are about 4.3 h compared to 23 d of IgG-digoxin complexes.

The following example clearly demonstrates the effectiveness of FAB in the treatment of massive digoxin overdosage. A patient who swallowed 22.5 mg of digoxin was treated at a stage when his serum potassium level was 8.7 mmol/l and the QRS complex on ECG widened. He received 1100 FAB per intravenous infusion over a 2-h period, and within 10 min after discontinuation of the infusion, his heart was in stable sinus rhythm of 75/min, and serum potassium level rapidly decreased to normal range (4 mmol/l).[51,54,59,65,66,68,70,75]

3. Beta-Adrenergic Blocking Agents[1,3,58,72,74]

Propranolol and other beta-adrenergic blocking agents which have more specific cardiac effects can be used to control arrhythmias, ventricular tachycardia, and ventricular ectopic rhythms arising from cardiac glycoside toxicity. Due to its beta-adrenergic nerve blockade, propranolol effectively reduces the heart rate, reduces atrioventricular conduction time, and reduces contractility. It also has a quinidine-like effect which abolishes ectopic ventricular arrhythmias. Contraindications will be cases where partial or complete heart block or bradycardia occurs.

4. Phenytoin (Diphenylhydantoin)[1,3,58,72,74]

It has an almost specific antagonism to ectopic ventricular rhythms induced by cardiac glycosides. It is also used to counter paroxysmal ventricular tachycardias. Contraindications are atrial flutter or fibrillation, ventricular fibrillation, and ventricular bradycardia.

5. Procainamide[1,3,58,72,74]

It effectively raises the threshold of both atrium and ventricle to impulse flow. Arrhythmias are counteracted by this effect, and slowed conduction time in both atria and ventricles is established. Tachycardias of atria and ventricles can be controlled.

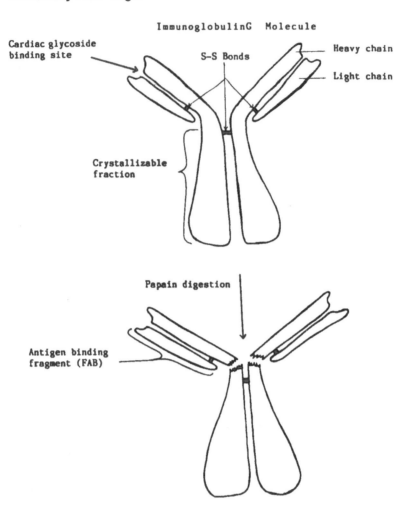

FIGURE 8.   The separation of antigen binding fragments (FAB) from an immunoglobulin G (IgG) molecule.

Contraindications — Partial or complete heart block cannot be treated with procainamide as it may result in asystole.

6. Atropine[1,3,58]

It abolishes vagal tone and thus increases heart rate where bradycardia occurs and can be beneficial where partial or total heart block occurs.

Contraindications — Atrial and ventricular tachycardias will be exacerbated by atropine.

VIII. TOXICOSES OF ANIMALS

A. Plants of Importance

Most of the genera of cardiac-glycoside-containing poisonous plants which cause stock losses on a large scale grow in the southern half of Africa.[2,4,9] Naudé mentions that this group of plants causes one of the six most important plant poisoning syndromes in South Africa.[2]

1. Crassulaceae

Among the crassulaceae are *Cotyledon* spp., *Adromischus* spp., *Kalanchoe* spp., and

Tylecodon spp. with *C. ventricosa*, *T. wallichii*, *C. orbiculata*, *C. decussata*, and *K. rotundifolia* as some of the more important species.[2,4,9,16] Botha[16] describes nenta poisoning or krimpsiekte caused by the Crassulaceae in the southern Karoo as one of the main factors limiting the expansion of the Angora goat and mohair industry. He found, in a survey made over 3 years in four districts covering about 302,000 ha, that the average yearly losses of Angora goats to nenta poisoning are about 1200.[16]

2. Iridaceae

Two genera are of great importance in this family, namely, *Homeria* and *Moraea*. All of the 37 species of *Homeria* are endemic to southern Africa, as are 61 of the 90 species of *Moraea*.[12] All of them are regarded as poisonous, with the following as the economically most important ones: *Homeria breyniana*, *H. miniata*, *H. pallida* (= *H. glauca*), *Moraea polystachya*, *M. polyanthos*, *M. spathulata*.[2,41] Farmers in southern Africa annually lose stock to tulp poisoning (poisoning by these Iridaceae), and it can be illustrated by the following example.

Cattle were moved from southwest Africa to the Cape Province where 42 out of 60 died of tulp poisoning as they were unaccustomed to tulp-infested grazing.[2] Cardiac glycosides have been isolated from *H. glauca* (which is now described as *H. pallida*), *M. polystachya* and *M. polyanthos*.[2]

3. Liliaceae

The well-known *Urginea maritima* of the Mediterranean countries is cultivated for the production of proscillaridin and scillirosidine, used in human medicine.[1-4] In southern Africa the economically important ones are *U. saguinea* (was known as *U. burkei*), *U. altissima*, *U. pusilla*, *U. capitata*, *U. macrocentra*, *U. rubella*, *Ornithoglossum viride*, and *Pseudogaltonia clavata*. They are poisonous throughout the year but cause most stock losses in spring when their green flower heads may be the only green vegetation available.[2,4,9,25,27,39-41] The leaves appear later, and if no alternative grazing is found, it will be eaten by stock. Out of a herd of 144 cattle, 44 died when they ate the leaves of *U. capitata* shortly after they were introduced to a new camp.[40] *Bowiea volubilis* more often causes deaths in humans overdosed by African herbalists.[2]

4. Santalaceae

Thesium lineatum and *T. namaquense* are known poisonous plants, and thesiuside, a cardiac glycoside, was isolated from *T. lineatum*. Large stock losses among sheep, Angora goats, and cattle occur annually in the dry areas of the Cape Province owing to these poisonous plants, while in the same area stock are often seen well grazed and with no associated incidents of poisoning.[4,43]

5. Asclepiadaceae

Asclepias species have a worldwide distribution and are seen as weeds in many countries. They are only eaten when no other grazing is available such as during drought periods or with gross overgrazing. *A. eriocarpa* and *A. labriformis* are noted for the large numbers of stock poisonings in some of the western states of the U.S.[13] As many as 108 species are described, and cardiac glycosides of a number of them have been isolated.[4,7,10,13]

B. Species Susceptible.

Cardiac-glycoside-plant poisoning occurs most often in cattle, sheep, and goats, and a few incidents of poisoning may occur in horses, dogs, cats, and birds.[2,4,9] Wild herbivores normally avoid poisonous plants, but where deer, for instance, are farmed, circumstances may induce them to eat some of these plants.[7] Laboratory tests with cardiac glycosides are done with rabbits, guinea pigs, rats, mice, and cats.[4,9,77-79] Humans who mistakenly thought the corms of *Homeria* spp. were edible bulbs, died from cardiac glycoside poisonging after eating them.[12]

Cattle are more prone to eat the Iridaceae as the leaves are often growing intertwined with grass, and cattle do not discriminate as well as sheep and horses when grazing.[2,4,9,41] Goats more often succumb to the Crassulaceae because goat farming is practiced in the hilly and mountainous areas where these plants are abundant.[2,16] Cattle, sheep, and goats are equally exposed to the Liliaceae and *Asclepias* spp.[4,13,41] Horses are exposed to Iridaceae and *Asclepias* spp. in hay as both can be cut with grass and remain toxic when dry.[4,9,10,12,13,41]

Species not susceptible to cardiac-glycoside-plant poisoning include the spring hare (*Pedetes capensis*) and cape porcupine (*Hystrix africae-australis*). They naturally feed on (among other food) corms of Iridaceae and bulbs of Liliaceae.[41] With laboratory experiments, the European hedgehog was found to be resistant to cardiac-glycoside-containing grasshoppers.[10]

C. Predisposing Factors

1. Factors Associated with the Plants

Most of these poisonous plants are eaten under adverse grazing conditions caused by drought, overgrazing, and dry winter conditions. It is particularly true for unpalatable plants such as the *Asclepias* spp.[9,12,13] Under these conditions, the poisonous plants may be the only green vegetation available.[4,9,12,13] *Urginea* spp. sprout their green flower heads in the dry springtime, when they are very attractive and poisonous to stock.[2,4,9] The long green leaves of *Homeria* and *Moraea* spp. are often the only green vegetation on harvested maize (corn) lands which are used for winter grazing of cattle and sheep on the high veld of South Africa.[1,2] On the banks of streams and rivers, the leaves of these two genera are intertwined with grass where it can be eaten with the grass or included in hay cut from the grassy banks.[1,2] All of these poisonous plants are more toxic and more attractive (to stock) in the flowering stage.[4,9,12,41]

2. Factors Associated with the Animals

Animals raised on a farm seem to "know" the poisonous plants on that farm and avoid them. This knowledge may be obtained when they eat a nonlethal amount of the plant, and the bitter taste and bad effects of it are afterward associated with the plant.[9,12,16,41] Young animals, perhaps because they feed indiscriminately, and old animals are more prone to eat poisonous plants and to succumb to them.[9,12,16,41]

Newly introduced stock, stable-fed animals, and hungry animals, especially after being driven fast over a long distance, will readily eat these poisonous plants.[4,9,12,16,38,41] Symptoms of poisoning will be enhanced by drinking water after poisonous plants are eaten. If these animals are driven after they consume poisonous plants, symptoms will be seen sooner, and they will be more acute.[4,9,12]

D. Symptomatology

1. General Symptoms and Syndromes

Animals can die peracutely due to cardiac-glycoside-plant poisoning, as Button and co-workers demonstrated when they dosed sheep at a rate of 3 g/kg body mass with finely milled, air-dried *Homeria glauca* plant material. One sheep died within 2.2 h from dosing.[80,81]

Acute to subacute syndromes are frequently seen with many of the wide variety of cardiac-glycoside-containing poisonous plants. In cattle, sheep, and goats the symptomatology are about the same. The animal is restless; there is grinding of the teeth, forced and labored breathing, ruminal atony (sometimes with bloat), cyanotic mucous membranes, frequent urination, and defecation, feces becoming fluid if the animal lives long enough; rapid heart rate occurs, sometimes arrhythmic with ectopic ventricular beats or atrioventricular dissociation; sometimes bradycardia may occur; and terminally, there is ventricular fibrillation.[2,4,7-13,15,16,18,38,41,82]

Homeria and *Moraea* spp. in subacute to chronic cases can cause severe gastroenteritis with bloat and profuse diarrhea. Nervous involvement is shown as nervousness, muscular fasciculations, progressive ataxia, and posterior paresis. Heart involvement can be bradycardia or tachycardia or arrhythmias. The animal dies in a state of severe depression, paresis, and dehydration.[2,4,9,12,38,41,80,81]

Some of the Crassulaceae, for example, *Tylecodon wallichii* and *Cotyledon ventricosa*, cause a subacute and chronic syndrome with more pronounced nervous involvement called krimpsiekte or nenta poisoning. In the Angora goat, typical symptoms may start off with an acute attack where typical signs described above can be noted plus a drooping lower jaw, with salivation and accumulation of unchewed food in the back of the mouth. If the animal survives, more chronic symptoms appear. This chronic form can arise without an acute phase if such an animal ate small amounts of nenta (as these plants are called) over a number of days. The chronically sick animal stands with an arched back, is very thin and tucked-up in appearance. The head is held lower than the back line, and the lower jaw droops with saliva dribbling out. The tongue is frequently wiggled and sometimes protrudes, while the lower lip sags and trembles. Breathing is labored when the animal walks, and the head may dangle loosely as if there were no control over the muscles of the neck. With sufficient rest, the animal may appear normal, but if it is driven hard, it will soon collapse, and these symptoms can appear again. Some may develop a torticollis which is rather permanent, while the other symptoms can disappear. Suckling kids can develop the wry neck syndrome, and with the head twisted sideways, they cannot suckle and die of starvation. Horses and dogs can develop torticollis when they ingest bufadienolides of the Crassulaceae.[2,4,8,9,41,82] Acute cardiac-glycoside poisoning in the dog is characterized by prolonged exhaustive vomiting and death following cardiac arrest. Monkeys show symptoms of vomiting, muscular spasms, and dyspnea before death.[4]

2. Physiopathology

Acute poisoning of sheep with *Homeria glauca* results terminally in tachycardia, raised systolic blood pressure, progressive hypoxemia, hypercarbia, acidemia, hemoconcentration, hyperkalemia, hypochloremia, elevated serum creatinine and glucose, and progressive increases in plasma cortisol, adrenaline, noradrenaline, and dopamine.[80,81]

3. Electrocardiographic Changes

The following ECG changes were recorded in sheep experimentally poisoned with *Urginea sanguinea* and *Homeria glauca*. Widening of QRS-wave, ST depression and ectopic foci, abnormal enlargement of P-wave, QRS-wave, or T-wave, sinus arrhythmia, tachycardia, and ventricular fibrillation.[80,81,83-86]

4. Macroscopic Pathology

There are no typical lesions, and the following are usually seen in acute and subacute cases: There is general venous congestion, congestion and edema of the lungs, subepi- and subendocardial hemorrhages, gastroenteritis (which may be hemorrhagic in some cases). Remnants of the leaves of the offending plant may be recognized in the ruminal contents, and dehydration can be marked.[4,7-13,15,22,23,38-41] In chronic cases, less venous congestion is seen and less marked gastroenteritis, but dehydration, emanciation, and hemorrhages in the myocardium can occur.[4,8,9,82]

5. Microscopic Pathology

Mild to moderate, acute to subacute, multifocal cardiomyopathy can be seen. It is suggested that the altered transfer of ions across the sarcolemma renders the myocardium more susceptible to injury.[22,23,87,88]

E. Diagnosis
1. General Methods of Diagnosis

Unlike in human medicine, cardiac-glycoside-plant poisoning in animals cannot be diagnosed by RIAs as they are developed for specific cardiac glycosides such as digoxin, digitoxin, and ouabain.[59,60,66,67,69] Where poisoning with foxglove (digoxin and digitoxin) or *Strophanthus* spp. and *Acokanthera* spp. (ouabain) are suspected, these tests can be useful. Normally, a case history can indicate the involvement of a cardiac-glycoside-containing plant. This may be substantiated by finding the suspected plants, and it is more definite if there are signs of it having been eaten by animals.[9,12]

Clinical examination of sick animals, taking note of symptoms displayed and post-mortem examination of dead animals can give more evidence to indicate the cause of poisoning.[2,4,9]

ECG examination can give additional evidence.[80-81,83-86] For ruminant animals, Schultz et al.[85] propose that reproducible results with ECG recording can be obtained if the Einthoven triangle is moved to the sagittal plane. This is achieved by shifting the right forelimb lead to the head between the ears, the left forelimb lead to the top of the sacrum, the left hindlimb lead to the median sternal region, the earth electrode to the left hind leg just above the hock, and using subcutaneous needle electrodes. Physiopathological parameters can be measured in critically ill animals and compared to those indicating cardiac-glycoside poisoning.[80,81] In addition, the ratio of potassium to sodium ions of red cells of the living animal can be determined. Cardiac glycosides cause a significant increase in sodium and decrease in potassium ions, and by determining these concentrations, cardiac-glycoside poisoning can be diagnosed.[3]

2. The Rubidium Test[89]

With the rubidium test the presence of any unknown cardiac glycoside, in toxic quantities, can be determined in fresh specimens of liver, kidney, and ruminal contents. This is an indirect test which is difficult to perform, but nonspecific, as it measures the effects of cardiac glycosides and can indicate the presence of all cardiac glycosides affecting an animal. Tissues of dead animals can be taken as specimens for this test.

For this method, red blood cells are made use of as models of cells which possess the $Na+,K+ -ATPase$. If a cardiac glycoside is added to the physiological fluid surrounding red cells, it adheres to this enzyme and inhibits active uptake of potassium by the red cells. In this method, rubidium is used instead of potassium as it is an element with similar characteristics as potassium, and very little of it occurs normally in tissues of man and animals.

To test a specimen, red cells are suspended in a physiological fluid which contains no potassium. An extract of the specimen is added to the red-cell suspension and incubated (to allow cardiac glycoside activity, if present). Then a quantity of rubidium is added, and reincubation takes place. After a sufficient time lapse, the rubidium concentration is measured in the physiological fluid. If there was a cardiac glycoside in the specimen, rubidium uptake by the red cells would have been inhibited. To measure inhibition of rubidium uptake, a standard rubidium-uptake graph in the absence of cardiac glycosides must be determined as well as a graph of standard inhibition with graded concentrations of, for instance, digoxin. The test result is then compared to these graphs in order to determine any inhibition of rubidium uptake, which indicates the presence of a cardiac glycoside in the specimen.

F. Treatment
1. Activated Charcoal

This old and well-known remedy is still in use today because it can effectively adsorb cardiac glycosides and many other substances and harmlessly remove them. A large number of acutely poisoned animals can be saved by dosing the charcoal at a rate of 2 g/kg body mass after shaking it into water. The activated charcoal adsorbs all the cardiac glycosides still remaining in the rumen and a few in the small intestine where some circulating cardiac glycosides can go owing to enterohepatic recirculation. In most cases a single dose is sufficient.[15,52,78,82-84,90]

2. *Potassium Chloride*
With more chronic cases such as the paretic syndrome (nenta or krimpsiekte) of the Crassulaceae, it appears that potassium chloride at a dose rate of 1 g/kg body mass can be used beneficially in addition to activated charcoal. Hypokalemia is more likely to occur in chronic poisoning, and that explains why potassium chloride can be used in these cases.[1,3,58,71,82] No definite cardiac therapy is described in veterinary literature, but a few remedies used in human medicine can be considered.

3. *Atropine*
Atropine has been used with visible improvement in Angora goat rams poisoned with *Cotyledon orbiculata*. The dose rate was 10 mg/g atropine sulfate subcutaneously initially, and again 5 mg on the following day. Atropine can abolish vagal tone and, thus, increase the heart rate where bradycardia occurs. It can also be useful where partial or total heart block occurs.[1,3,15]

4. *Beta-Adrenergic Blocking Agents*
Button et al.[80,81] state that these agents can be beneficial where there are ventricular dysrhythmias. They can aid in restoring normal cardiac function in the critically poisoned animal. Contraindications are partial and total heart block.[1,3,58,80,81]

5. *Miscellaneous Supporting Drugs*
Other antiarrhythmic drugs can be used, but they can also aggravate the condition where heart block occurs.[1,3,58] Epanutin (diphenylhydantoin) was not of practical use in experimentally poisoned animals as its therapeutic and toxic levels almost overlap in sheep and calves.[93] Oxygen and antacids can be beneficial in acute cases of poisoning and can be used to support treatment with activated charcoal and beta-adrenergic blocking drugs or atropine.[80,81]

Hemoperfusion with an activated-charcoal or chloestyramine filter is described for use in laboratory animals where it is effective. It may be successfully applied to treat a single valuable domestic animal.[3,53,55-57,76-78] Adsorption of cardiac glycosides to resins such as chloestyramine is another effective method available. Like activated charcoal, the cholestyramine is dosed to the animal in order to remove cardiac glycosides from the rumen and small intestine.[1,3,53,58,76]

6. *No Treatment*
In many cases, acutely poisoned animals are kept quietly without any treatment, and only water, food, and shelter are provided. A large number of them can survive with this "treatment". The reason is that some of the wilder poisoned animals will struggle fiercely when caught and held to be treated. Their wild attempts to escape overburden the heart, and they can die while being treated, or shortly thereafter.[4,8,9,41]

G. Prevention
There is no existing preventative medicine or vaccine available. Prevention can only be achieved by removing the stock from pastures during potentially dangerous periods, for

instance, during spring when the Iridaceae are flowering, or by eradication of the plants mechanically or with herbicides. Areas where these plants grow abundantly can be fenced off, and camps infested with poisonous plants should be used for mature stock only.[4,9,39,41] Other measures taken can lessen the losses.

First, newly introduced stock must be well fed before they enter a camp infested with these poisonous plants. Young stock and newly acquired stock must be placed in such a camp with stock which are used to the poisonous plants, and, hopefully, the new ones will follow the example of the accustomed ones to avoid the poisonous plants.[49]

Second, many South African farmers maintain that cattle and sheep can be "taught" to avoid eating the Iridaceae by predosing them with tulp (*Homeria* or *Moraea* spp.) before they are introduced to a tulp-infested camp. Strydom and Joubert tried to simulate this with a group of calves which they predosed with chopped tulp leaves, a boiled extract of tulp leaves and charred tulp leaves. They found that the various predoses gave no protection when compared to a control group and that all of the calves learned to avoid tulp by the fourth day in a tulp-infested camp.[42] Attempts to produce vaccines against cardiac glycoside plant poisoning are unsuccessful so far. Antigens used are conjugates of proteins (e.g., BSA) or "naked" bacteria and the commercially available cardiac glycosides. With these antigens, specific immunity is elicited against the cardiac glycoside in use. No cross-immunity to afford protection against the many known and unknown cardenolides and bufadienolides is achieved.[91,92]

IX. FUTURE PERSPECTIVES

Many cardiac glycosides of plant origin have been isolated and many others still await chemical investigation. Continued work in this field may yield more knowledge about cardiac glycosides and may add useful commercially extractable ones for cardiac therapy.[1,3,12] At the same time, chemical synthesis of cardiac glycosides or, at least, their unsaturated lactone rings can open new doors in veterinary medicine. With the aid of various conjugates of such lactone rings to naked bacteria, it may become possible to produce a vaccine and hyperimmune serum to some of the groups of cardiac-glycoside-containing poisonous plants.[91,92] For human medicine, a cardiac glycoside or synthesized cardiac glycoside with a wide safety margin is sought.[1,3] Grazing conditions can be improved by judicious rangeland management, and, additional to that, edible adapted plants can be planted where poisonous plants are eradicated.[4,9,41]

REFERENCES

1. **Moe, G. K. and Farah, A. E.,** Digitalis and allied cardiac glycosides, in *The Pharmacological Basis of Therapeutics,* 6th ed., Goodman, L. S. and Gilman, A., Eds., Macmillan, New York, 1980, chap. 30.
2. **Naudé, T. W.,** The occurrence and significance of South African cardiac glycosides, *J. S. Afr. Biol. Soc.,* 18, 7, 1977.
3. **Reynolds, J. E. F.,** Ed., Digoxin and other cardiac glycosides, in *Martindale The Extra Pharmacopoeia,* 28th ed., The Pharmaceutical Press, London, 1982, 531.
4. **Watt, J. M. and Breyer-Brandwijk, M. G.,** *The Medicinal and Poisonous Plants of Southern and Eastern Africa,* E. & S. Livingstone, Edinburgh, 1962, 63.
5. **Anderson, L. A. and Koekemoer, J. M.,** The chemistry of *Melianthus comosus* Vahl. II. The isolation of hellebrigenin 3-acetate and other unidentified bufadienolides from the rootbark of *Melianthus comosus* Vahl., *J. S. Afr. Chem. Inst.,* 21, 155, 1968.
6. **Anderson, L. A. P.,** Chemistry of plant poisons, *Bienn. Rep. Dir. Vet. Res. Inst. Onderstepoort,* 11, 1985/86.

7. **Corrigall, W., Moody, R. R., and Forbes, J. C.,** Foxglove *(Digitalis purpurea)* poisoning in farmed red deer *(Cervus elaphus),* Vet. Rec., 119, February 11, 1978.

8. **Henning, M. W.,** Krimpsiekte, *Rep. Vet. Res. Un. S. Afr.,* 11/12, 331, 1926.

9. **Steyn, D. G.,** *Vergiftiging van Mens en Dier,* J. L. Van Schaik Bpk, Pretoria, 1949, 114.

10. **Seiber, J. N., Lee, S. M., and Benson, J. M.,** Cardiac glycosides (Cardenolides) in species of Asclepias (Asclepiadaceae), *Handbook of Natural Toxins,* Vol 1., Keeler, R. F. and Tu, A. T., Eds., Marcel Dekker, New York, 1983, chap. 2.

10a. **Rothschild, M., Von Euw, J., and Reichstein, T.,** Cardiac glycosides in the oleander aphid, *Aphis nerii, J. Insect Physiol.,* 16, 1141.

11. **Everist, S. J.,** *Poisonous Plants of Australia,* Angus and Robertson, Sydney, Australia, 1974.

12. **Naudé, T. W. and Potgieter, D. J. J.,** Studies on South African cardiac glycosides. I. Isolation of toxic principles of *Homeria glauca* (W. & E.) N.E. BR. and observations on their chemical and pharamcological properties, *Onderstepoort J. Vet. Res.,* 4, 255, 1971.

13. **Benson, J. M., Seiber, J. N., Keeler, R. F., and Johnson, A. E.,** Studies on the toxic principle of *Asclepias eriocarpa* and *Asclepias labriformis,* in *Effects of Poisonous Plants on Livestock,* Keeler, R. F., Van Kampen, K. R., and James, L. F., Eds., Academic Press, New York, 1978, 273.

14. **Withering, W.,** An account of the foxglove and some of its medicinal uses: with practical remarks on dropsy and other diseases, C.G.J. & J. Robinson, London, 1785, Reprinted in *Med. Class.,* 1937, 2, 305, in *The Pharmacological Basis of Therapeutics,* 6th ed., Goodman, L. S. and Gilman, A., Eds., Macmillan, New York, 1980.

15. **Tustin, R. S., Thornton, D. J., and Kleu, C. B.,** An outbreak of *Cotyledon orbiculata* poisoning in a flock of Angora goat rams, *J. S. Afr. Vet. Assoc.,* 55, 181, 1984.

16. **Botha, P.,** Incidence and distribution of nenta poisoning ("krimpsiekte"), *Angora Goat Mohair J.,* July, 57, 1983.

17. **Cheeke, P. R. and Shull, L. R.,** Natural toxicants in feeds and poisonous plants, AVI Publishing, Westport, CT, 1985, 190.

18. **Benson, J. M., Seiber, J. N., Bagley, C. V., Keeler, R. F., Johnson, A. E., and Young, S.,** Effects on sheep of the milkweeds *Asclepias eriocarpa* and *A. labriformis* and of cardiac glycoside-containing derivative material, *Toxicon,* 17, 155, 1979.

19. **Anderson, L. A. P., Steyn, P. S., and Van Heerden, F. R.,** The characterization of two novel bufadienolides, lanceotoxins A and B from *Kalanchoe lanceolata* (Forssk) Pers., *J. Chem. Soc. Perkin Trans.* 1, 1573, 1984.

20. **Steyn, P. S., Van Heerden, F. R., and Van Wyk, A. J.,** The structure of cotyledoside, a novel toxic bufadienolide glycoside from *Tylecodon wallichii* (Harv) Toelken, *J. Chem. Soc. Perkin Trans. 1,* 965, 1984.

21. **Steyn, P. S., Van Heerden, F. R., Vleggaar, R., and Anderson, L. A. P.,** Bufadienolide glycosides of the Crassulaceae. Structure and sterochemistry of orbicusides A—C, novel toxic metabolites of *Cotyledon orbiculata, J. Chem. Soc. Perkin Trans., 1,* 1633, 1986.

22. **Anderson, L. A. P., Joubert, J. P. J., Prozesky, L., Kellerman, T. S., Schultz, R. A., Procos, J., and Olivier, P. M.,** The experimental production of krimpsiekte in sheep with *Tylecodon grandiflorus* (Burn. F.) Toelken and some of its bufadienolides, *Onderstepoort J. Vet. Res.,* 50, 301, 1983.

23. **Anderson, L. A. P., Schultz, R. A., Joubert, J. P. J., Prozesky, L., Kellerman, T. S., Erasmus, G. L., and Procos, J.,** Krimpsiekte and acute cardiac glycoside poisoning in sheep caused by bufadienolides from the plant *Kalanchoe lanceolata* Forsk., *Onderstepoort J. Vet. Res.,* 50, 295, 1983.

24. **Naudé, T. W. and Schultz, R. A.,** Studies on South African cardiac glycosides. II. Observations on the clinical and haemodynamic effects of cotyledoside, in *Onderstepoort J. Vet. Res.,* 49, 247, 1982.

25. **Louw, P. G. J.,** The cardiac glycoside from *Urginea rubella* Baker. I. Isolation and properties of rubellin, in *Onderstepoort J. Vet. Sci. Anim. Ind.,* 22, 313, 1949.

26. **Steyn, P. S., Van Heerden, F. R., Vleggaar, R., and Anderson, L. A. P.,** Structure elucidation and absolute configuration of the tyledosides, bufadienolide glycosides from *Tylecodon grandiflorus, J. Chem. Soc. Perkin Trans.,* 1, 429, 1986.

27. **Louw, P. G. J.,** Transvaalin, a cardiac glycoside isolated from *Urginea burkei* BKR (Transvaal slangkop), *Onderstepoort J. Vet. Res.,* 25, 123, 1952.

28. **Veldsman, D. P.,** On venenatin, the cardiac glycoside from the bark and wood of *Acokanthera venenata* G. Don. I. Historical introduction, isolation, physical and chemical properties, *S. Afr. Ind. Chem.,* August, 145, 1949.

29. **Katz, A.** Über die glykoside von *Bowiea volubilis* Harvey. Glykoside und aglykone, *Helv. Chim. Acta,* 36, 1417, 1953.

30. **Stoll, A.,** The cardio active glycosides, *J. Pharm. Pharmacol.,* 1, 849, 1949.

31. **Schnell, Von R., Euw, J. V., Richter, R., and Reichstein, T.,** Die glykoside der samen von *Strophnthus sarmentosus* A.P.DC. Glykoside und aglykone, *Pharm. Acta Helv.,* 28, 289, 1953.

32. **Sapeika, N.,** The action of venenatin, a glycoside from *Acokanthera venenata, S. Afr. J. Med. Sci.,* 14, 87, 1949.
33. **Van Wyk, A. J. and Enslin, P. R.,** Bufadienolides from Iridaceae, *J. S. Afr. Chem. Inst.,* 22, 566, 1969.
34. **Mohr, K. and Reichstein, T.,** Isolierung von digitalinum verum aus den samen von *Digitalis purpurea* L. und *Digitalis lanata* Ehrh. Glykoside und aglykone, *Pharm. Acta Helv.,* 24, 246, 1949.
35. **Enslin, P. R., Naudé, T. W., Potgieter, D. J. J., and Van Wyk, A. J.,** 1 α, 2 α-Epoxyscillirosidine, the main toxic principle of *Homeria glauca* (Wood and Evans) N.E.Br., *Tetrahedron,* 22, 3213, 1966.
36. **Zoller, P. and Tamm, Ch.,** Die konstitution dses transvaalins, *Helv. Chim. Acta,* 36, 1744, 1953.
37. **Soldin, S. J.,** Digoxin—issues and controversies, *Clin. Chem.,* 33, 5, 1986.
38. **Hutcheon, D.,** The poisoning of stock by tulp, *Agric. J. Cape Good Hope,* 84, 1900.
39. **Mitchell, D. T.,** *Urginea macrocentra* (Baker): Its toxic effects on ruminants, in 11th and 12th Rep. Dir. Vet. Ed. Res., 303, 1927.
40. **Mitchell, D. T.,** *Urginea capitata* (Baker) the berg slangkop: Its toxic effect on Ruminants, *Onderstepoort J. Vet. Sci. Anim. Ind.,* 2, 681, 1934.
41. **Vahrmeijer, J.,** *Poisonous Plants of Southern Africa,* Tafelberg, Cape Town, 1981, chap. 5.
42. **Strydom, J. A. and Joubert, J. P. J.,** The effect of pre-dosing *Homeria pallida* Bak. to cattle to prevent tulp poisoning, *J. S. Afr. Vet. Assoc.,* 54, 201, 1983.
43. **Anderson, L. A. P., Joubert, J. P. J., Kellerman, T. S., Schultz, R. A., and Pienaar, B. J.,** Experimental evidence that the active principle of the poisonous plant *Thesium lineatum* L.f. (Santalaceae) is a bufadienolide, *Onderstepoort J. Vet. Res.,* in press.
44. **Joubert, J. P. J.,** Die effek van hartglikosiede en elektroliete op die selmembraan se Natruim, Kalium-adenosientrifosfatase, *J. S. Afr. Vet. Assoc.,* 52, 225, 1981.
45. **Glyn, I. M.,** The action of cardiac glycosides on ion movements, in *Pharmacol. Rev.,* 16, 381, 1964.
46. **Schwartz, A., Lindenmayer, G. E., and Allen, J. C.,** The sodium-potassium adenosine triphosphatase, in *Pharmacol. Rev.,* 27, 3, 1975.
47. **Skou, J. C.,** Enzymatic basis for active transport of Na^+ and K^+ across the cell membrane, in *Physiol. Rev.,* 45, 596, 1965.
48. **Hajdu, S. and Leonard, E.,** The cellular basis of cardiac glycoside action, *Pharmacol. Rev.* 11, 173, 1959.
49. **Otani, A., Palumbo, N., and Read, G.,** Pharmacodynamics and treatment of mammals poisoned by *Bufo marinus* toxin, *Am. J. Vet. Res.,* 30, 1865, 1969.
50. **Chen, K. K. and Kovarikova, A.,** Pharmacology and toxicology of toad venom, *J. Pharm. Sci.,* 56, 1535, 1967.
51. **Schmidt, D. H. and Butler, V. P., Jr.,** Immunological protection against digoxin toxicity, *J. Clin. Invest.,* 50, 866, 1971.
52. **Neuvonen, P. J., Elfving, S. M., and Elonen, E.,** Reduction of absorption of digoxin, phenytoin and aspirin by activated charcoal in man, *Eur. J. Clin. Pharmacol.,* 13, 213, 1978.
53. **Caldwell, J. H. and Greenberger, N. T.,** Interruption of the enterohepatic circulation of digoxin by cholestyramine. I. Protection against lethal intoxication, *J. Clin. Invest.,* 50, 2626, 1971.
54. **Ochs, H. R. and Smith, T. W.,** Reversal of advanced digitoxin toxicity and modification of pharmacokinetics by specific antibodies and fab fragments, *J. Clin. Invest.,* 60, 1303, 1977.
55. **Vale, J. A., Rees, A. J., Widdop, B., and Goulding, R.,** Use of charcoal haemoperfusion in the management of severely poisoned patients, in *Br. Med. J.,* 1, 5, 1975.
56. **Smiley, J. W., March, N. M., and Del Guercio, E. T.,** Hemoperfusion in the management of digoxin toxicity, *JAMA,* 240, 2736, 1978.
57. **Gilfrich, H. J., Kasper, W., Meinertz, T., Okonek, S., and Bork, R.,** Treatment of massive digitoxin overdose by charcoal haemoperfusion and cholestyramine, *Lancet,* 505, March 1978.
58. **Chung, E. K.,** Guide to managing digitalis intoxication, *Postgrad. Med.,* 49, 99, 1971.
59. **Butler, V. P., Jr., Schmidt, D. H., Smith, T. W., Haber, E., Raynor, B. D., and Demartini, P.,** Effects of sheep digoxin-specific antibodies and their fab fragments on digoxin pharmacokinetics in dogs, *J. Clin. Invest.,* 59, 345, 1977.
60. **Oliver, G. C., Jr., Parker, B. M., Brasfield, D. L., and Parker, C. W.,** The measurement of digitoxin in human serum by radioimmunoassay, *J. Clin. Invest.,* 47, 1035, 1968.
61. **Watson, J. F. and Butler, V. P., Jr.,** Biologic activity of digoxin—specific antisera, *J. Clin. Invest.,* 51, 638, 1972.
62. **Hinds, J. A., Pincombe, C. F., Morris, H., and Duffy, P.,** Competitive solid-phase enzyme immunoassay for measuring digoxin in serum, *Clin. Chem.,* 32, 16, 1986.
63. **Beyers, A. D., Spruyt, L. L., Seifart, H. I., Kriegler, A., Parkin, D. P., and Van Jaarsveld, P. P.,** Endogenous immunoreactive digitalis-like substance in neonatal serum and placental extracts, *S. Afr. Med. J.,* 65, 878, 1984.

64. **Scherrmann, J.-M., Sandouk, P., and Guedeney, X.,** Specific interaction between antidigoxin antibodies and digoxin-like immunoreactive substances in cord serum, *Clin. Chem.,* 32, 97, 1986.
65. **Smith, T. W., Butler, V. P., Jr., and Haber, E.,** Characterization of antibodies of high affinity and specificity for the digitalis glycoside digoxin, *Biochemistry,* 9, 331, 1970.
66. **Smith, T. W.,** Ouabain—specific antibodies: immunochemical properties and reversal of Na$^+$, K$^+$- activated adenosine triphosphatase inhibition, *J. Clin. Invest.,* 51, 1583, 1972.
67. **Freytag, J. W. and Litchfield, W. J.,** Liposome-mediated immunoassays for small haptens (digoxin) independent of complement, *J. Immunol. Methods,* 70, 133, 1984.
68. **Gardner, J. D., Kiino, D. R., Swartz, T. J., and Butler, V. P., Jr.,** Effects of digoxin—specific antibodies on accumulation and binding of digoxin by human erythrocytes, *J. Clin. Invest.,* 52, 1820, 1973.
69. **Smith, T. W.,** Radioimmunoassay for serumdigitoxin concentration: methodology and clinical experience, *J. Pharmacol. Exp. Ther.,* 175, 352, 1970.
70. **Butler, V. P., Jr., Smith, T. W., Schmidt, D. H., and Haber, E.,** Immunological reversal of the effects of digoxin, in *Fed. Proc. Fed. Am. Soc. Exp. Biol.,* 36, 2235, 1977.
71. **Follath, F.,** Digitalisintoxikation: Häufigkeit, symptome und behandlung, *Schweiz. Rundsch. Med. (Praxis),* 67, 415, 1978.
72. **Bismuth, C., Gaultier, M., Conso, F., and Efthymiou, M. L.,** Hyperkalaemia in acute digitalis poisoning: Prognostic significance and therapeutic implications, *Clin. Toxicol.,* 6, 153, 1973.
73. **Pilati, C. F. and Paradise, N. F.,** Ouabain-induced mechanical toxicity. Aberrations in left ventricular function, calcium concentration and ultrastructure, *Proc. Soc. Exp. Biol. Med.,* 175, 342, 1984.
74. **Lyon, A. F. and De Graff, A. C.,** Reappraisal of digitalis. X. Treatment of digitalis toxicity, *Am. Heart J.,* 73, 835, 1967.
75. **Gold, H. K. and Smith, T. W.,** Reversal of ouabain and acetyl strophanthidin effects in normal and failing cardiac muscle by specific antibody, *J. Clin. Invertebr.,* 53, 1655, 1974.
76. **Caldwell, J. H., Bush, C. A., and Greenberger, N. J.,** Interruption of the enterohepatic circulation of digitoxin by cholestyramine. II. Effect on metabolic disposition of tritium-labeled digitoxin and cardiac systolic intervals in man, *J. Clin. Invest.,* 50, 2638, 1971.
77. **Galloway, E. J. and Liu, C. T.,** Use of activated charcoal for hemoperfusion in Dutch rabbits, *Am. J. Vet. Res.,* 42, 542, 1981.
78. **Caldwell, J. H., Caldwell, P. B., Murphy, J. W., and Beachler, C. W.,** Intestinal secretion of digoxin in the rat, in *Naunyn Schmiedebergs Arch. Pharmakol.,* 312, 271, 1980.
79. **Vick, R. L., Kahn, J. B., Jr., and Acheson, G. H.,** Effects of dihydroouabain, dihydrodigoxin and dihydrodigitoxin on the heart-lung preparation of the dog, *J. Pharmacol., Exp. Ther.,* 121, 330, 1957.
80. **Button, C., Reyers, F., Meltzer, D. G. A., Mülders, M. S. G., and Killeen, V. M.,** Some physio-pathological features of experimental *Homeria glauca* (Wood and Evans) N.E. BR. poisoning in Merino sheep, *Onderstepoort J. Vet. Res.,* 50, 191, 1983.
81. **Button, C. and Mülders, M. S. G.,** Further physiopathological features of experimental *Homeria glauca* (Wood and Evans) N.E. BR. poisoning in Merino sheep, *Onderstepoort J. Vet. Res.,* 51, 95, 1984.
82. **Vermeulen, S. O. and Joubert, J. P. J.,** Die eksperimentele verwekking en behandeling van chroniese krimpsiekte by Angora bokke, *Karoo Agric.,* 3, 16, 1984.
83. **Joubert, J. P. J. and Schultz, R. A.,** The treatment of *Urginea sanguinea* Schinz poisoning in sheep with activated charcoal and potassium chloride, *J. S. Afr. Vet. Assoc.,* 53, 25, 1982.
84. **Joubert, J. P. J. and Schultz, R. A.,** The treatment of *Moraea polystachya* (Thunb) KER-GAWL (cardiac glycoside) poisoning in sheep and cattle with activated charcoal and potassium chloride, *J. S. Afr. Vet. Assoc.,* 53, 249, 1982.
85. **Schultz, R. A., Pretorius, P. J., and Terblanche, M.,** An electrocardiographic study of normal sheep using a modified technique, *Onderstepoort J. Vet. Res.,* 39, 97, 1972.
86. **Pretorius, P. J., Van der Walt, J. J., and Kruger, J. M.,** A comparison between haemodynamic and cellular micro-electrode experiments with the cardiac aglycone epoxyscillirosidin, *S. Afr. Med. J.,* 43, 1360, 1969.
87. **Newsholme, S. J., Van Ark, H., and Howerth, E.,** Measurements of mass, length and valve diameters from normal formalinfixed ovine hearts, *Onderstepoort J. Vet. Res.,* 51, 103, 1984.
88. **Newsholme, S. J. and Coetzer, J. A. W.,** Myocardial pathology of domestic ruminants in Southern Africa, *J. S. Afr. Vet. Assoc.,* 55, 89, 1984.
89. **Bourdon, Par. R. and Mercier, M.,** Dosages des hétérosides cardiotoniques dans les liquides biologiques par spectrophotométrie D'absorption atomique, *Ann. Biol., Clin. (Paris),* 27, 651, 1969.
90. **Joubert, J. P. J. and Schultz, R. A.,** The minimal effective dose of activated charcoal in the treatment of sheep poisoned with the cardiac glycoside containing plant *Moraea polystachya* (Thunb) KER-GAWL, *J. S. Afr. Vet. Assoc.,* 53, 265, 1982.
91. **Joubert, J. P. J.,** Voorlopige bevindinge in die ontwikkeling van 'n entstof teen krimpsiekte, *Karoo Agric.,* 3, 10, 1985.

92. **Bellstedt, D. U., Kruger, J. S., Polson, A., and Van der Merwe, K. J.,** Naked bacteria — an intigen carrier and adjuvant in one, *S. Afr. J. Sci.* 81, 57, 1985.
93. **Joubert, J. P. J.,** Unpublished data, 1981.

Chapter 4

SAPONINS

David Oakenfull and Gurcharn S. Sidhu

TABLE OF CONTENTS

I. INTRODUCTION

Saponins, as the name implies, are a group of naturally occurring compounds which have properties resembling soaps and detergents. They are a complex and chemically diverse group of compounds, mainly of plant origin,[1-5] but also occurring in a number of marine animals.[6,7] Their physiological effects are as diverse as their chemical structures and properties. They are by no means all as toxic as their inclusion in this series might imply. Some saponins are indeed toxic to both mammals and cold-blooded animals.[8,9] Others, however, occur in significant amount in commonly used food and forage plants, such as clover, alfalfa (lucerne), soybeans, and chick-peas.[3,4] Their effects may actually be beneficial. In human diets, for example, it has been suggested that saponins may lower plasma cholesterol concentrations[4] and reduce the risk of heart disease. Saponins are an ingredient in a recently patented potion to prevent snoring.[10]

Long before the first saponin was identified by Schmeideberg over 100 years ago,[11] peoples of cultures as diverse as Medieval European and Australian Aboriginal[12] had used green twigs from certain plants to poison and harvest fish from water holes and ponds. Saponin-rich plants are prominent among traditional herbal remedies in Europe, China, and India.[13]

In this chapter we review the physical, chemical, and physiological properties of the saponins occurring in food and forage plants. The subject has also been reviewed recently by Price, Johnson, and Fenwick.[3] Their main emphasis is the structural chemistry of saponins, and the reader's attention is drawn to this excellent article, particularly for this type of information.

II. CLASSES OF SAPONINS AND THEIR DISTRIBUTION IN PLANTS

In chemical terms, a saponin is a triterpene or steroid linked to one or more sugar groups.[5,14-16] There are thus two broad subdivisions of saponins: those with triterpenoid aglycones and those with steroid aglycones. Almost all the saponins in food and forage plants are of the triterpene class.

More than a hundred different saponins have been isolated from plants, identified, and, to greater or lesser extents, characterized. As well as different plant species producing many different steroid or triterpene aglycones, these can be linked to different sugars producing a proliferation of possible structures. The saponins present in soybeans are an apt example because they have been thoroughly investigated owing to the economic importance of soybeans. Five different aglycones have been identified (Figure 1), and these are linked to three monosaccharides which may be any three from galactose, arabinose, rhamnose, glucose, xylose, or glucuronic acid.[17] Thus, there are $5 \times {}^6P_3 = 600$ possible structures! Not all of these, of course, need be present, but there seems little doubt that there are considerably more than the ten saponins which have so far been separated chromatographically from soybeans.

Despite their diverse chemistry, saponins have a number of common and characteristic properties which include:

1. Bitter taste
2. Formation of stable foams in aqueous solution
3. Hemolysis of red blood cells
4. Toxicity to cold-blooded animals such as fish, snails, insects, etc.
5. An ability to interact with bile acids, cholesterol, or other 3-β-hydroxy steroids in aqueous or alcoholic solutions to form mixed micelles or coprecipitates

Saponins are present in many hundreds of plant species, but only a few of these are used as food by man or domestic animals. A review which appeared in 1980 identified 28 plant species widely used for human consumption as containing saponins.[4] However, this list now needs to be considerably extended. In Tables 1 to 3 we have listed those food plants used for human or animal consumption which are now known to contain significant levels of saponins. Many of these plants are herbs, spices, or medicinal plants and consumed only in trivial quantities. However, some are staple items of diet for a large part of the population of the world. Among the major ones are chick-peas in the Middle East, India, and Pakistan, soybeans in much of Southeast Asia, and peanuts in central Africa. Another is quinua (*Chenopodium quinoa* Willd.) which is a food crop grown in the Andes at altitudes of 2500 to 4000 m. It produces seeds in sorghum-like clusters which can be used in soup or ground

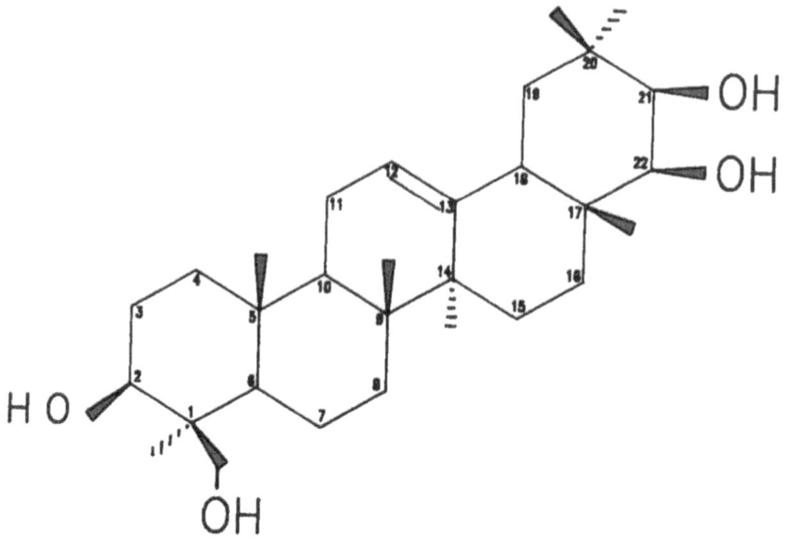

A

FIGURE 1.   Structures of the soysapogenins. That for soysapogen A shows the standard numbering for the triterpenoid ring system.

into flour for bread or cake.[38] Yams (*Dioscorea* species) are a food staple in parts of Southeast Asia and the Pacific region. The tubers of many members of the yam family contain saponins[53] which can be present at relatively high concentrations. *D. composita*, which contains about 13% saponin, is cultivated as a source of diosgenin and used in the manufacture of synthetic steroid-based hormones.[3] Only *Dioscoria* with low saponin contents are used for human consumption. Those with high levels of saponin are bitter and often toxic.[3] It has recently been reported that blackberries contain saponins[39] though the evidence is at present somewhat tenuous.

There are fewer significant animal feed or forage plants which contain saponins (Table 3). The major ones are alfalfa (lucerne), some clovers, guar, sunflowers, and lupine.

Where possible we have given the level of saponins present. Remarkably high concentrations are found in alfalfa (lucerne), particularly in the the immature sprouts, and in chick-peas. In some species there is considerable variation in the saponin content among different varieties and cultivars. This is particularly so for alfalfa for which a more than tenfold difference has been reported between low-saponin and high-saponin varieties.[54] Fenwick and Oakenfull[41] have noted a substantial difference in the saponin contents of navy beans (*Phaseoulus vulgaris*) from different sources (Canada, U.S., and Australia), and recently, a substantial difference has been reported between two named cultivars: Kerman, 6.4 g saponin per kilogram dry weight and Gallaroy, 19 g per kilogram dry weight.[57] In contrast, the different varieties of lentils (*Lens culinaris*) have much the same saponin content, despite very obvious differences in color and size of the seeds.[41] Different cultivars of chick-pea and mung bean have also been found to have only small differences in saponin content.[56] In general, there seems to be no special preponderance of saponin-containing plants in any particular genus. However, the vast majority of those used for food or forage and appearing in Tables 1 to 3 are from legumes. Table 4 shows the per capita consumption of some of these foods. They represent a not inconsiderable dietary intake of saponins.

Within plants, the saponins appear to be most concentrated in the roots and rapidly growing shoots. High concentrations of saponin have been found in alfalfa shoots[41,54] and in the

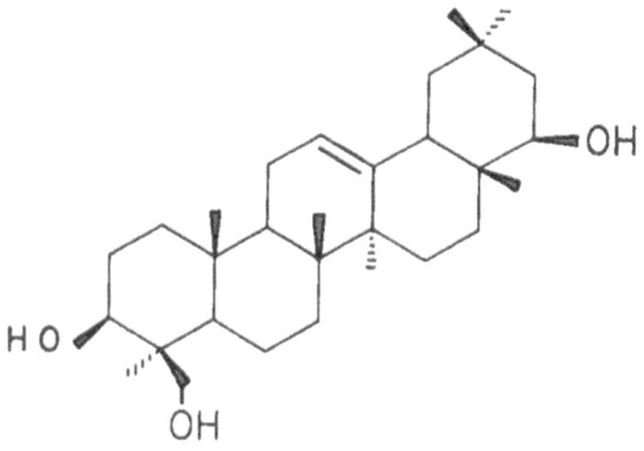

FIGURE 1 (continued)

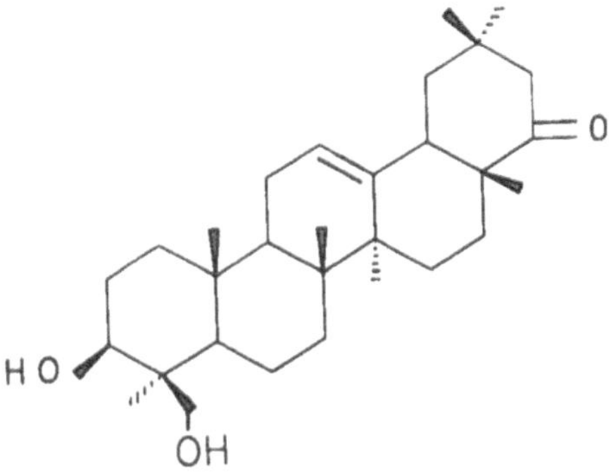

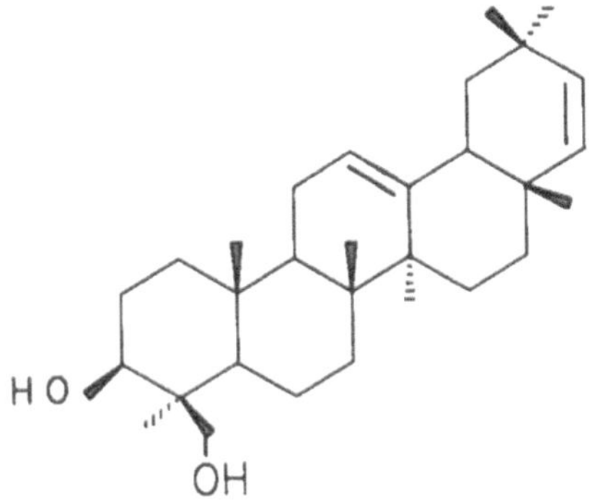

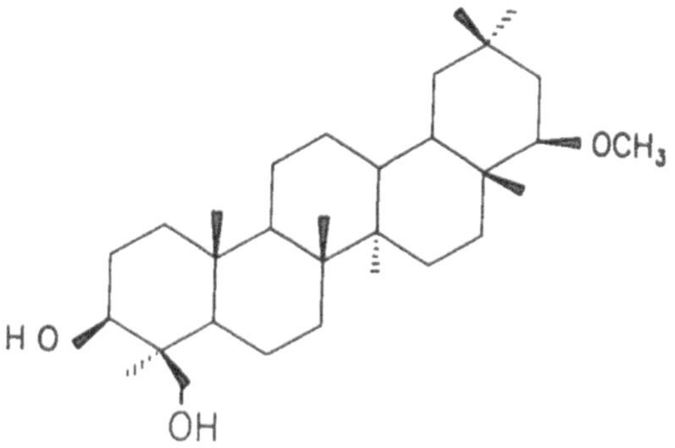

sprouts of chick-peas and mung beans.[41] Also, some trees, such as *Quillaia saponaria* and the Australian melaleucas,[57] have high levels of saponin in the bark. This distribution is consistent with the idea that the role of saponins is to protect the plant from insect or fungal attack,[18] with the saponins being concentrated in the most vulnerable parts.

III. METHODS OF EXTRACTION AND ANALYSIS

A. Extraction and Isolation

Saponins can be isolated from plant materials by extraction with organic solvents. The dried plant material is first extracted with acetone or hexane, preferably using a Soxhlet® extractor, to remove lipids, pigments, etc. The residue is then further extracted with methanol which removes the saponins, along with many other compounds such as simple sugars, oligosaccharides, and flavonoids. Various schemes have been devised for isolating the saponins. For example, the methanol extract can be partitioned between a 1:1 mixture of *n*-butanol and water.[58,59] The saponins pass into the *n*-butanol layer which is evaporated to dryness, the residue taken up in methanol, and the saponin precipitated by adding a large amount of diethyl ether. A method using liquid flash chromatography has recently been described by Price and Fenwick.[60] They isolated soyasaponin I from peas (*Pisum sativum*). They applied their methanol extract (redissolved in water) to a flash chromatography column containing a reverse-phase octadecylsilane bonded to silica gel. The column was washed with water and then eluted with methanol. The crude saponin fraction was then purified by using normal phase silica gel and eluting with chloroform:methanol (1:1 by volume).

Table 1
PLANT FOOD STAPLES WHICH CONTAIN
SIGNIFICANT LEVELS OF SAPONIN

Plant	Saponin content (g/kg)	Ref.
Chick-pea (*Cicer arietinum*)	2.3[a]—60[b]	18—20
Soybean (*Glycine Max*)	5.6[a]—56[b]	17,21
Navy bean (*Phaseolus vulgaris*)	4.5—21[b]	20,22
Kidney bean (*P. vulgaris*)	2[c],3.5[a],16[b]	20,22
Mung bean (*P. mungo*)	0.5[a],5.7[b]	23
Mung bean shoots	27[b]	
Broad bean (*Vicia faba*)	3.5[b]	24
Green pea (*Pisum sativum*)	1.8[a],1.1[b]	18,25
Lentil (*Lens culinaris*)	1.1[a],5.1[b]	18,19
Azuki bean (*Vigna angularis*)	—	26—28
Peanut (*Arachis hypogyea*)	16[a],0.05[b]	18,29
Onion (*Allium cepa*)	—	30
Garlic (*Allium sativum*)	—	30
Leek (*Allium porrum*)	—	30
Asparagus (*Asparagus officinalis*)	15[b]	31,32
Spinach (*Spinacea oleracea*)	47[b]	33
Silver beet (*Beta vulgaris*)	58[b]	34
Eggplant (*Solanum melongena*)	—	35
Sunflower (*Helianthus annuus*)	—	36
Sesame seed (*Sesamum indicum*)	3[b]	20
Oats (*Avena sativa*)	1[b],1.3[d]	37
Quinua (*Chenopodium quinoa*)	—	38
Blackberry (*Rubus ssp. Hyb.*)	—	39

[a] Reference 40.
[b] Reference 41.
[c] Reference 42.
[d] Reference 43.

Table 2
SOME SAPONIN-CONTAINING
PLANTS COMMONLY USED AS
FLAVORINGS, HERBS, OR
SPICES

Plant	Ref.
Ginseng (*Panax ginseng*)	43
Sarsaparilla (*Smilax aristolochiifolia*)	1
Licorice (*Glycyrrhiza macedonica*)	1
Nutmeg (*Myristica fragrans*)	44
Fenugreek (*Trigonella foenum-graecum*)	45
Thyme (*Thymus vulgaris*)	1
Sage (*Salvia officinalis*)	1
Linden (*Tilia europaea*)	1
Tea (*Thea sinensis*)	1
Quillaia bark (*Quillaia saponaria*)	46

Table 3
SAPONIN-CONTAINING
PLANTS COMMONLY USED
AS LIVESTOCK FEED

Plant	Ref.
Alfalfa (lucerne; *Medicago sativa*)	47
White sweet clover (*Melilotus alba*)	48
Red clover (*Trifolium pratense*)	49
Labino clover (*Trifolium repens*)	50
Burr clover (*Medicago hispida*)	50
Guar (*Cyamopsis tetragonobla*)	48
Lupine (*Lupinus* spp.)	52

Table 4
AVERAGE DAILY PER CAPITA
CONSUMPTION (g) IN THREE
COUNTRIES OF SAPONIN-
CONTAINING FOOD PLANTS[a]

	Syria	Hong Kong	U.K.
Chick-pea	15	—	—
Lentil	17	—	1
Dry peas/beans[b]	7	—	7
Sesame seed	3	1	—
Soybean	—	9	—
Peanuts	5	4	4
Eggplant	36	1	—
Tomato	19	7	14
Onions/garlic	4	3	16
Green peas/ beans[b]	30	2	29

[a] Food Balance Sheets, Food and Agriculture Organization of the United Nations, Rome, 1980.
[b] Not otherwise specified.

B. Identification of Saponins

1. Foam Characteristics

Aqueous solutions of saponins form very stable foams, and this has long been used as a first indication of the presence of saponins in a plant extract,[1,61] although it is important to be aware that there are many other potentially surface-active compounds present in plants which could also cause the formation of stable foams.

2. Hemolytic Activity

Another property characteristic of saponins which has long been recognized as diagnostic is their ability to lyse red blood cells.[1] This property, which is discussed in more detail in Section VII, accounts for their extreme toxicity when given intravenously. Again, it is important to note that other surfactants also have hemolytic activity and that not all saponins, for example, some of those from soybeans, are necessarily hemolytically active. Thus, hemolytic activity from a plant extract, like the ability for form stable foams, is suggestive, but not in itself an unequivocal indication that saponins are present.

3. Chromatography

Chromatographic methods provide the most reliable means of detecting the presence of saponins in plant extracts. Most of these can be used quantitatively and are described in the following section, but thin-layer chromatography (TLC) still appears to be the most rapid and reliable method for quantitative detection.[20,41]

Several solvent systems have been described as suitable for developing TLC. Examples are chloroform/methanol/water (65:35:10)[43] or n-butanol/ethanol/15 M ammonia (7:2:5).[62] A variety of spray reagents can be used which give characteristic colors with saponins, for example, vanillin in phosphoric acid[63] or anisaldehyde in sulfuric acid.[63] The reactions are characteristic of steroids or triterpenes in general and so are not specific for saponins. Another useful procedure for detecting saponins on TLC plates exploits their hemolytic activity. If the plate is coated with a suspension of erythrocytes in isotonic buffer and gelatin, saponins appear as white spots on a red background.[64] Oakenfull and Fenwick[20,41] have suggested that positive results from at least three of these tests are necessary for reliable identification of saponins.

4. Interaction With Other Steroid Molecules

Some saponins form insoluble complexes with sterols. For example, the interaction of digitonin with cholesterol and related sterols is well known and was extensively exploited in various isolation and identification procedures in the days before chromatography.[1] It has been found that specific structures are required for complex formation. For example, in the case of the triterpene saponins from alfalfa, complex formation requires the presence of an intact steroid-ring system having the conformation of cholesterol to which a side chain characteristic of cholesterol or phytosterols is attached. On the aglycone of saponin, unsubstituted carboxyl groups are required at carbons 23 and 28 since blocking these removes the ability of saponin to form complexes.[65]

C. Quantitative Methods

1. Early Methods

Reliable quantitative estimation of the saponin content of plant material has proved very difficult in the past, and any data obtained before 1980 are at best only approximate. Here these early methods will be only briefly outlined.

A useful indication of the saponin content of plant material, albeit only a lower limit, can be obtained by determining the yield of purified saponin gravimetrically.[54,66] Other simple methods exploit the characteristic physical properties of saponins, such as their foaming power[61] or their hemolytic activity.[67] Spectrophotometric methods have been reported which exploit the colors produced by the reactions of saponins with vanillin[68] or anisaldehyde.[69] Although very sensitive, these methods are not suitable for estimating saponins in crude plant extracts because the reactions are not specific, and colored compounds can be produced from other phytosterols and compounds such as flavonoids which are likely to accompany saponins in such preparations. The vanillin method has, though, been used to estimate the quantity of ginseng saponins in candies and jellies.[70] Bioassay methods have also been used extensively and include toxicity to fish,[71] inhibition of germination of seeds,[72] and inhibition of growth of the mold *Trichoderma viride*.[73] This last method has been developed specifically for alfalfa saponins.

More reliable methods have been developed in recent years, mostly based on TLC or high performance liquid chromatography (HPLC).

2. Thin-Layer Chromatography

Quantitative TLC is a well-established technique[74] which has been applied to saponin estimation by several workers.[19,20,25,41] The essence of the technique is to spot a TLC plate with a crude saponin extract, develop the plate with a suitable solvent system, and use one of a number of methods for estimating the quantity of saponin on the plate. The solvent systems used most frequently for developing plates have been *n*-butanol:ethanol:concentrated ammonia (7:2:5)[41] and chloroform:methonol:water (13:7:2).[75] A scanning densitometer is the quickest and easiest means of estimating the intensity of the saponin spots which can be visualized by spraying the plate with 10% sulfuric acid in ethanol and then heating it at 110° for 10 min. The method has the great advantage that there is no risk of losing any material in elaborate purification procedures. A major disadvantage, though, is the large number of different spots produced on a chromatogram of a crude extract. Only a few of these will be saponins as many other compounds, particularly sugars, are likely to be present. Fenwick and Oakenfull[20,41] have discussed this problem in some detail. They advocated making a number of chromatograms of the particular extract and visualizing the spots in three different ways:

1. Spray the chromatogram with anisaldehyde (1 ml) and sulfuric acid (2 ml) in acetic acid (97 ml) and heat at 110° for 10 minutes to give characteristic purple-mauve colors with steroids or triterpenes.
2. Treat another chromatogram with silver nitrate followed by sodium hydroxide to visualize reducing sugars.
3. Coat a third plate with a suspension of erythrocytes in isotonic buffer with gelatin. Saponins then slowly appear as white spots against a red background.

Only spots appearing with all three treatments should be attributed to saponins. Despite these precautions, it is possible that some of Fenwick and Oakenfull's measurements of levels of saponins in food plants could be considerably overestimated.[3,40]

Curl, Price, and Fenwick[25] have recently introduced flash chromatography as a purification step preceding quantitative TLC. The methanolic extract (in aqueous solution) is applied to a flash chromatography column containing reversed phase octadecylsilane bonded to silica gel. The column is then eluted with methanol and the eluate evaporated, redissolved quantitatively, and used for TLC. Using this method, Curl et al.[25] found considerably lower levels of saponin in soybeans than those found by Fenwick and Oakenfull (Table 1). The problem remains unresolved, but it is possible that the lower values of Curl et al. result from loss of material during the flash chromatography step.

3. Gas-Liquid Chromatography (GLC)

GLC is of very limited applicability because it can only be used for the separation and analysis of the nonpolar aglycone part of the saponin molecule. Thus, there are potential losses of material during hydrolysis and derivatization (the trimethylsilyl ether has been used[76]). The method is a useful source of structural information but less useful for quantitative analysis. It has, though, been used for analysis of the saponins in ginseng[77] and soybeans.[78]

4. High-Performance Liquid Chromatography

HPLC effectively separates nonpolar compounds and has been used for the separation and analysis of both aglycones and intact saponins.[75] Ireland and Dziedzic[79] have described a method for rapid analysis of soybean sapogenins by HPLC which can give a quantitative estimate of saponins by assuming that the carbohydrate to sapogenin ratio is 1:1 (by weight). Defatted soy flour is boiled under reflux with 1.5 *M* sulfuric acid in dioxane-water (1:3). This hydrolyzes the saponins, and the sapogenins can then be extracted with ethyl acetate. HPLC is then carried out with a commercial silica column, eluting with light petroleum-ethanol with a gradient technique, increasing the proportion of ethanol. More recently, the same authors have described separation of intact saponins by HPLC and suggest that the method could be made quantitative once suitable standards have been obtained.[80]

Oleanane saponins have also been separated by HPLC, and it has been suggested that the method could be made quantitative and generally applicable to other saponins.[81]

IV. CHEMISTRY OF SAPONINS

Saponins have been found in many hundreds of plants, and there is an abundance of information about their chemistry and general properties. Here we restrict ourselves to the relatively few food saponins whose chemistry has been studied in any detail.

A. Saponins in Plants Used for Human Consumption

1. Soybeans

Soy saponins are the most thoroughly investigated of the saponins occurring in food plants. The sapogenins were first identified and characterized in 1964,[83] and more recently, the structures of many of the saponins have been elucidated, particularly by the group of Kitagawa et al.[21,83] Recent work has cast some doubt on the earlier identifications of the sapogenins, and the current situation is that there are four of these compounds whose structures are well established (Figure 1).[89] Several more are less certainly identified. Confusion has been caused by the production of additional sapogenin-like compounds during hydrolysis of the crude saponins. Kitagawa et al.[83] have concluded that what was called soysapogenol D is such an artifact, and Jurzysta[84] has shown that soysapogenols C and F are also artifacts produced by hydrolysis of soysapogenol B.

The sugars present on the intact saponins are galactose, arabinose, rhamnose, glucuronic acid, xylose, and glucose.[85] Kitagawa et al.[21] have succeeded in isolating, purifying, and characterizing five of the saponins: soysaponins A_1 and A_2 (which have soysapogenol A as the aglycone) and soysaponins I, II, and III (which have soysapogenol B as the aglycone). The disaccaride analogue of II, soysaponin IV, is a minor constituent of this mixture.

2. Peas and Beans

Many of the edible beans are varieties of *Phaseolus vulgaris* (examples of these are navy beans, kidney beans, green beans). Soysaponin I (Figure 2) has been identified in all of these and also in the closely related runner bean (*P. aureus*),[19,86] butter bean (*P. lunatus*),[19,86] and scarlet runner bean (*P. coccineus*).[19,86] Soysapogenols B and C have been identified as sapogenols present in these species. Soysapogenol B has been identified as an aglycone present in the saponins from faba beans (*Vicia faba*),[19] chick-peas (*Cicer arietinum*),[19] and lentils (*Lens culinaris*).[19] These species, as well as garden peas (*Pisum sativa*) have also been found to contain soysaponin I.[60] The saponins from azuki beans (*Vigna angularis*) have recently been carefully investigated by Kitagawa et al.[26,28] They found four sapogenols: soysapogenol B, gypsogenic acid, sophoradiol, and the previously unknown compound azukisapogenol (Figure 3).

3. Silver Beet and Sugar Beet

These are varieties of *Beta vulgaris*, the first, eaten as a green vegetable and the second, used for sugar production. The most abundant sapogenin is oleanolic acid (Figure 4), but there are at least five others.[87] A structural formula has been suggested for one of the saponins.[87]

4. Spinach (Spinacea oleracea)

There are two major sapogenins, oleanolic acid and hederagenin (Figure 4),[33] and a third has recently been identified as olean-12-en-2,3-diol-23-one-28-oic acid.[3] Structural formulas have been reported for the two most abundant saponins.[33]

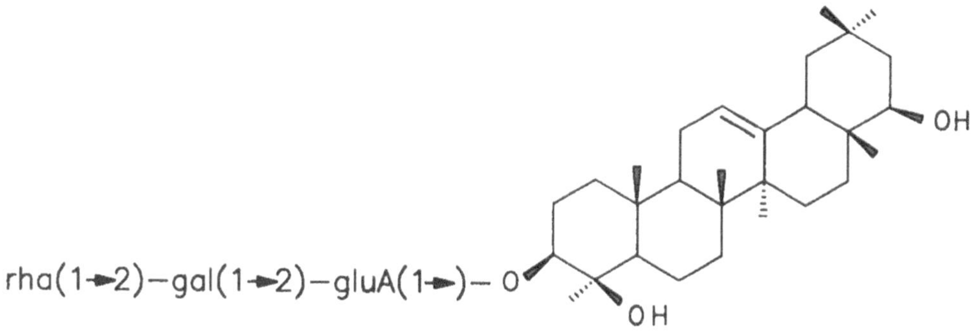

FIGURE 2.   Soysaponin I

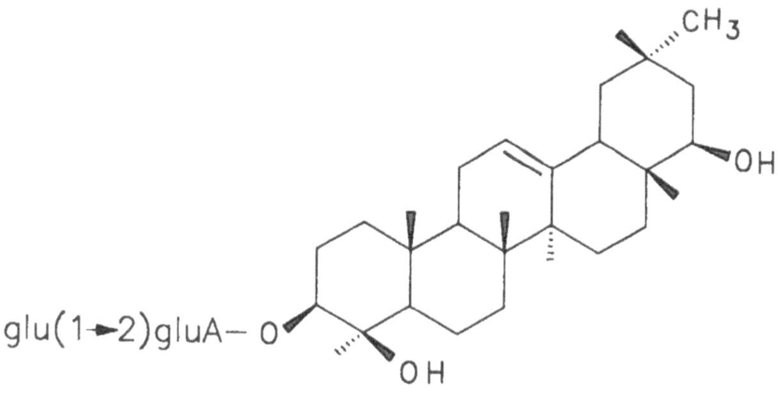

FIGURE 3.   Azukisapogenol

5. Asparagus (Asparagus officinalis)

Asparagus saponins seem to be concentrated in the roots and white shoots.[32] They are one of the few groups of steroidal saponins known to occur in food or forage plants. Two aglycones, sarsapogenin and yamogenin (Figure 5),[31,32] have been identified. There are at least five saponins and structural formulas for two of them have been reported.[31,32]

6. Oats (Avena sativa)

Interestingly, oats are the only cereal known to contain saponins, and they also have steroidal aglycones.[88] Four have been identified: avenestergins A-1, A-2, B-1, B-2; and recently four of the saponins have been isolated by HPLC and their structures identified.[89-91]

7. Potatoes and Tomatoes (Solanum spp.)

The saponin-like toxic alkaloids, α-chaconine and α-solanine (Figure 6), are present in potatoes, particularly any parts that are green through exposure to light.[92] Tomatoes contain the similar compound α-tomatine (Figure 6).[93] The seeds contain a saponin with the structure shown in Figure 7.[89] Saponins have been isolated and identified from the related species eggplant (*S. melongena*),[35] but may not be present in all cultivars because none could be detected in a variety grown in Australia.[41]

8. Onions, Garlic, and Leeks (Allium) spp.

Five distinct saponins have been found in garlic and leeks and three in onions.[30] Garlic

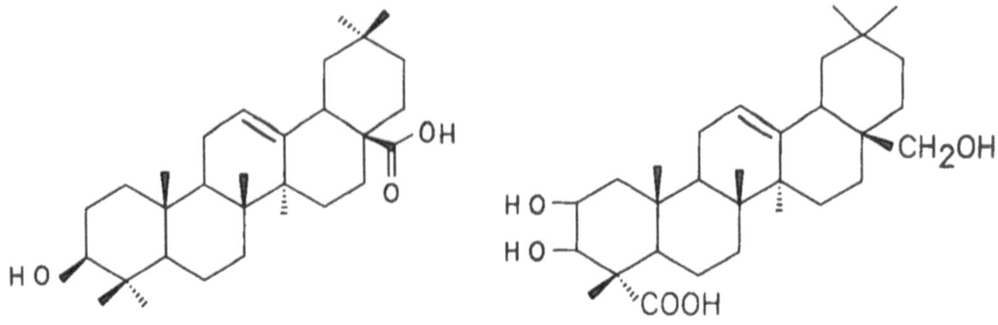

FIGURE 4.   Structures of (A) oleanolic acid and (B) hederagenin.

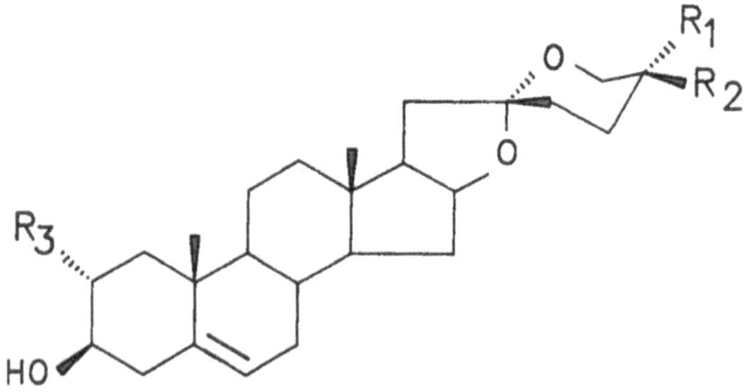

	R_1	R_2	R_3
(a):	$-CH_3$	H	H
(b):	H	$-CH_3$	$-OH$

FIGURE 5.   Structures of (a) sarsapogenin and (b) yamogenin.

saponins appear to be based on only one aglycone, sitosterol.[30] Onion saponins have three: sitosterol, oleanic acid,[30] and β-amyrin.[30] Leeks have been reported to contain oleanolic acid and gitogenin together with three unknown sapogenins.[30] None of the saponins have been identified.

9. Peanuts (Arachis hypogaea)

Four different aglycones have been isolated from peanut saponins, and one of these has been shown to be identical to soysapogenol B (Figure 1).[29] The structures of the others remain unknown, although recently one of the saponins has been identified as soysaponin I (Figure 2).[94]

10. Quinua (Chenopodium quinoa)

Several saponins have been separated by column chromatography. The sapogenins have been analysed by GLC and found to be 79.5% oleanolic acid.[95]

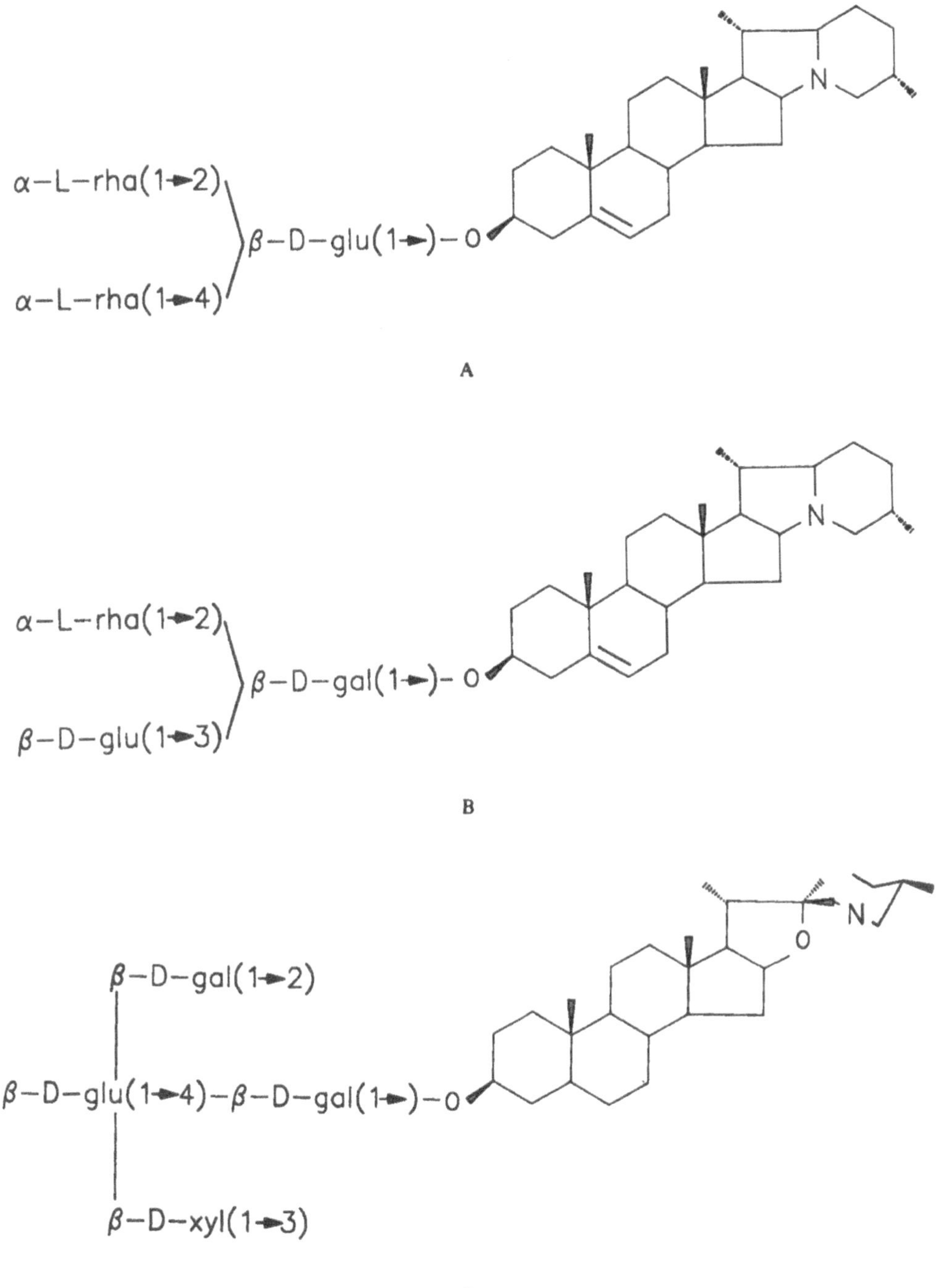

α–L–rha(1→2)
α–L–rha(1→4) $\Big\}$ β–D–glu(1→)–O

A

α–L–rha(1→2)
β–D–glu(1→3) $\Big\}$ β–D–gal(1→)–O

B

β–D–gal(1→2)
|
β–D–glu(1→4)–β–D–gal(1→)–o
|
β–D–xyl(1→3)

C

FIGURE 6. Structures of (A) α-chaconine, (B) α-solanine, and (C) α-tomatine.

11. Tea (Thea sinensis)

Five sapogenins have been isolated from tea: theasapogenols A to E (Figure 8).[96,97] Tschesche et al.[98] have found that the saponin contains 45% theasapogenol E, 30% A, 15% B, and traces of C and D. One of the intact saponins has been characterized. There were four sugars: glucuronic acid, galactose, xylose, and arabinose.[99]

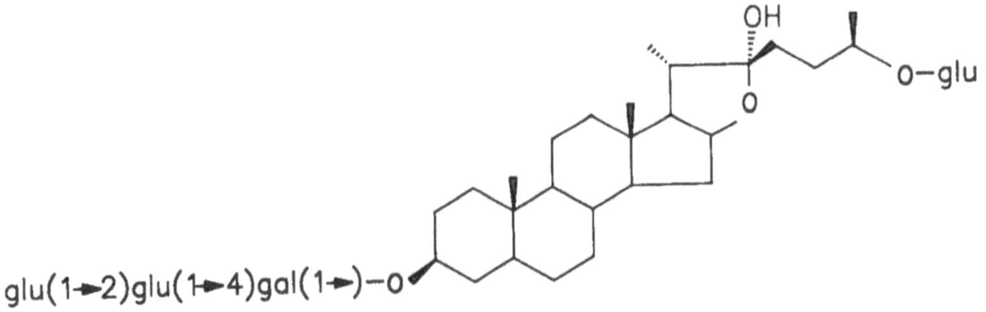

FIGURE 7.   A tomato saponin.

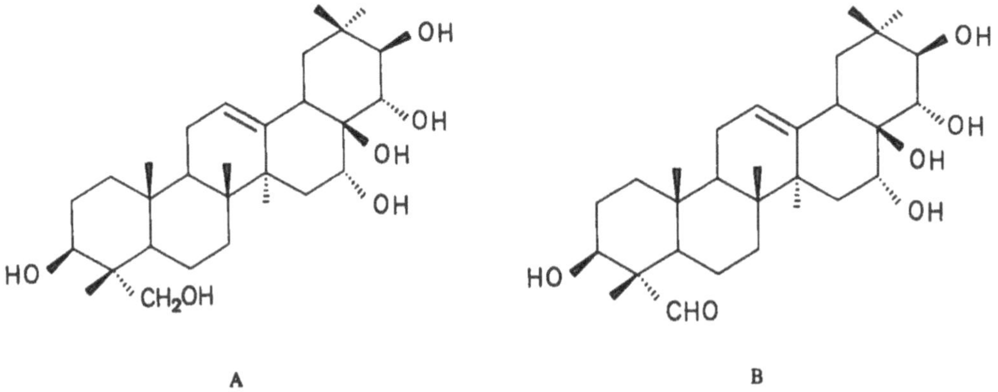

A

B

FIGURE 8.   Structures of the theasapogenols.

12. Fenugreek (Trigonella foenum-graecum)

The green leaves of this plant are used in India as a vegetable, but in the West it is the seeds that are used, primarily as a spice, but also for medicinal purposes. Gangrade and Kaushal[100] have separated five saponins, fenugrins A to E, by a combination of column and thin-layer chromatography. They found that the major aglycone was diosgenin (Figure 9) and that the sugars were glucose, arabinose, and rhamnose. Another group, Varshney et al.,[101] have reported separating three saponins which they called graecunins A, B, and C. Again, the aglycone found was diosgenin with graecunin B containing glucose, xylose, and rhamnose, and graecunin C containing glucose and rhamnose. Two furostanol glycosides have recently been isolated from seeds by droplet countercurrent chromatography and their structures characterized.[102,103] The saponin content is relatively high, and there is consequently some interest in using fenugreek as a source of diosgenin for synthesis of steroid-based hormones.[104] Sapogenin levels of 0.6 to 1% have been reported, and the actual saponin content would be about double that.

13. Licorice (Glycyrrhiza glabra)

Licorice has a long history as a medicinal plant, and it has been known to contain saponins since the last century.[1] The major sapogenin is glycyrrhetinic acid (Figure 9), but a large number of minor ones have been isolated.[3] Few of the saponins have yet been characterized, but one of them appears to be a polyglucuronic derivative of glycyrrhetinic acid, a combination which makes a very unusual saponin in that it is sweet — reputedly 50-times sweeter than sucrose![105]

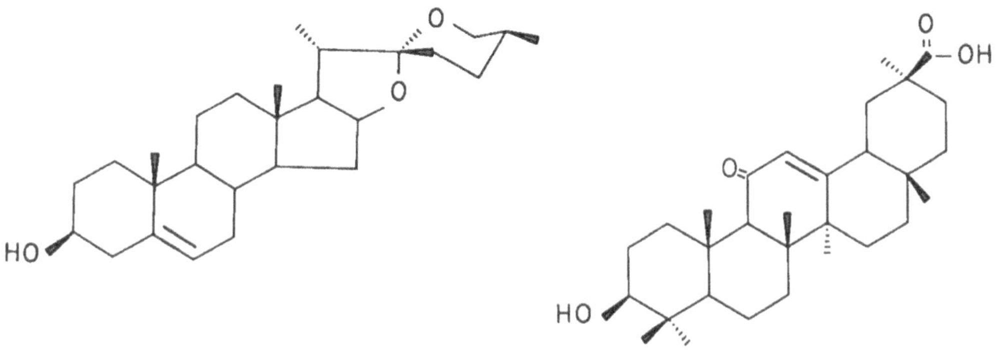

FIGURE 9.   Structures of (A) diosgenin and (B) glycyrrhetinic acid.

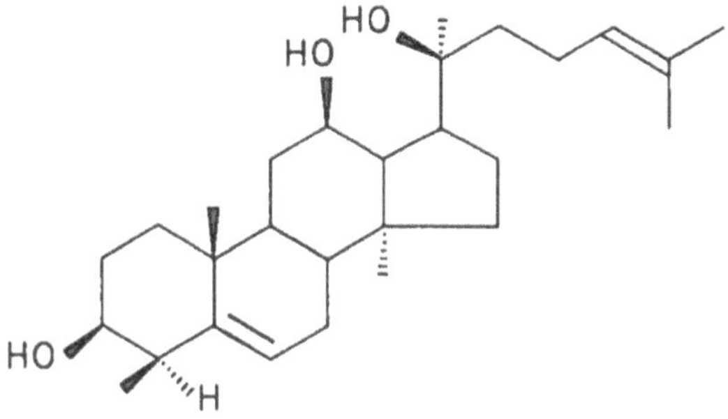

FIGURE 10.   Ginseng 20 (*S*)-protopanaxdiol.

14. Ginseng (Panax spp.)

There is great interest in the use as medicinal plants of the several species of plant known as ginseng. Ginseng root has long been a part of Chinese medicine and is becoming popular in the West as a stimulant and tonic[106,107] (with undertones of aphrodisiac powers). *Panax ginseng* is the traditional ginseng, but other species, particularly, *P. quinquefolium* (American ginseng), are used as cheaper alternatives. As a consequence of this interest, many investigations have been carried out on the chemistry of the ginseng saponins. A large number of them have been separated. The major sapogenin is 20(*S*)-protopanaxdiol (Figure 10) with the sugars glucose, arabinose, and xylose.

15. Quillaia (Quillaia saponaria)

The bark of the quillaia tree contains about 10% saponin and is a major commercial source.[108] It is included here because of its use by the food industry as a foaming agent, particularly in soft drinks such as ginger beer.[108] Quillaic acid (Figure 11) is the major aglycone, but gypsogenic acid (Figure 11) has recently been shown to be present also.[46] The sugars present are rhamnose, xylose, arabinose, galactose, glucuronic acid, and fucose. Kartnig and Ri[109] have isolated one of the saponins, quillinin A, by column chromatography. They found it to be quillaic acid linked to glucuronic acid, galactose, glucose, arabinose, xylose, fucose, and rhamnose. More recently, Higuchi et al.,[110] as a first step in structure determination, have elucidated the structure of two deacylsaponins obtained by mild alkaline hydrolysis of quillaia saponin. The structures of these are shown in Figure 12.

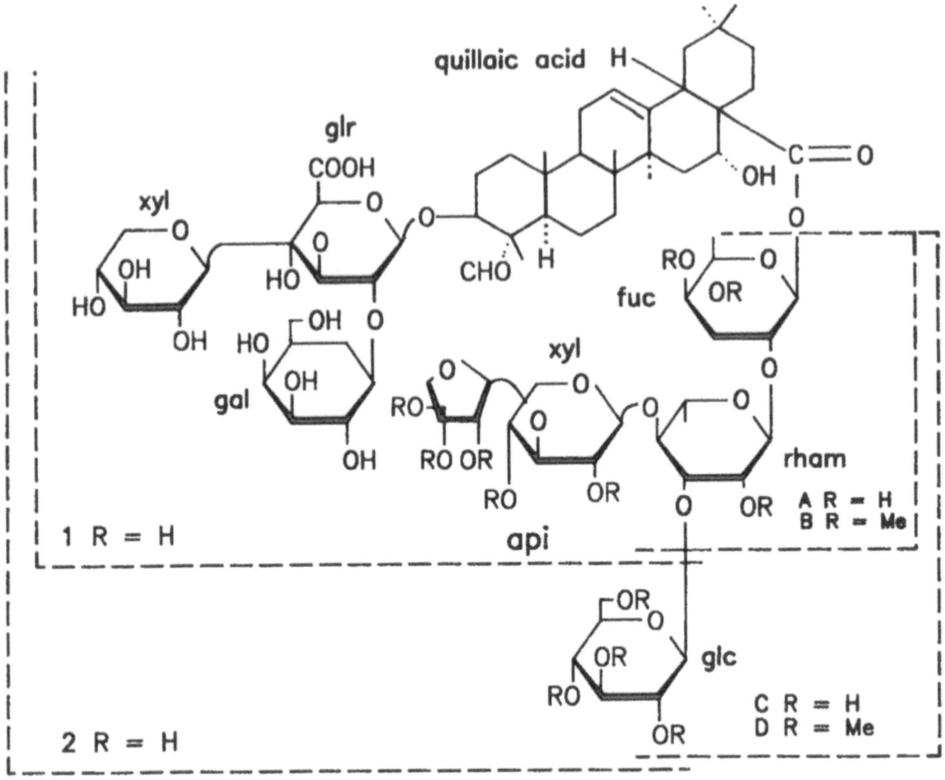

FIGURE 11.   Structures of two quillaia saponins.

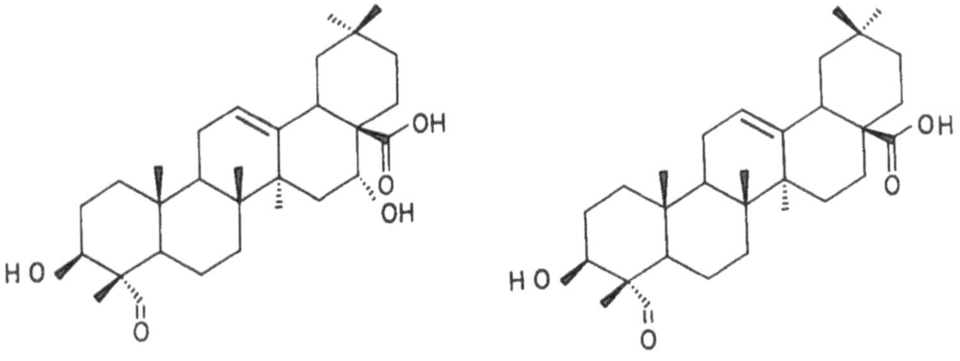

FIGURE 12.   Structures of (A) quillaic acid and (B) gypsogenic acid.

B. Saponins in Plants Used for Animal Feed

1. Alfalfa (Lucerne; Medicago sativa)

More seems to be known about alfalfa saponins than those from any other source.[47] This probably results from the economic importance of this crop as animal feed in many parts of the world[111] and the fact that antinutritional properties have been attributed to saponins and that they have been suggested as a cause of ruminant bloat,[112] although recent work has cast some doubt on this.[113]

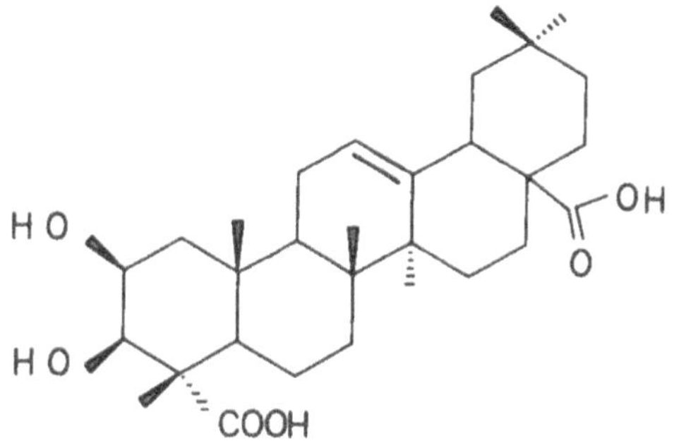

FIGURE 13.   Medicagenic acid.

The saponins of alfalfa seem to be a particularly complex mixture.[114] Hederagenin (Figure 4) and medicagenic acid (Figure 13) are the major aglycones,[115] but soysapogenols A to F have also been detected as minor components.[116,117] Chemical fractionation and TLC have revealed the presence of 33 saponins in one cultivar of alfalfa (Du Puits) and 27 in another (Lahontan).[116] The structures of three of the saponins have been determined.[118,119]

2. Other Forage Crops

Many of the clovers contain saponins, and there has been considerable interest in these because, as in alfalfa, it was once thought that they were responsible for ruminant bloat.[112] A saponin, melilotin, from white sweet clover (*Melilotus alba*) has been isolated and identified.[48] Two saponins have been isolated from red clover (*Trifolium pratense*) which are based on soysapogenins B and F,[49] and saponins based on soysapogenins B and C have been detected in labino clover (*T. repens*).[50] A saponin isolated from burr clover (*Medicago hispida*) has been shown to be hederagenin (Figure 4) linked to arabinose and rhamnose.[50]

3. Sunflowers (Helianthus annuus)

Hiller and Voigt[15] have isolated and characterized three saponins, which they call helianthosides A, B, and C, from sunflower petals. These are derived from echinocystic acid.

4. Guar (Cyamopsis tetragonobola)

As the cluster bean, guar is used extensively as a stock feed in the Indian subcontinent.[120] There is also interest in using guar meal as poultry feed.[121] Guar meal has been reported to contain about 12% saponin.[121] Curl et al.[51] have recently isolated two saponins and established that one of them has the structure shown in Figure 14. The other is an acetyl derivative of the same compound, the site of derivatization as yet unspecified.

5. Lupine (Lupinus spp.)

Hudson and El-Difrawi[52] have reported finding saponins in four species of lupine: *L. albus, L. augustrifolius, L. luteus,* and *L. mutabilis.* They found indications that the sapogenins were similar to those from soy, and recently, soysapogenol B has been found in another species, *L. verbasciformis.*[122]

V. PHYSICAL PROPERTIES OF SAPONINS

A. Solubility

Saponins generally are readily soluble in water, methanol, and aqueous ethanol. They are, however, virtually insoluble in such solvents as absolute ethanol, acetone, or diethyl ether. The solubility of gypsophilia saponin in water is 74 g/l at 30°, increasing to 180 g/l at 70°.[123]

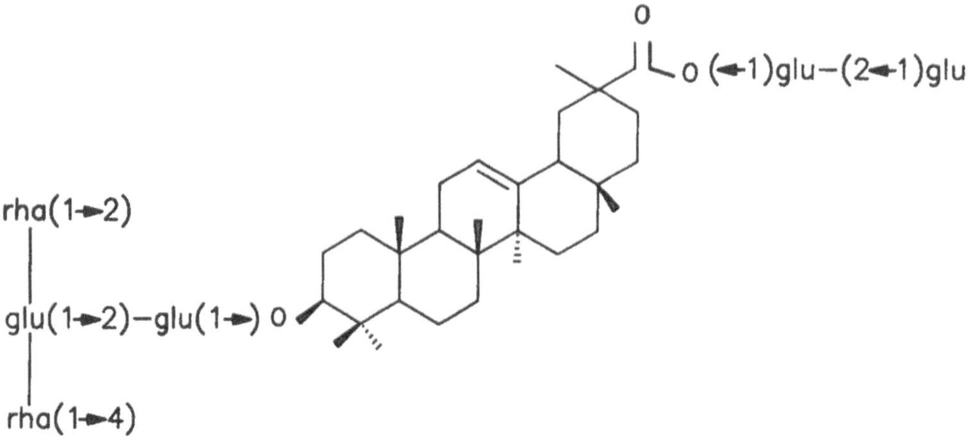

FIGURE 14.   Structure of a guar saponin.

B. Aggregation

Because saponins are amphiphilic, they form micelle-like aggregates in water.[124] Like many other amphiphiles,[125] they appear to have critical micelle concentrations (CMC). Below this concentration the molecules remain unassociated, and the transition to the micellar state is marked by an abrupt change in some physical property of the solution such as surface tension or the spectral characteristics of a solubilized dye.[125] The CMC of gypsophila saponin at 25° is about 0.14 g/l,[126] and that for saponin white (from *Saponaria officinalis*) is 0.6 g/l at 25° in 0.02 *M* phosphate buffer at (pH 7.12).[124] The CMC of ginseng saponin has been found from measurements of osmotic pressure and electrical conductivity to be ≈20 g/l.[127] The micelles formed by saponins seem generally to be very small (like those formed by the bile acids[128]). Soy saponins and saponin white associate only to the extent of forming dimers.[124] Gypsophila saponin forms larger aggregates of about 10 molecules which appear from X-ray scattering data to be roughly globular.[126] Quillaia saponin (from *Quillaia saponaria*) forms somewhat larger aggregates of about 50 molecules.[124] The saponins appear to aggregate by hydrophobic interaction of their aglycones, leaving the hydrophilic sugar groups exposed to the water as shown in Figure 15A.

C. Formation of Mixed Micelles with Bile Acids

A physical property of saponins which appears to have particularly significant physiological consequences for animals which ingest them (see Section VII) is their ability to form large mixed micelles with bile acids.[129] The bile acids themselves are surface-active steroids (Figure 16) which form small micelles similar to those of saponins,[128] their hydrophobic steroid groups stacking together like small piles of coins (Figure 15B). When the two types of compound are mixed, their hydrophobic groups interleave with each other in their stacks (Figure 15C).[124] The size of the micelles formed by saponins alone is limited by steric effects of the bulky sugar groups; the size of bile salt micelles is limited by electrostatic repulsion of the charged acid groups. In the mixed micelles the steric and electrostatic constraints are relieved and the stacks become greatly extended, incorporating many hundreds of molecules. This is graphically illustrated by the electron micrographs (shown in Figure 17) of the mixed micelles formed by saponin white and saponins from soy or quillaia with cholate.[124]

The sizes and detailed structures of these mixed micelles depend on the detailed chemical structure of the saponin molecules.[124] Some characteristics of the mixed micelles formed by three different saponins with the bile acid sodium cholate are shown in Table 5. The micelles

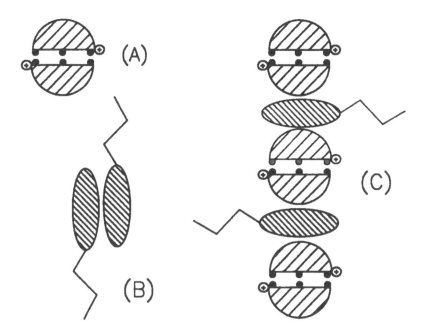

FIGURE 15. Schematic diagram of the aggregation of (A) bile acids, (B) saponins, and (C) saponins and bile acids.

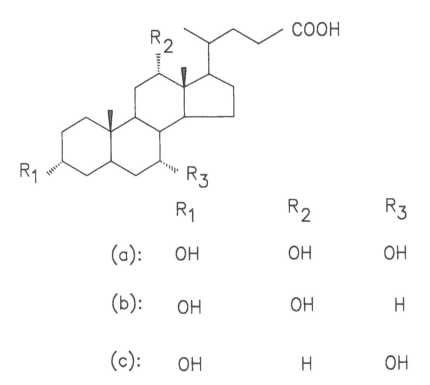

	R_1	R_2	R_3
(a):	OH	OH	OH
(b):	OH	OH	H
(c):	OH	H	OH

FIGURE 16. The structures of the three commonest bile acids: (a) cholate, (b) deoxycholate, and (c) chenodeoxycholate.

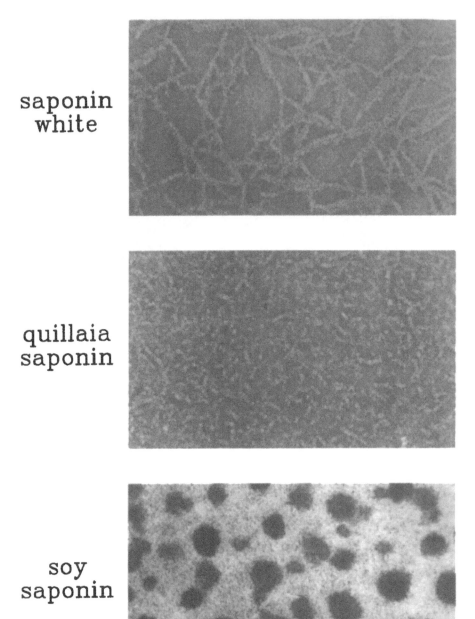

saponin
white

quillaia
saponin

soy
saponin

FIGURE 17.   Electron micrographs obtained by negative staining of the mixed micelles of cholate with three saponins: saponin white, quillaia saponin, and soy saponin.

formed by cholate with saponin white or soy saponin are very large but have different geometries (also indicated by the electronmicrographs in Figure 17); the micelles with quillaia saponin are smaller but resemble those formed with saponin white in structure. The very different structure of the micelles with soy saponin is hard to explain. Spectroscopic studies indicate a loose internal structure. The micelles are too large for the hydrophilic sugar groups

Table 5
CHARACTERISTICS OF THE MIXED
MICELLES FORMED BY THREE
SAPONINS WITH SODIUM CHOLATE

Saponin	Micellar weight	Shape	Axial ratio	Mean length (nm)
Saponin white	1.1×10^6	rod	43	170
Quillaia	3.5×10^5	rod	11	56
Soy	2.4×10^6	sphere	≈ 1	73[a]

[a] Diameter.

to be on the surface in a structure like that of the classical nonionic detergent micelle. They are also highly hydrated, which is again consistent with the sugar groups being in the core as well as on the surface. Soy saponin differs from saponin white and quillaia saponin in that all five of its aglycones have no carboxylic acid or other charged groups. Consequently, it is possible that there might be fewer constraints on how these might pack together. Other factors being equal, loose globular micelles would be entropically more stable than stacks in which the molecules have lost considerable freedom of motion.

A fourth saponin, that from gypsophila, also appears to form large mixed micelles with bile acids. Ljusberg-Wahren[130] has studied aqueous mixtures of this saponin with sodium cholate with the polarizing microscope and found a birefringent phase indicating the presence of extensive ordered structures.

D. Saponins as Emulsifiers

Saponins are amphiphilic, strongly surface active, and might, therefore, reasonably be expected to be good emulsifiers. A saponin of unspecified origin has been shown to emulsify kerosene/water and xylene/water systems better than detergents such as sodium dodecyl-sulfate, lecithin, gelatin, or plant gums.[131] Barla et al. have constructed the phase diagram for the system gypsophila saponin/sunflower oil monoglycerides/water.[132] The saponin effectively stabilizes the oil/water interface, forming various micellar and lamellar structures depending on the composition of the mixture.

VI. BIOSYNTHESIS OF SAPONINS

Only a brief outline of the biosynthesis of saponins is appropriate here. The area has been reviewed by Heftman, and Reference 133 should be consulted for detailed information.

The key consideration is Ruzicka's "biogenic isoprene rule", the essence of which is that squalene is the acyclic precursor of animal and plant sterols and triterpenes.[134] Many saponins have an O-substituent at C_3 which has been shown[135] to result from the intermediate 2,3-epoxysqualene. Nowacki et al.[136] have investigated the biosynthesis of alfalfa saponins by following the uptake of ^{14}C-acetate. Their results were explained in terms of initial synthesis of medicagenic acid from which the other aglycones were subsequently derived. Similar experiments by Peri et al.[137] on germinating soybeans incubated with ^{14}C-mevalonate also suggested that medicagenic acid is the primary aglycone.

A few experiments have been carried out to determine the major sites of biosynthesis. In oats saponins are formed during leaf development and extension, with maximum content of saponin being reached when the leaf blade is fully developed.[138] With licorice, a series of different isotopic labelling experiments has shown glycyrrhinic acid is only synthesized in the root tissue.[139] This seems consistent with the idea that saponin biosynthesis is most active in the tissues most vulnerable to mold attack or insect predation.

VII. PHYSIOLOGICAL AND PHARMACOLOGICAL PROPERTIES OF SAPONINS

The wide range of chemical and physical properties of saponins is matched by the extent and range of their physiological and pharmacological effects. The toxic nature of many saponins has been known and exploited for many centuries, particularly their toxicity to cold-blooded animals such as mollusks, amphibians, and fish. There are already several excellent reviews covering this area;[1-4,9,47,140] here we outline the more important physiological and pharmacological properties of saponins, giving most emphasis to recently published work.

We must emphasize again, at this point, that saponins as studied are almost invariably complex mixtures. Purified single compounds have been unavailable until very recently and even now are only available in very small quantities. Consequently, almost all the studies we describe have been carried out on crude extracts or simply saponin-rich plant materials such as quillaia bark.

A. Interaction with Biological Membranes

An ability to lyse red blood cells has long been thought to be a characteristic property of saponins (see Section I). Different saponins have very different hemolytic activities, and there is a wide variation among different animal species in the susceptibility of their erythrocytes to hemolysis by saponins. Some of the soy saponins have little hemolytic activity[141,142] whereas digitonin will cause 50% hemolysis in a 2%-suspension of bovine erythrocytes at a concentration as low as 2.5 mg/l.[143] Among different animal species the order of susceptibility is guinea pig and horse > dog, rat, and rabbit > goat, sheep, and cattle.[143]

The mechanism by which saponins cause hemolysis has been much investigated. Attempts have been made to correlate hemolytic activity with capacity of the saponin to lower the surface tension of water or produce stable foams,[112] but with no conclusive or convincing results. More successfully, extensive investigations[144,146] have suggested that chemical structure — the positions of attachment of the sugars and other polar groups on the aglycone — might be the determining factor. These results are summarized in Figure 18. However, these structural requirements are by no means absolute. Further support for this idea has come from a series of experiments in which it was shown that hemolytic activity is altered by small changes in the chemical struture of the aglycone. Ginseng saponins containing 20-*S*-protopanaxtriol are strongly hemolytic whereas those containing 20-*S*-protopanaxdiol have no hemolytic activity and even have a protective action.[147]

This all suggests that the hemolytic activity of saponins cannot be ascribed to their detergent-like properties but must be related to an ability to interact with some constituent (or constituents) of the cell membrane.[148,149] The fact that digitonin has a high hemolytic activity and also readily forms molecular complexes with cholesterol immediately suggests that it is this interaction with cholesterol which causes the cell membranes to leak. Supporting this idea, Gestetner et al.[65] have found that the antifungal and hemolytic properties of alfalfa and soy saponins are indeed closely related to their ability to form complexes with cholesterol,[141] but again, there are exceptions. Some strongly hemolytic saponins do not form complexes with cholesterol. Also, studies of effects of saponins on liposomes formed by egg lecithin and cholesterol showed that cholesterol depletion enhanced the effects of some saponins on the membrane integrity, but diminished the effects of others.[150]

Current evidence suggests that the changes to cell membranes brought about by saponins proceed by the following steps:[148,149]

1.   The hydrophobic aglycone of the saponin molecule penetrates the lipid bilayer and may specifically interact with other membrane components such as cholesterol.

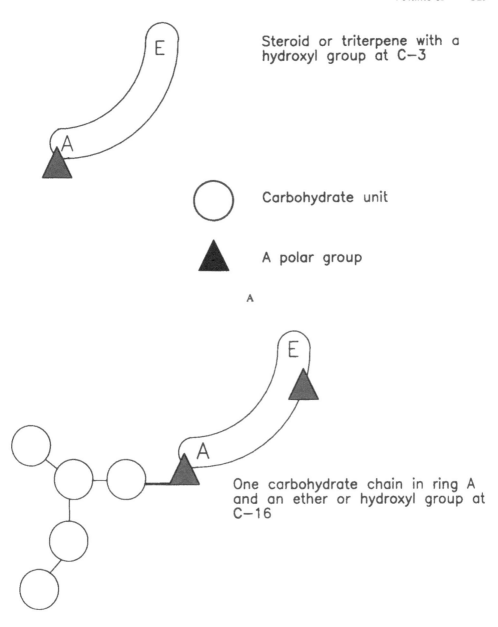

Steroid or triterpene with a
hydroxyl group at C−3

Carbohydrate unit

A polar group

A

One carbohydrate chain in ring A
and an ether or hydroxyl group at
C−16

Strongly hemolytic saponins

B

FIGURE 18. Schematic diagram of the structural requirements for hemolytic activity.

2. The presence of the saponin induces cholesterol-rich domains, producing conducting channels and making the membrane leaky.
3. This increased cell permeability allows efflux of K^+ and influx of water, Na^+, and other ions.
4. Complete lysis occurs with leakage of large molecules such as hemoglobin into the exterior medium.

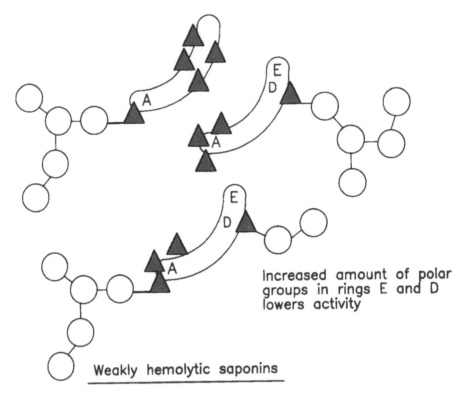

Increased amount of polar
groups in rings E and D
lowers activity

Weakly hemolytic saponins

C

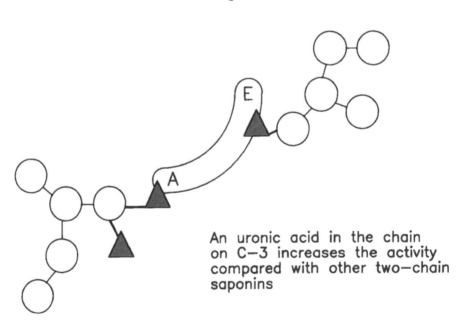

An uronic acid in the chain
on C—3 increases the activity
compared with other two—chain
saponins

Moderately hemolytic saponins

D

FIGURE 18 (continued)

Arguments in favor of this scheme include the fact that membranes rich in cholesterol are much more susceptible to damage by saponins than those with little or no cholesterol.[151] NMR studies have shown that saponin can interact with cholesterol within the bilayer.[151] Also, the formation of conducting channels has been observed by freeze-etching techniques and electron microscopy.[152,153] These show holes in the membrane surface bounded by hexagonal arrays of saponin, cholesterol, and phospholipid molecules. Some of these holes are permeable to molecules as large as ferritin.[154] Model studies of phospholipid bilayers and liposomes, using such techniques as NMR, also tend to confirm the role of the saponin aglycone and cholesterol in the damage to cell membranes caused by saponins.[149,151,155]

It has been suggested that interactions of ginseng saponins with appropriate membranes explain their effects on the brain;[156] they inhibit uptake of neurotransmitters by brain synaptosomes. Saponins also modify the permeability of the mucosal epithelial cells of the small intestine to allow entry of intact protein molecules and affect absorption of drugs and also various compounds more normally present in the digesta.[157] Agglutination of erythrocytes and phospholipid vesicles is another property of certain saponins which is related to their interactions with biological membranes. A number of ophiopogonins and ginsengosides, with little direct hemolytic activity, cause hemagglutination of erythrocytes from humans, rabbits, and sheep. They also appear to cause an analogous clumping of phospholipid vesicles.[158]

B. Effects on Enzymes and Metabolism

Despite the fact that saponins have such powerful metabolic effects in many organisms, few studies have been made of their direct effects on enzyme activities. Alfalfa saponins have been shown to inhibit respiration in the isolated rat diaphragm which can be explained by inhibition of succinate oxidase in the liver.[159] Alfalfa saponins also cause inhibition of trypsin and chymotrypsin *in vitro,* but the physiological significance of this *in vivo* has been questioned.[160] Ginseng saponins have been reported to inhibit Na^+, K^+-ATPase from dog cardiac sarcolemma, apparently by direct interaction with the membrane-bound enzyme.[161] Gypsophila saponin and nonionic detergents such as Triton® X-100 increase the activity of the enzyme, apparently by disrupting the membrane and making the enzyme more accessible to the substrate.[161]

The saponins from licorice (particularly glycyrrhizin) potentiate the action of the glucocorticoids, and this appears to be related to inactivation of the hormones by certain liver enzymes. Glycyrrhizin inhibits Δ^4-5-B-reductase which reduces the Δ^4-3-keto groups of cortisol, deoxycorticosterone, progesterone, and aldosterone. The saponin also inhibits side-chain degradation of cortisol.[162]

Turning to more general metabolic effects, a number of triterpene glycosides have been extensively investigated. Crude ginseng root saponins have been found to have powerful effects on carbohydrate and lipid metabolism.[163] The saponins have been claimed to reduce muscle fatigue from exercise, apparently by conserving endogenous carbohydrate stores with lower production of lactate and pyruvate and increased oxidation of fatty acids. Ginseng saponins injected intravenously have also been reported to increase protein synthesis in the liver, along with increased synthesis of RNA and increased RNA-polymerase activity.[164,165] Another reported effect is that ginseng saponins increase mitotic activity in the bone marrow and, thus, can protect animals (mice were used) against radiation injury.[166] They may also protect against liver damage from carbon tetrachloride, galactosamine, or ethanol.[166] In contrast, there are other saponins such as cauloside A and C, plasticodoside C, and hederagenin which strongly inhibit protein synthesis in bone marrow.[167]

C. Effects on Fungi and Microorganisms

Some saponins show a strong fungistatic and antimicrobial activity. Much of the early work in this area was reviewed only 15 years ago by Cucu and Grecu,[168] and their work, together with that of Wolters,[169] should be consulted for details. Inhibition of the growth of *Trichoderma* has been used extensively as a bioassay for alfalfa saponins. Alfalfa saponins have also been shown to inhibit growth of *Scleroticum rolfisii*, mycoplasma, and various species of *Pythium*.[170,171] Many other saponins have been shown to have similar effects. As with hemolytic activity, the magnitude of the effect depends on the detailed chemical structure of the saponin and is presumably related to the ability of saponins to interact with cell membranes. Spirostanolic glycosides, such as agaoroside, funkioside, and tomatine, added to the culture media arrest the growth of *Phytopthora infestans*, while the chemically related furostanolic glycosides show little or no activity.[172] Camellidin I and II isolated from various species of camellia inhibit the germination of *Candidia* and the growth of the hyphae of *Piticularia oryzae*, *Cochlibolus miyabeanus*, *Pestalotia logiseta*, *P. theae*, and other fungi.[173] Formulations containing tomatine (25 µg/ml) have been used to protect crops against fungi.[174]

Anisimov et al.[175-177] have tested the antimicrobial activity of 49 different triterpenoid saponins using *Saccharomyces carlsbergensis* as the test organism. While glycosides of oleanolic acid or hederagenin, with free -COOH groups at C_{25} or C_{28}, showed the highest activity, glycosides of meristropic acid, lupanine, and mecedonic acid showed little activity, even at relatively high concentrations in the culture medium. That a membrane phenomenon is involved is further indicated by a number of other observations. The antifungal activity of alfalfa saponins is overcome by adding cholesterol to the culture medium.[171] Saponins show less toxicity to fungi lacking cholesterol in their cell membranes. Saponins which complex strongly with cholesterol are the most strongly antifungal. Saponins induce alterations in the selective permeability of plasma membranes of sea urchin embryos resulting in suppression of nucleoside and amino acid transport into the cell.[177]

These effects may explain why so many plants contain saponins — as a protection against fungal and microbial attack.

D. Effect on Cold-Blooded Animals
1. Insects

The toxicity of saponins to insects has led to the suggestion that they might also provide protection from insect predation.[18,178] Saponins and their aglycones have been implicated in providing resistance of certain ancient wood in temples to attack from termites.[179] It is probably not a coincidence that in some trees the bark, a particularly inert and therefore vulnerable tissue, is rich in saponins. Examples are *Quillaia saponaria* from South America and *Melaleuca* spp. from Australia.

Extensive investigations have been reported[180] of the effects of the saponins from soy and alfalfa roots and shoots on the peach aphid (*Myzus persicae*). The results were consistent with the idea that it is the saponins in these legumes which provide resistance to the insect. It has also been shown that the bruchid or azuki beetle (*Callobrachus chinensis*) does not survive in the saponin-containing seeds of soybeans, navy beans, peanuts, and garden peas.[178] The saponins of alfalfa (but not red or white clover) are toxic to the larvae of the grass grub (*Costelytra zealandica*).[181] Several species of locust have shown increased mortality when fed on alfalfa or other saponin-containing legumes. Their larvae developed more slowly and the emerging adults were smaller than when they were fed saponin-free herbage.[180] Toxicity seems to result from inhibition of water resorption from the hind gut, decrease in hemolymph volume, and associated dehydration. Again this is a membrane-related phenomenon, presumably involving the epithelium of the hind gut. Cholesterol and some phytosterols overcome the deleterious effects when added to growth media.

As might be expected, not all saponins are equally toxic. In the case of the azuki beetle, the saponins of chick-peas (*Cicer arietinum*) and lentils (*Lens culinaris*) are much less toxic than those from alfalfa or soybeans.[178] Also, saponins from the same species fractionated by calcium oxide precipitation showed quite different levels of toxicity. Alfalfa root saponins rich in medicagenic acid are more toxic to the flour beetle (*Tribolium castraneum*) than soy saponins.[178] While hydrolytic removal of one sugar group from medicagenic acid increased its toxicity, the same process made soy saponins nontoxic.

Not all insects are equally susceptible to saponins. Some have evolved strategies for overcoming the saponin defenses of the plants on which they prey. Several alfalfa pests, such as alfalfa weevil (*Hypera postica*), spotted aphid (*Thereoaphid maculata*), clover root curculio (*Stona hispidulus*), and seed chalcid (*Bruchophagus roddi*) are hardly affected by the saponin content of their diet.[180] It has even been suggested that the alfalfa weevil may have taken positive advantage of the saponins in its diet for its development. However, an apparent correlation between the total saponin content of a plant and resistance to insect predation must be considered with caution.[181] Aphids feed on sieve tubes and may remain unaffected by saponins present in other parts of the leaves or in the roots of alfalfa.[181] It has recently been shown that alfalfa saponins in the diet neither inhibit feed intake nor have any repellent effect on the green pea aphid *Acyrthosiphon pisum*; yet, ingested alfalfa saponins are definitely toxic (radioactive tracer techniques were used).[181,182]

Thus, it seems there is no direct link between the presence in a plant of saponins and the plant's resistance to insect predation. Some saponins are vastly more toxic than others; some insects have developed strategies for dealing with the toxic properties of saponins.

2. Fish, Tadpoles, and Snails

Toxicity to fish is a characteristic property of saponins which has been known and exploited for centuries.[1] Saponins added to the water rapidly cause paralysis and death. Small fish, such as minnows or guppies, and tadpoles and freshwater snails have been used to bioassay saponin toxicity. Saponins extracted from the African medicinal plant *Phytolacca dodecandra* have been suggested as a means of controlling the water snail *Biomphalaria glabrata* which transmits the endemic disease schistosomiasis.[183,184] Saponin-containing tubers of another African plant, *Talinum tenuissimum*, have also been reported to be effective in this way.[185]

Attempts have been made to relate the toxicity of saponins to fish to their ability to lower surface tension, their ability to increase the permeability of cell membranes, or to their hemolytic power.[186] None of these correlations, though, have been particularly convincing. The effect seems to be directly related to damage caused by the saponins to the delicate membranes of the gills. Scanning electron microscopy has revealed the extensive damage that occurs in the gills of the climbing perch (*Anabas testidineus*) exposed to the saponins from *Mollugo pentaphylla* in solution at 50 ppm.[187] The saponin caused swelling of the lamellar and interlamellar epithelium with some dissociation of the epithelium and development of microridges in the minute water spaces between the lamellae. These changes caused asphyxiation and death of the fish.

E. Effects on Birds and Mammals

There has been a tendency in the past to treat saponins exclusively as antinutritional or toxic constituents of plants when consumed by animals or humans. Recent work has brought to light more positive, beneficial effects from dietary saponins, and in this section we shall discuss both these aspects.

1. Digestion and Absorption of Saponins

Saponins are 10 to 1000 times less toxic orally than when given by intravenous injection. They are absorbed very slowly, if at all, from the gastrointestinal tract.[143] Gestetner et al.[188] have studied the digestion and absorption of saponins in chicks, rats, and mice. The researchers fed the animals diets containing soy saponins and failed to detect any saponins in the blood. The enzymes of the gastrointestinal tract had no effect on the saponins, but they were hydrolyzed to their respective sapogenins by the cecal microflora.

Because of the importance of ginseng (*Panax ginseng*) in Chinese medicine, several investigations have been made of the fate of ginseng saponins in the gut. Chen et al.[189] failed to detect Rg_1, one of the main 20(*S*)-protopanaxtriol saponins, in the plasma or urine of rabbits after oral administration. However, Odani et al.,[190] using what they claimed were more sensitive techniques, found various metabolic products arising from oral or intravenous administration of the saponin. They found that 2 to 20% of an oral dose (100 mg/kg body weight) of Rg_1 was absorbed from the upper digestive tract. Peak levels of 3.5 ± 2.0 and 2.6 ± 1.5 µg/g in the liver and kidney, respectively, were reached 1.5 h after ingestion. Peak levels of 0.9 µg/ml were reached in the plasma 30 min after injection. When injected intravenously (5 mg/kg), Rg_1 was rapidly excreted in the urine (23.5%) and bile (57.2%). However, Rb_1, one of the main 20(*S*)-protopanaxdiol saponins, showed very different pharmokinetics.[191] Very little, if any, was absorbed from the gastrointestinal tract. Injected Rb_1 was strongly bound to plasma proteins and very little taken up by the liver and other tissues. Some partial hydrolysis of the saponins occurred in the stomach, similar to mild acid hydrolysis *in vitro*.

Certain saponins, such as those from horse chestnut (*Aesculus hippocastanum*), cause violent inflammation of the gastrointestinal tract. This can increase the permeability of the gut sufficiently to allow saponins to pass into the bloodstream and cause hemolytic damage.[8,9,143] Such violent reaction seems, however, quite rare, and from the evidence available it can be concluded that generally saponins are absorbed only to a very small extent, if at all, although the fact that licorice has definite pharmacological effects suggests[192] that glycyrrhizin might be to some extent absorbed. It should be borne in mind that there appears to be considerable variation among saponins in this respect.

2. Digestion and Absorption of Other Substances

There is evidence that saponins can influence the digestion and absorption of other components of the digesta in a number of different ways:

1. Saponins may interact chemically or physically with other components of the digesta; for example, the formation of mixed micelles with the bile acids.
2. Saponins may interact with the mucosal cell membranes causing permeability changes or loss of activity of membrane-bound enzymes.
3. Long-term, sustained ingestion may cause trophic effects on the physiology of the gastrointestinal tract.

It has long been known that saponins can reduce the absorption of cholesterol. Interaction between dietary cholesterol and saponins was indicated by the early work of Peterson, who showed in 1950 that the suppression of growth of chicks by alfalfa meal or quillaia saponin could be overcome by including cholesterol in the diet.[193] Cayen[194] later found that dietary tomatine lowered plasma cholesterol and inhibited the absorption of cholesterol in the rat by forming an insoluble complex in the gut. After injection of ^{3}H-cholesterol, dietary tomatine greatly increased fecal excretion of labeled cholesterol (but not labeled bile acids) and increased the rate of cholesterol synthesis by the liver. Morgan et al.[195] have reported contrary results, but from experiments with different saponins. They fed chick digitonin or gypsophila saponins and again found lower plasma cholesterol, but they found no evidence for increased fecal excretion of cholesterol. They suggested that increased fecal excretion of bile acids was responsible for the lower plasma cholesterol concentrations.

Recent work with different animal species and different saponins has provided conclusive evidence that some saponins (but by no means all) interact with cholesterol and directly interfere with its absorption. An extensive series of studies by Malinow et al.[196-199] with rats and nonhuman primates has provided some clear evidence about the role of saponins in cholesterol absorption. Saponins from alfalfa tops or roots were given intragastrically to rats (5 to 20 mg per animal) which at the same time were given oral or intravenous doses of ring-labeled cholesterol. Three methods of estimating cholesterol absorption were used with allowances being made for intestinal degradation of cholesterol and secretion of endogenous cholesterol into the lumen. In each case it was convincingly shown that alfalfa saponins reduced cholesterol absorption. Malinow et al.[199] also found that partial hydrolysis of the alfalfa saponins reduced cholesterol absorption still further. Digitonin, on the other hand, was more effective in the unhydrolyzed form. These results, combined with some studies *in vitro*, led Malinow et al.[199] to conclude that the mechanism of the effect was not simply that of formation of saponin-cholesterol complexes. They suggested that saponins perturb the mixed micelles of cholesterol, partially hydrolyzed lipids, bile acids, etc. from which cholesterol is absorbed into the mucosal cells from the lumen.[200] Cayen and Dvornik[201] have similarly studied the effects of diosgenin on lipid metabolism in rats. They found that diosgenin reduced the absorption of both exogenous and endogenous cholesterol and as a consequence increased the rate of synthesis of cholesterol by the liver and intestine. Diosgenin also increased secretion of cholesterol into the bile and excretion of other neutral sterols but without increasing biliary or fecal excretion of bile acids.

Thus, it seems that the chemical structure of the aglycone moiety of the saponin molecule is important in its interactions with cholesterol, as it is in the interactions of saponins with cell membranes discussed earlier. The glycosidic moiety provides the solubility. To illustrate this point, the triterpenes tigogenin or disogenin themselves have only a very small effect on cholesterol absorption, but when converted into their β-glucosides or β-cellobiosides, absorption is greatly inhibited.[200] These synthetic saponins produce an inhibition in cholesterol absorption similar to that produced by diosgenin.

Sidhu et al.[202] have studied the effect of tigogenin β-cellobioside on cholesterol absorption from perfused loops of small intestine *in vivo* in the rat. They compared the synthetic saponin with soy saponin and found that the former was the more effective in inhibiting cholesterol absorption by a substantial margin. The concentrations required for 50% inhibition of absorption were 2.09 g/l for soy, but only 0.27 g/l for the synthetic saponin. They also found that the saponins had no permanent effect on the mucosal cells: the full absorption rate was restored when the loops of intestine were washed free of saponin and subsequently reperfused with micellar cholesterol solutions.

Saponins can also inhibit absorption of bile acids, which are intimately related to cholesterol metabolism as discussed in the following section. Some saponins, as discussed in Section IV, form large mixed micelles with bile acids, and this effectively blocks the reabsorption process. Sidhu and Oakenfull[203] have shown that saponin white and soy saponin inhibit uptake of bile acids from perfused loops of small intestine in the rat. These same saponins also lower plasma cholesterol concentrations and increase fecal excretion of bile acids in rats.[204]

Saponin white and quillaia saponins also inhibit absorption from perfused loops of small intestine of glucose and the amino acids L-leucine and L-lysine.[205] Dialysis of glucose and the amino acids through cellulose membranes is also inhibited by the saponins. Close examination of the electron micrographs[124] of the mixed micelles of sodium cholate and saponin white or quillaia saponin (Figure 17) indicate helical substructures which could provide a matrix with which small molecules like glucose or amino acids could interact. Soy saponins also inhibit glucose uptake from perfused loops of intestine, but the synthetic saponin trigogenin β-cellobioside is without any effect.[202]

These interactions could explain the lower growth rates often observed in response to dietary saponins.[206] The slower rate of glucose absorption could flatten the glucose response curve as have been observed following ingestion of some forms of dietary fiber.[207] Slower absorption rates would distribute the sites of nutrient absorption more evenly over the entire length of the small intestine, instead of absorption occurring mostly in the upper section. Interactions of saponins with bile acids could restrict lipid absorption through limiting the availability of bile acids for emulsification and micellization. In the long term such effects are likely to produce significant changes to the gut wall.

3. Trophic Effects

Trophic effects from sustained ingestion of saponins have received little attention. In general, three factors most influence the trophic response of the mucosa:[206] the physical nature of the digesta, enteric hormones, and the pancreatic and biliary secretions. Other, but minor, factors which influence this response are mucosal blood flow, neural effects, epidermal growth factor, and the gut microflora. In the case of saponins, the permeability changes brought about by interactions of saponins with the membranes of the mucosal cells are also likely to be of some significance. Hiai[209] has shown that saponins can stimulate the production or release of neuropeptide hormones from the hypothalamus or pituitary, ginseng saponin Rd being particularly active. Ginseng saponins have been found to have binding affinity for steroid receptors specific for glucocorticoid, mineralocorticoid, and progestin hormones in the rat uterus.[210] When administered orally or by injection, saponins affect the levels of cAMP in the plasma, brain, adipocytes, and adrenal glands of rats.[211-213] Forskolin, a diterpene from the roots of *Coleus forskohlii*, has been found to reversibly stimulate adenylate cyclase both intact cells and cell membranes prepared from small intestine and other tissues. Forskolin also potentiated their response to amines and vasoactive intestinal peptides.

The deleterious effects on the gastrointestinal tract caused by cholera toxin are due to irreversible activation of the cAMP system. Thus, modulation of the cAMP system by dietary saponins could affect the activity of putative paracrine or hormonal messengers in the enterocrine cells.[213] In the gastrointestinal tract, these messengers control gastric emptying, motility, absorption and secretion of water and electrolytes, and absorption of nutrients. Ultimately, these processes control appetite, feed intake, and body growth. Thus, there are many potential causes for trophic effects.

Few studies have been made of the long-term implications of trophic effects in relation to possible toxicity of dietary saponins. Such effects could explain how saponins, which are poorly absorbed, if at all, influence tissues remote from the gastrointestinal tract. There is a particular need to fill the gaps in our knowledge in this area.

VIII. TOXICITY OF SAPONINS

As we have already seen, the physiological and pharmacological effects of saponins depend intimately on detailed chemical structure. It is, therefore, not surprising that saponins from different plant sources vary widely in their toxicity.[9] Different animal species also differ in their susceptibility to saponins. The single lethal dose can vary from 1 to 6000 mg/kg body weight.[9] It also, of course, makes a big difference whether the saponin is administered orally or intravenously. The LD_{50} in the rat for saponin from *Gypsophila paniculata* is 1.9 mg/kg body weight when given intravenously and 50 mg/kg when given orally.[9]

Saponins normally remain within the gastrointestinal tract,[34] but massive doses can possibly cause intestinal lesions allowing the saponins to enter the blood stream. The result is liver damage, hemolysis of red cells, respiratory failure, convulsions, and coma.[8,9] Irritation of the gastrointestinal tract by some other cause may also increase an animal's susceptibility to saponin poisoning, and it has been suggested that continued ingestion of sublethal doses of saponin can cause "corrosion of the intestine"[143] so that saponins eventually enter the bloodstream. Ewart[143] advised that people using purgatives or having inflamed intestines or cirrhotic livers should avoid all foods containing saponins. Poisoning during periods of drought of Australian sheep and cattle consuming various bush plants not normally eaten has been attributed to saponins. A more recent study[214] has similarly shown toxicity to cattle of the saponin-rich seeds of the palo santo tree (*Bulnesia sarmicientii*), a native of Paraguay. These seeds contain about 3.5% saponins, and cattle eating them displayed such symptoms as bradycardia, distended jugular vein, tympanic bloat, higher than usual frequencies of defecation and urination, and abnormal neurological responses. The LD_{50} in hamsters of the crude palo santo saponins was 860 mg/kg body weight. Such toxicity however, seems to be exceptional. Ishaaya et al.,[215] for example, were unable to detect any adverse effects in chicks, rats, or mice fed high levels of soy saponins. A recent study on purified ginseng saponins has shown that for rats the oral LD_{50} is greater than 5000 mg/kg body weight. Malinow et al.[216] have found alfalfa saponin not to be toxic to monkeys when fed at levels of 10 g/kg diet for 6 months. The saponin from *Quillaia saponaria* was found not to be toxic to mice when fed at levels of up to 0.5% of the diet for a period of 84 weeks.[217] Saponin extract from *Yucca mohavensis* was similarly well tolerated by rats when fed at levels of up to 2.16 g/kg diet over a period of 12 weeks.[218] Saponins may even protect against the toxic effects of other materials. Soysaponins have been reported to prevent liver damage from consumption of peroxidized corn oil.[219] Simultaneous consumption of saponin with tannins by herbivores appears to inhibit the absorption of the toxic tannins from the gastrointestinal tract.[220]

However, when monogastric animals consume diets containing saponins at subacute levels, there often occurs a reduction in feed intake and growth.[221] This became obvious when attempts were first made to utilize protein-rich alfalfa meal in feed for pigs or poultry.[222] The problem is exacerbated if leaf protein concentrate is used because there is a tendency for the saponins to be retained and concentrated in the press cake.[223] Vigorous efforts have therefore been made to develop alfalfa cultivars with low levels of saponin. Cheeke et al.[224] have established that pigs fed low saponin alfalfa meal showed superior growth, and researchers found a convincing negative correlation between saponin content of the alfalfa meal and growth rates. It has been suggested that growth retardation is caused by low feed intake due to the bitterness of saponins, but against this is the often repeated observation that growth retardation is reversed if cholesterol is added to the feed. An alternative explanation is that saponins inhibit absorption of various micronutrients as described in Section VII.

IX. A ROLE FOR SAPONINS IN THE PREVENTION OF CARDIOVASCULAR DISEASE?

In the context of human nutrition, one of the most interesting and potentially beneficial properties of saponins is their ability to lower the concentrations of cholesterol in blood plasma and in organs such as liver.[225]

Saponins can inhibit absorption of cholesterol or bile acids (or both) from the small intestine (Section VII). The potential value of saponins in cholesterol-lowering diets or in producing low cholesterol meat or eggs has been recognized for about 30 years,[195,226] but until recently little research was done in this area.

A. Plasma and Egg Cholesterol in Poultry

Early experiments with chickens showed that including saponin-rich alfalfa meal or leaf protein concentrate or quillaia saponins in their feed lowered serum and liver cholesterol levels.[226,227] At the same time there was reduced growth or egg production, effects which could be abolished by including cholesterol in the feed. These studies were extended further by Morgan et al.,[195] who used semisynthetic diets, with or without cholesterol, to show that digitonin or gypsophila saponin also lowered plasma cholesterol.

Possible effects of ginseng saponins on poultry have been extensively investigated by

Qureshi et al.[228] *Panax ginseng* and *P. quinquefolium* were extracted serially with petroleum ether, methanol, and water, and the solvent-free extracts fed to chicks at levels not exceeding 0.1% of the diet. All fractions lowered total and LDL cholesterol and were associated with reduction in the activities of two key enzymes in cholesterol synthesis, β-hydroxy-β-methylglutaryl-CoA reductase and cholesterol-7-α-hydroxylase. Qureshi et al. have concluded that it is not in this case the saponins that are responsible for these effects, because their results seemed inconsistent with the stimulation by ginseng saponins of cholesterol and lipid synthesis reported by other workers.

With quail fed alfalfa top or root saponins (which have different proportions of midicagenic acid), there was no significant effect on plasma or liver cholesterol. There was, however, a marked increase in cholesterol and total lipid synthesis as measured by incorporation of ^{14}C-acetate. Quail, thus, seem to differ from chickens in their response to dietary saponins.[229]

Attempts to produce low cholesterol eggs by feeding saponins have met with only limited success. Small reductions in response to feeding alfalfa meal have been obtained,[230] but high dietary levels of saponins seem to be required. Another group, Nakaue et al.,[231] reported finding no effect from alfalfa meal at 10% of the diet. Sims et al.[232] have fed isolated saponins from the cactus *Yucca schottii* to laying hens. The saponins reduced the feed consumption and egg production, but again there was no effect on the level of cholesterol in the eggs. When they also fed ^{14}C-labeled cholesterol, they found that the saponins increased fecal excretion of cholesterol but decreased its deposition in eggs and decreased its level in plasma. They concluded that saponins did indeed reduce absorption of dietary cholesterol and its transfer to egg yolk, but that this deficiency was compensated for by endogenous synthesis in the ovaries. It is much more difficult to lower egg cholesterol by dietary manipulation than to increase it by feeding cholesterol.

B. Plasma Cholesterol in Mammals

The cholesterol-lowering activity of saponins first observed in poultry has since been observed in a number of mammalian species, particularly rats and monkeys. Kritchevsky et al.[233] were among the first to suggest that the saponins in alfalfa were the cause of the lower plasma cholesterol levels in rats fed alfalfa-based diets. Malinow et al.[196-200] have since provided conclusive evidence that the saponins in alfalfa are indeed responsible for the lower plasma cholesterol concentrations and that alfalfa fiber in the absence of saponins is not responsible for this activity. This group has demonstrated the cholesterol-lowering effect of isolated alfalfa saponins in rats,[196] monkeys (*Macaca fascicularis*),[199] and rabbits,[234] and in human subjects given autoclaved alfalfa seeds.[198]

Saponins from other plant species have also been investigated. Malinow et al. have demonstrated the effect in monkeys fed digitonin.[197] Oakenfull et al. using rats have found significant reductions in plasma and liver cholesterol concentrations in response to feeding commercial saponin white[235] or isolated saponins from quillaia bark,[204] soybeans,[204] navy beans (*Phaseolus vulgaris*),[55] or chick-peas (*Cicer arietinum*).[236] In these experiments the animals were fed diets containing sufficient cholesterol to induce hypercholesterolemia. In rats there is generally no significant cholesterol-lowering effect in the absence of dietary cholesterol. Sautier et al.,[237] for example, reported finding no effect from saponin white in rats in the absence of dietary cholesterol, and Gibney et al.[238] found no effect from the same saponin in rats or hamsters.

The only reported experiment in which human subjects were given saponins directly is that described by Bingham et al.[239] They gave a group of 174 arthritic patients tablets of a saponin-rich extract from *Yucca schidigen* in a treatment which also included controlled diets, exercise, and physiotherapy. A control group was given a placebo, with the rest of the treatment the same. Substantial reductions in plasma cholesterol were observed in response to the saponin, particularly, in individuals with higher levels initially. There were also significant falls in plasma triglycerides and blood pressure.

Foods containing saponins as a significant (but naturally occurring) component have also been shown to be effective in lowering plasma cholesterol in hyperlipidemic humans. Mathur[240] found that giving hypercholesterolemic patients chick-peas as an isocaloric substitute for wheat flour and other cereals lowered their mean plasma cholesterol concentrations from 206 mg/100 ml to 160 mg/100 ml. Other legumes have been found to be similarly effective,[241] and it has been suggested that the saponins remaining attached to isolated soy protein could be responsible for the cholesterol-lowering activity of soy products.[242]

The mechanism of this effect seems to be, at least in part, that saponins inhibit absorption from the small intestine of cholesterol or bile acids as described in the previous section. Which of these is dominant depends on the chemical structure of the saponin. In a comparative experiment, Oakenfull et al.[204] found that saponin white and isolated saponins from soybeans gave a relatively small increase in fecal excretion of cholesterol and other neutral sterols, but greatly increased fecal excretion of bile acids. In contrast, quillaia saponin caused a large increase in fecal excretion of neutral sterols, but only a small increase in fecal bile acids. Bile acid and cholesterol metabolism are closely interrelated because bile acids are synthesized from cholesterol in the liver, cholesterol being their immediate precursor.[243] Bile acids and cholesterol (along with phospholipids) are secreted by the liver as bile to promote the digestion and absorption of lipids. Bile acids are normally very efficiently recycled, and after they have completed their function as detergents in lipid digestion and absorption, they return to the liver in the portal bloodstream (the enterohepatic cycle). The ability of some saponins to form mixed micelles with bile acids (as described in Section V) interrupts this cycle. Bile acid molecules incorporated into mixed micelles are not available for reabsorption and are consequently carried over into the large bowel. Similarly, the formation of molecular complexes of saponins with cholesterol (Section VII) prevents cholesterol absorption. The net result is a higher loss by fecal excretion of cholesterol and bile acids than can be met from endogenous or exogenous sources, with consequent reduction in the cholesterol levels of blood plasma, liver, and other tissues.

Increased fecal excretion of bile acids in response to dietary saponins, but without any significant lowering of plasma or tissue cholesterol, has been found in animals and human subjects with plasma cholesterol levels already low before dietary intervention. Pigs fed low-lipid diets had increased fecal bile acids when given saponin white in their drinking water.[244] Human volunteers with normal plasma cholesterol concentrations given soy flour with high or low levels of cholesterol also showed increased fecal excretion of bile acids (and neutral sterols) in response to the saponin.[245] This probably explains why there have been a few isolated claims that saponins have no effect on plasma cholesterol levels.[238]

Thus, it appears that saponins act as a natural alternative to cholestyramine. Cholestyramine is an ion exchange resin which has a powerful ability to bind bile acids, and is a routine treatment for hypercholesterolemia. But because it can also interact with other components of the digesta, affecting absorption of minerals, lipids, and other nutrients, abdominal discomfort and other side effects are not uncommon. Compliance by patients is therefore not good in the long term. Saponins appear to offer a potentially more acceptable alternative. There seems every indication that foods rich in saponins could usefully be included in cholesterol-lowering diets.

X. USEFUL APPLICATIONS OF SAPONINS

A. The Food and Drink Industries

In terms of human nutrition, the most useful potential application of saponins could be in the prevention of cardiovascular disease.[196-200,225] Numerous animal feeding trials have now established that saponins lower plasma cholesterol concentrations through increasing fecal excretion of cholesterol directly or indirectly as bile acids. Because of the occurrence of saponins in a number of staple foods, vegetarians are likely to consume up to 10 g of saponins daily.[4] Saponins consumed in this way appear to have no injurious effects, and the vegetarian diet appears to have positive health benefits to which saponins may be a contributing factor.

Purified saponins or concentrated extracts are added to manufactured foods and drinks primarily as foaming agents or emulsion stabilizers. In addition, soysaponins have been reported to have antioxidant properties and to prevent liver injury caused by highly peroxidized fats.[219,246] Saponins from certain species of Cucurbitaceau (gourds) have been patented as antioxidants for food use.[246] The alleged ability of Korean ginseng to retard aging has been attributed to the antioxidant properties of its saponins.[247] Some "health foods" popular in Japan contain saponin as ingredients.[248] Saponins have also been reported useful in the preparation of spray-dried powders containing vitamin E for enrichment of foods, drinks, and animal feeds.[249,250] Saponin from *Saponaria officinalis* is used traditionally as an emulsifier in the preparation of halvah, a confectionery product from Greece and the Middle East. Saponin-rich fenugreek seeds are used in many Indian pickles and chutneys to inhibit fungal growth.

There seems to be considerable confusion as to the regulatory status of saponins as food components. Crude extracts of licorice and sasparilla are permitted as food flavorings in Australia,[251] Britain,[9] and the U.S.[9] Quillaia saponin and saponin isolated from *Yucca mohavensis* are generally recognized as safe (GRAS) in the U.S., but quillaia saponin is prohibited as a food additive in West Germany, and all saponins are prohibited for food use in Spain and Morrocco.[9]

B. Pharmaceutical Applications

The pharmaceutical uses of saponins are as varied as their origins and chemical structures. It is impossible here to cover the entire range of applications; we mention here some of the major ones.

First, a number of saponins or saponin-rich mixtures have found use as antiinflammatory, antidiuretic, antipyretic, and analgesic agents, central-nervous-system (CNS) depressants, and as treatment for ulcers.[192] These effects seem to be related to the stimulatory effects of saponins on the hypothalamus or pituitary glands.[209] Antiulcerogenic activity is reputed to be related to effects of saponin on the synthesis of prostaglandins in the intestinal mucosa. Aescin saponins have been used in France topically to treat hemorrhoids and venous congestion.[192] Saponins from *Buffa operculata*, primrose, and tea seeds act as expectorants and promote secretion of mucus from the respiratory tract.[142]

Second, saponins appear to influence blood pressure, but the direction of the effect varies with plant source. Saponins from *Glycyrrhiza* species have been shown to be hypertensive,[252] and those from *Yucca* to be hypotensive.[239]

Third, ginseng has a central place in Chinese medicine.[252] The literal translation of ginseng from Chinese is "spirit of man", and according to the Shin-Nung Pen T'sao Ching written 2000 years ago, ginseng "effectively calms the mental condition, stabilizes the spirit, stimulates the mind, extends the memory, makes the sight clear and removes ill-feeling". At least some of these virtues seem to be due to the ginseng saponins. Opposing pharmacological effects have been found with the two groups of ginseng saponins. Rg_1 has been found to stimulate the CNS and to have antifatigue activity, whereas Rb_1 depressed the CNS and had a tranquilizing action.

Fourth, cimicifugocide saponins[253] possess strong immunosuppressive activity which is preferentially directed towards β-cell function, but which in large doses also affects T-cell function.[253] Licorice extracts also have immunosuppressive activity,[254] and triterpenes from *Maytenus diversifolia* have been found to inhibit growth *in vivo* of leukemic lymphocytes.[255]

Fifth, a mixture of saponins isolated from *A. aspera* acted as a heart stimulant when tested on isolated hearts from frogs, guinea pigs, and rabbits.[256] Antispasmogenic effects have been found with a saponin from *Gardenia turgid* which has also been claimed to have potential as an antiasthmatic drug.[257] Crude saponin extracts from *Albizzia lebbek* have been shown to protect mast cells against antigenic shock.[258] Soy saponins counteract the effects of endotoxins or thrombin in rats by inhibiting the formation of fibrin thrombi and increasing fibrin degradation.[259] Similar effects have been reported for saponins from *Ruscus* spp.[260]

In general, it is the triterpene saponins which appear to have the actual or potential direct pharmaceutical uses. The steroidal saponins appear more useful as starting materials for chemical synthesis of steroid hormones and related compounds for pharmaceutical use.

C. General Industrial Applications

Saponins possess excellent emollient properties and have found extensive use in cosmetics.[261,262] They also have numerous applications as surfactants.[263] This is because they have the advantage of being nonionic surfactants which are almost unaffected by the presence of salt or acid or alkaline conditions. (Those with acid groups on the aglycone are somewhat more sensitive to pH in their surfactant properties.) Shampoos, other hair preparations, and various cleaning agents are manufactured with saponin levels ranging from 0.001 to 10%.[264-266] Diazo copying materials usually contain saponins to improve the spreadability and stability of coating layers.[267,268] In biological waste-treatment plants, addition of saponins can improve oxygen transfer efficiency resulting in increased growth of bacteria and increased production of biogas. The saponins also appear to counteract adverse effects of pH or heavy metals on bacterial growth.[269] Saponins have found use as impregnating agents for cigarette filters.[270] They have also been used extensively in fire extinguishers, photographic emulsions and in the preparation of light weight concrete.

XI. CONCLUSIONS AND RESEARCH NEEDS

Saponins occur in many food and forage plants. They are extraordinarily physiologically active compounds, but their reputation for toxicity is in general unfounded. Some saponins are indeed highly toxic, but most are not. Ingested saponin mostly remains within the gastrointestinal tract, and it is only when saponins enter the bloodstream that their hemolytic activity causes damage. Their physiological activity appears to arise from two main causes: (1) their powerful physical interactions, as surface-active agents, with other components of the digesta and (2) their ability to interact with the membranes of the mucosal cells, causing sometimes profound changes to the associated membrane biochemistry.

Their interactions with various components of the digesta explain why saponins sometimes have antinutritional properties. Through inhibiting nutrient uptake they inhibit growth. Although of obvious economic disadvantage with domestic animals, this property would appear to have advantages for obese Western man (and woman).

Another potential benefit from saponins in foods for human consumption could be in reducing the risk of cardiovascular disease. Dietary saponins have been repeatedly shown to have the ability to lower plasma cholesterol concentrations in animals. There are indications that diets based on food rich in saponins would produce similar effects in hypercholesterolemic humans, but no well-controlled clinical trials have yet been carried out. These are urgently needed. There is also an urgent need to use modern techniques such as immunoassay and protein binding to investigate the metabolic effects of the interactions of saponins with cell membranes. A start has been made, but much remains to be learned in this area.

Purified saponins have only very recently become available in gram quantities, and this has greatly restricted the scope of investigations of their properties. Now that purified saponins are available, the more significant results previously obtained with mixtures need to be confirmed with pure compounds.

In saponins, nature has provided a vast stock of pharmacologically active and industrially useful compounds. They have many actual and potential uses in improving the lot of mankind.

REFERENCES

1. Kofler, I., *Der Saponine*, Springer-Verlag, Vienna, 1927.
2. Tschesche, von R. and Wulff, G., Chemie und Biologie der Saponine, in *Progress in the Chemistry of Organic Natural Products*, Vo. 30, Herz, W., Giesbach, H., and Kirby, G. W., Eds., Springer-Verlag, Vienna, 1973, 461.
3. Price, K. R., Johnson, I. T., and Fenwick, G. R., The chemistry and biological significance of saponins in foods and feeding stuffs, *Crit. Rev. Food Sci. Nutr.*, in press.
4. Oakenfull, D., Saponins in food: a review, *Food Chem.*, 6, 19, 1981.
5. Rodd, E. H., *Chemistry of Carbon Compounds*, Vol. 2 (Part B), Elsevier, Amsterdam, 1953, 1035.
6. Turner, A. B., Smith, D. S. H., and Mackie, A. M., Characterization of the principal steroidal saponins of the starfish *Marthasterias glacialis*: structures of the algycones, *Nature (London)*, 233, 209, 1971.
7. Kitagawa, I. and Kobayashi, M., On the structure of the major saponin from the star fish *Acanthaster planci*, *Tetrahedron Lett.*, p. 859, 1977.
8. Martindale, W., *The Extra Pharmacopoeia*, 26th ed., Pharmaceutical Press, London, 1972, 393.
9. George, A. J., Legal status and toxicity of saponins, *Food Cosmet. Toxicol.*, 3, 85, 1965.
10. Reichert, D., Antisnoring agent, German Patent 3,317,530, 1985; *Chem. Abstr.*, 102, 84406, 1985.
11. Schmeideberg, H., cited by Rodd, E. H., *Chemistry of Carbon Compounds*, Vol. 4 (Part C), Elsevier, Amsterdam, 1960.
12. Bondi, A., Birk, Y., and Gestetner, B., in *Toxic Constituents of Plant Foodstuffs*, Liener, I. E., Ed., Academic Press, New York, 1969, 169.
13. Essman, E. J., The medicinal use of herbs, *Fitoterapia*, 55, 279, 1984.
14. Mahato, S. B., Gangley, A. N., and Sahu, N. P., Steroid saponins, *Phytochemistry*, 21, 959, 1982.
15. Hiller, K. and Volgt, G., Neue Ergebniße in der Eforschung der Triterpensaponine, *Pharmazie*, 7, 365, 1977.
16. Das, M. C. and Mahato, S. B., Triterpenoids, *Phytochemistry*, 22, 1071, 1983.
17. Willner, D., Gestetner, B., Lavie, D., Birk, Y., and Bondi, A., Soya bean saponins. V. Soyasapogenol E, *J. Chem. Soc. Suppl.*, 1, 5885, 1964.
18. Applebaum, S. W., Marco, S., and Birk, Y., Saponins as possible factors of resistance of legume seeds to attack of insects, *J. Agric. Food Chem.*, 17, 618, 1969.
19. Price, K. R., Curl, C. L., and Fenwick, G. R., The saponin content and sapogenol composition of the seed of thirteen varieties of legume, *J. Sci. Food Agric.*, 37, 1185, 1986.
20. Fenwick, D. E. and Oakenfull, D. G., Saponin content of soya beans and some commercial soya products, *J. Sci. Food Agric.*, 32, 273, 1981.
21. Kitagawa, I., Yoshikawa, M., and Yoshioka, I., Saponins and sapogenol. XIII. Structures of three soybean saponins, soyasaponin I, soyasaponin II, and soyasaponin III, *Chem. Pharm. Bull.*, 24, 121, 1976.
22. Kretsu, L. G., Mtya, P. K., and Chirva, V. Y., Structure of *Phaseolus vulgaris* saponins, *Khim. Prir. Soedin.*, p. 618, 1972; *Chem. Abstr.*, 79, 97962, 1973.
23. Basu, N. and Rastogi, R. P., Triterpenoid saponins and sapogenins, *Phytochemistry*, 6, 1249, 1967.
24. Vinkler, G. N. and Smyslova, G. I., The discovery of a natural auto-antimutagenic system, *Genetika*, 52, 1967.
25. Curl, C. L., Price, K. R., and Fenwick, G. R., The quantitative estimation of saponin in pea and soya, *Food Chem.*, 18, 241, 1985.
26. Kitagawa, I., Wang, H. K., Saito, M., and Yoshikawa, M., Saponin and sapogenol. XXXI. Chemical constituents of the seeds of *Vigua angularis* (Willd.) Ohwi et Ohashi, (1) Triterpenoidal sapogenols and 3-furanmethanol-β-D-glucopyranoside, *Chem. Pharm. Bull.*, 31, 664, 1983.
27. Kitagawa, I., Wang, H. K., Saito, M., and Yoshikawa, M., Saponin and sapogenol. XXXII. Chemical constituents of the seeds of *Vigua angularis* (Willd.) Ohwi et Ohashi, (2) Azukisaponins I, II, III and IV, *Chem. Pharm. Bull.*, 31, 674, 1983.
28. Kitagawa, I., Wang, H. K., Saito, M., and Yoshikawa, M., Saponin and sapogenol. XXXIII. Chemical constituents of the seeds of *Vigua angularis* (Willd.) Ohwi et Ohashi, (3) Azukisaponins V and VI, *Chem. Pharm. Bull.*, 31, 683, 1983.
29. Dieckert, J. W., Morris, N. J., and Mason, A. F., Saponins of the peanut: isolation of some peanut sapogenins and their comparison with soysapogenols by glass paper chromatography, *Arch. Biochem. Biophys.*, 82, 220, 1959.

30. **Smoczkiewiczowa, M. A., Nitschke, D., and Wieladek, H.,** Plants of the species *Allium* (onion, garlic, leek) as sources of saponin glycosides, in *Proc. Int. Congr. Commodity Science of Natural Resources,* Trieste, 1978, 407.

31. **Kawano, K., Sakai, K., Sato, H., and Sakamura, S.,** A bitter principle of asparagus: isolation and structure of furostanol saponin in asparagus storage root, *Agric. Biol. Chem.,* 39, 1999, 1975.

32. **Kawano, K., Sato, H., and Sakamura, S.,** Isolation and structure of furostanol saponin in asparagus edible shoots, *Agric. Biol. Chem.,* 41, 1, 1977.

33. **Tschesche, von R. and Wulff, G.,** Triterpenes. XXVII. Saponins of spinach *(Spinacea oleracea), Justus Liebigs Ann. Chem.,* 726, 125, 1969.

34. **Birk, Y.,** Saponins, in *Toxic Constituents of Plant Foodstuffs,* Liener, I. E., Ed., Academic Press, New York, 1969, 169.

35. **Kintia, P. K. and Schvets, S. A.,** Melongosides N, O and P: steroidal saponins from the seeds of *Solanum melongena, Phytochemistry,* 24, 1567, 1985.

36. **Chirva, V. Y., Chebov, L., and Lazur'evskii, G. V.,** Sunflower saponins, *Khim. Prir. Soedin.,* 4, 140, 1968; *Chem. Abstr.,* 69, 41746, 1969.

37. **Tschesche, von R. and Schmidt, W.,** Zwei neue Saponine der oberirdischen Teile des Hafers *(Avena sativa)* mit Nuatigenin als Aglykon, *Z. Naturforsch.,* 218, 896, 1966.

38. **White, P. L., Avistur, E., Dias, C., Vinas, E., White, H. S., and Collazos, C.,** Nutrient content and protein quality of quinua and canihua, edible seed products of the Andes mountains, *J. Agric. Food Chem.,* 3, 531, 1955.

39. **Yoon, K.-R. and Wrolsted, R. E.,** Investigation of marion blackberry, strawberry, and plum fruit for the presence of saponins, *J. Agric. Food Chem.,* 32, 691, 1984.

40. **Price, K. R., Curl, C. L., and Fenwick, G. R.,** The saponin content and sapogenol composition of the seed of thirteen varieties of legume, *J. Sci. Food Agric.,* in press.

41. **Fenwick, D. E. and Oakenfull, D.,** Saponin content of food plants and some prepared foods, *J. Sci. Food Agric.,* 34, 186, 1983.

42. **Ireland, P. A., and Dziedzic, S. Z.,** Saponins and sapogenins of chickpea, haricot bean and red kidney bean, *Food. Chem.,* 23, 105, 1987.

43. **Tschesche, von R. and Wulff, G.,** Chemie und Biologie der Saponine, *Fortschr. Chem. Org. Naturst.,* 30, 461, 1973.

44. **Varshney, I. P. and Sharma, S. C.,** Saponins and sapogenins. XXII. Chemical investigation of seeds of *Myristica fragrans, Indian J. Chem.,* 6, 474, 1968.

45. **Valoshina, D. A., Kiselev, V. P., and Rumyantseva, G. N.,** Content of sapogenins and their yield from *Trigonella foenum-graecum* seeds, *Rastit. Resur.,* 13, 655, 1977; *Chem. Abstr.,* 88, 34506, 1978.

46. **Varshney, I. P., Beg, M. F. A., and Sankaram, A. V. B.,** Saponins and sapogenins from *Quillaia saponaria, Fitoterapia,* 56, 254, 1985.

47. **Birk, Y. and Peri, I.,** Saponins, in *Toxic Constituents of Plant Foodstuffs,* Liener, I. E., Ed., Academic Press, New York, 1980, chap. 6.

48. **Nicollier, G. and Thompson, A. C.,** A new triterpenoid saponin from the flowers of *Melilotus alba,* white sweet clover, *J. Nat. Prod.,* 46, 183, 1983.

49. **Oleszek, W. and Jurzysta, M.,** Isolation, chemical characterization and biological activity of red clover *(Trifolium pratense* L.) root saponins, *Acta Soc. Bot. Pol.,* 55, 247, 1986.

50. **Walter, E. D.,** Isolation of a saponin, hederin and its sapogenin, hederagenin, from Burr clover *(Medicago hispida), J. Am. Pharm. Assoc.,* 46, 466, 1957.

51. **Curl, C. L., Price, K. R., and Fenwick, G. R.,** Isolation and structural elucidation of a triterpenoid saponin from guar *(Cyamopsis tetragonoloba),* in press.

52. **Hudson, B. J. F. and El-Difrawi, E. A.,** The sapogenin of the seeds of four lupin species, *J. Plant Foods,* 3, 181, 1979.

53. **Chen, A. H.,** Applications of the saponins of *Dioscorea.* I. Studies on the saponins of *Dioscorea* in Taiwan, *Hua Hsueh,* 43, 79, 1985.

54. **van Atta, G. R., Guggolz, J., and Thompson, C. R.,** Determination of saponins in alfalfa, *J. Agric. Food Chem.,* 9, 77, 1961.

55. **Kozuharov, S., Oakenfull, D. G., and Sidhu, G. S.,** Navy beans and navy bean saponins lower plasma cholesterol concentrations in rats, *Proc. Nutr. Soc. Aust.,* 11, 162, 1986.

56. **Basu, N. and Rastogi, R. P.,** Triterpenoid saponins and sapogenins, *Phytochemistry,* 6, 1249, 1967.

57. **Simes, J. J. H., Tracey, J. G., Webb, L. J., and Dunstan, W. J.,** An Australian phytochemical survey. III. Saponins in Eastern Australian flowering plants, *CSIRO Bull.,* p. 281, 1959.

58. **Kitagawa, I., Saito, M., and Yoshikawa, M.,** Structure of soyasaponin A, a bisdesmoside of soyasapogenol A from soybean, the seeds of *Glycine max.* Merrill, *Chem. Pharm. Bull.,* 33, 1069, 1985.

59. **Kitagawa, I., Saito, M., and Yoshikawa, M.,** Saponin and sapogenol. XXXVIII. Structure of soysaponin A_2, a bisdesmoside of soyasapogenol A_2 from soybean, the seeds of *Glycine max.* Merrill, *Chem. Pharm. Bull.,* 33, 598, 1985.

60. Price, K. R. and Fenwick, D. R., Soyasaponin I, a compound possessing undesirable taste characteristics, isolated from the dried pea, (*Pisum sativum* L.), *J. Sci. Food Agric.*, 35, 887, 1984.
61. O'Dell, B. L., Regam, W. O., and Beach, T. J., Toxic principle in red clover, *Mo. Agric. Exp. Stn. Res. Bull.*, 702, 12, 1959.
62. Wolf, W. J. and Thomas, B. W., Thin layer and anion exchange chromatography of soybean saponins, *J. Am. Oil Chem. Soc.*, 47, 86, 1970.
63. Lisboa, B. P., Thin layer chromatography of sterols and steroids, in *Thin Layer Chromatographic Analysis*, Marinetti, G. V., Ed., Marcel Dekker, New York, 1976.
64. Smoczkiewica, M. A., Nitschke, D., and Stawinski, T. M., Mikrobestimmung von Spuren von Saponinglykosiden in 'nicht-saponin fuhren dem' pflanzen Material, *Mikrochim. Acta*, 11, 597, 1977.
65. Gestetner, B., Assa, Y., Henis, Y., Tencer, Y., Rotman, M., Bondi, A., and Birk, Y., Interaction of lucerne saponins with sterols, *Biochim. Biophys. Acta*, 275, 181, 1972.
66. Gestetner, B., Birk, Y., Bondi, A., and Tencer, Y., Soyabean saponin. VII. A method for the determination of sapogenin and saponin content of soya beans, *Phytochemistry*, 5, 803, 1966.
67. Buchi, J. and Dolder, R., Die bestimmung des Hemolytischem Index (HI) offizineller Arzneidrogen, *Pharm. Acta Helv.*, 25, 179, 1950.
68. Hiai, S., Oura, H., and Nakajima, T., Color reaction of some saponins and sapogenins with vanillin and sulfuric acid, *Planta Med.*, 29, 116, 1976.
69. Baccou, J. C., Lambert, F., and Sauvaire, Y., Spectrophotometric method for the determination of total steroidal sapogenin, *Analyst (London)*, 102, 458, 1977.
70. Hyong-Soo, K. and Hee-Ja, L., Determination of total saponin in ginseng jellies and candies, *Korean J. Food Sci. Technol.*, 10, 356, 1978.
71. Jones, M. and Elliot, F. C., Two rapid assays for saponin in individual alfalfa plants, *Crop Sci.*, 9, 688, 1969.
72. Pedersen, M. W., Zimmer, D. E., McAllister, D. R., Anderson, J. O., Wilding, M. D., Taylor, G. A., and McGuire, C. F., Comparative studies of the saponin of alfalfa varieties using chemical and biochemical assays, *Crop Sci.*, 7, 349, 1967.
73. Livingston, A. L., Whitehand, L. C., and Kohler, G. O., Microbiological assay for saponin in alfalfa products, *J. Assoc. Off. Anal. Chem.*, 60, 957, 1977.
74. Touchstone, J. C. and Dobbins, M. F., *Practice of Thin Layer Chromatography*, John Wiley & Sons, New York, 1978, 235.
75. Kitagawa, I., Yoshikawa, M., Hayashi, T., and Taniyama, T., Quantitative determination of soyasaponins in soybeans of various origins and soybean products by means of high performance liquid chromatography, *Yakugaku Zasshi*, 104, 275, 1984.
76. Myoga, K. and Shibata, F., Labino clover sapogenins and gas chromatographic assay method, *J. Jpn. Soc. Grass. Sci.*, 23, 67, 1977.
77. Bombarelli, E., Bonati, A., Gabetta, B., and Martinelli, E. M., Gas-liquid chromatographic determination of ginsengosides in *Panax ginseng*, *J. Chromatogr.*, 196, 121, 1980.
78. Kitagawa, J., Yoshikawa, M., Hayashi, T., and Tanayama, T., Characterization of saponin constituents in soybeans of various origins and quantitative analysis of soyasaponin by gas-liquid chromatography, *Yakugaku Zasshi*, 104, 162, 1984.
79. Ireland, P. A. and Dziedzic, S. Z., High performance liquid chromatography on silica phase with evaporative light scattering detection, *J. Chromatogr.*, 361, 410, 1986.
80. Ireland, P. A., Dziedzic, S. Z., and Kearsley, M. W., Saponin content of soya and some commercial soya products by means of high performance liquid chromatography of the sapogenins, *J. Sci. Food Agric.*, 37, 694, 1986.
81. Domon, B., Dorsaz, A.-C., and Hostettman, K., High performance liquid chromatography of oleanone saponins, *J. Chromatogr.*, 315, 441, 1984.
82. Willner, D., Gestetner, B., Lavie, D., Birk, Y., and Bondi, A., Soya bean saponins. V. Soyasapogenol E, *J. Chem. Soc. Suppl.*, 1, 5885, 1964.
83. Kitagawa, I., Yoshikawa, M., Wang, H. K., Saito, M., Tosirusnuk, V., Fujiwara, I., and Tomita K.-I., Revised structures of soyasapogenols A, B and E, oleanene-sapogenols from soybean. Structures of soyasaponins I, II and III, *Chem. Pharm. Bull.*, 30, 2294, 1982.
84. Jurzysta, M., Arising artefacts during hydrolysis of soyasaponins, in *Proc. 16th Meet. Biochemical Society Poland*, 1978, 285.
85. Gestetner, B., Birk, Y., and Bondi, A., Soybean saponins. VI. Composition of carbohydrate and aglycone moieties of soyabean saponin extract and of its fractions, *Phytochemistry*, 5, 799, 1966.
86. Ireland, P. A. and Dziedzic, S. Z., Saponins and sapogenins of chickpea, haricot bean and red kidney bean, *Food Chem.*, 23, 105, 1987.
87. Wagner, J. and Sternkopf, G., Chemical and physiological studies of the saponins of sugar beets, *Nahrung*, 2, 338, 1958.

88. **Crombie, L., Crombie, W. M. L., and Whiting, D. A.,** Isolation of avenacins A-1, A-2, B-1 and B-2 from oat roots. Structures of their aglycones, the avenestergenins, *J. Chem. Soc. Chem. Commun.,* p. 244, 1984.

89. **Tschesche, von R., Tauscher, M., Fehlhaber, H.-W., and Wulff, G.,** Avenacocid A, ein bisdesmosidisches Steroidsaponin aus *Avena sativa, Chem. Ber.,* 102, 2072, 1969.

90. **Tschesche, von R. and Lauven, P.,** Avenacocid B, ein zweites bisdesmosidisches Steroidsaponin aus *Avena sativa, Chem. Ber.,* 104, 3549, 1971.

91. **Crombie, L., Crombie, W. M. L., and Whiting, D. A.,** Structures of the four avenacins, oat root resistance factors to "take-all" disease, *J. Chem. Soc. Chem. Commun.,* p. 246, 1984.

92. **Jadhav, S. J., Sharma, R. P., and Salunkhe, D. K.,** Naturally occurring toxic alkaloids in foods, *Crit. Rev. Toxicol.,* 11, 21, 1981.

93. **Sata, H. and Sakamura, S.,** A bitter principle of tomato seeds. Isolation and structure of a new furostanol saponin, *Agric. Biol. Chem.,* 37, 225, 1973.

94. **Rao, V., Rao, P. S., Tomori, T., and Kizu, H.,** Soyasaponin I from the roots of *Arachis hypogaea, Planta Med.,* 51, 71, 1985.

95. **Augusto Ruiz, W.,** Chromatographic analysis of the saponins of quinoa *(Chenopodium quinoa Willd., var. Kancolla), Inf. An., Fac. Eng. Aliment. Agric., Univ. Estadual Campinas,* 8, 2, 1980.

96. **Yasioka, I., Nishimura, T., Matsuda, A., and Kitagawa, I.,** Seed sapogenols of *Thea sinensis* L. I. Barringtogenol C (=theasapogenol B), *Chem. Pharm. Bull.,* 18, 1610, 1970.

97. **Yasioka, I., Nishimura, T., Matsuda, A., and Kitagawa, I.,** Seed sapogenols of *Thea sinensis* L. II. Theasapogenol A, *Chem. Pharm. Bull.,* 18, 1621, 1970.

98. **Tschesche, von R., Weber, A., and Wulff, G.,** Uber die Struktur des 'theasaponins', eines Gemisches von Saponinen aus *Thea sinensis* L. mit stark antiexsuitiver Wirksamkeit, *Justus Liebigs Ann. Chem.,* 721, 209, 1969.

99. **Ueda, Y.,** Samen Saponin von *Thea sinensis* L. III. Uber die Konstitution des Thea-Sapogenols, *Chem. Pharm. Bull.,* 2, 175, 1954.

100. **Gangrade, H. and Kaushal, R.,** Fenugrin, B., a saponin from *Trigonella foenum-graecum* Linn., *Indian Drugs,* p. 149, 1979.

101. **Varshney, I. P., Jain, J. C., Srivastava, H. C., and Singh, P. P.,** Study of saponins from *Trigonella foenum-graecum* Linn. leaves, *J. Indian Chem. Soc.,* 54, 1135, 1977.

102. **Gupta, R. K., Jain, D. C., and Thakur, R. S.,** Furostanol glycosides from *Trigonella foenum-graecum* seeds, *Phytochemistry,* 23, 2605, 1984.

103. **Gupta, R. K., Jain, D. C., and Thakur, R. S.,** Furostanol glycosides from *Trigonella foenum-graecum* seeds, *Indian J. Chem.,* 248, 1215, 1985.

104. **Puri, H. S., Jefferies, T. M., and Hardman, R.,** Diosgenin and yamogenin levels in some Indian plant samples, *Planta Med.,* 30, 118, 1976.

105. **Yogoshi, T., Shidehara N., and Nakamura, M.,** Purification of glycyrrhetic acid as a sweetener, Japan Kokai, Tokyo Koho 61 37 798, 1986; *Chem. Abstr.,* 105, 77808p, 1986.

106. **Hu, S.-Y.,** The genus *Panax* (ginseng) in Chinese medicine, *Econ. Bot.,* 30, 11, 1976.

107. **Hu, S.-Y.,** A contribution to our knowledge of ginseng, *Am. J. Chin. Med.,* 5, 1, 1977.

108. **Leung, A. Y.,** *Encyclopedia of Common Natural Ingredients Used in Food, Drugs and Cosmetics,* John Wiley & Sons, New York, 1980, 276.

109. **Kartnig, Th. and Ri, C. Y.,** Dunnschichtchromatographische Untersuchungen an den Saponinen aus *Cortex Quillajae, Planta Med.,* 23, 269, 1973.

110. **Higuchi, R., Tokimitsu, Y., Fujioka, T., Komori, T., Kawasaki, T., and Oakenfull, D. G.,** Structure of deacylsaponin obtained from the bark of *Quillaja saponaria, Phytochemistry,* 26, 229, 1987.

111. **Hill, W. S.,** *The Culture of Lucerne,* Whitcombe and Tombs, Auckland, 1924.

112. **Lindahl, I. L., Davis, R. E., and Tertell, R. T.,** Alfalfa saponins: Studies of their chemical, pharmacological and biological properties in relation to ruminant bloat, *U.S. Dep. Agric. Tech. Bull.,* 1161, 1957.

113. **Majak, W., Howarth, R. E., Fesser, A. C., Goplen, B. P., and Pedersen, M. W.,** Relationship between ruminant bloat and the composition of alfalfa herbage. II. Saponins, *Can. J. Anim. Sci.,* 60, 679, 1980.

114. **Berrang, B., Davis, K. H., Wall, M. E., Hanson, C. H., and Pedersen, M. E.,** Saponins of two alfalfa cultivars, *Phytochemistry,* 13, 2253, 1974.

115. **Shany, S., Gestetner, B., Birk, Y., Bondi, A., and Kirson, I.,** Isolation of hederagenin and the saponin from alfalfa, *Isr. J. Chem.,* 10, 881, 1972.

116. **Bondi, A., Birk, Y., and Gestetner, B.,** Forage saponins, in *Chemistry and Biochemistry of Herbage,* Vol. 1, Butler, G. W. and Bailey, R. W., Eds., Academic Press, New York, 1973, 511.

117. **Oleszek, W. and Jurzysta, M.,** Isolation, chemical characterization and biological activity of alfalfa *(Medicago media* Pers.) root saponins, *Acta Sci. Bot. Pol.,* 55, 23, 1986.

118. **Morris, R. J., Dye, W. B., and Gisler, P. S.,** Isolation, purification and structural identity of alfalfa root saponin, *J. Org. Chem.,* 26, 1241, 1961.

119. **Gestetner, B.,** Structure of a saponin from lucerne *(Medicago sativa), Phytochemistry,* 10, 2221, 1971.
120. **Salunke, D. K. and Desai, D. B.,** *Postharvest Biotechnology of Vegetables,* Vol. 1, CRC Press, Boca Raton, FL, 1984, 154.
121. **Verma, S. R. S.,** The Nutritive Value of Guar Meal *(Cyamopsis tetragonoloba L.)* for Poultry, Ph.D. thesis, University of Edinburgh, 1977.
122. **Nakano, T., de Azcunes, B. C., and Spinelli, A. C.,** Chemical studies on the constituents of *Lupinus verbasciformis, Planta Med.,* 29, 241, 1976.
123. **Biran, M. and Baykut, S.,** Physico-chemical properties of gypsophila saponin, *Chim. Acta Turk.,* 3, 63, 1975.
124. **Oakenfull, D.,** Aggregation of saponins and bile acids in aqueous solution, *Aust. J. Chem.,* 39, 1671, 1986.
125. **Oakenfull, D. G. and Fisher, L. R.,** Micelles in aqueous solution, *Chem. Soc. Rev.,* 6, 25, 1977.
126. **Baria, P., Larsson, K., Ljusberg-Wahren, H., Norin, T., and Roberts, K.,** Phase equilibria in a ternary system saponin-sunflower oil-monoglycerides-water; interactions between aliphatic and alicyclic amphiphiles, *J. Sci. Food Agric.,* 30, 864, 1979.
127. **Joo, C. L. and Lee, J. J.,** Biochemical studies on ginseng saponins. IX. Determination of critical micelle concentrations of the saponin of Korean ginseng roots and its effect on lipid dispersion and enzyme reactions, *Han guk Saenghwahakhoe Chi,* 10, 59, 1977; *Chem. Abstr.,* 88, 100647, 1978.
128. **O'Connor, C. J. and Wallace, R. G.,** Bile salt micelles, *Adv. Colloid Interface Sci.,* 22, 1, 1985.
129. **Oakenfull, D. G. and Sidhu, G. S.,** A physico-chemical explanation for the effects of dietary saponins on cholesterol and bile salt metabolism, *Nutr. Rep. Int.,* 27, 1253, 1983.
130. **Ljusberg-Wahren, H.,** Amphiphiles with a rigid hydrocarbon part, a comparative study of aqueous systems of bile salts, saponins and resin acid soaps, Dissertation, Royal Institute of Technology, Stockholm, 1986.
131. **Shukla, S. D. and Vaish, A. K.,** Effect of medicinal plant gums and other surface active substances on the particle size distribution of emulsion stabilized by nicotine hydrogen tartarate, *J. Indian Chem. Soc.,* 57, 982, 1980.
132. **Baria, P., Larsson, K., Ljusberg-Wahren, H., Norin, T., and Roberts, K.,** Phase equilibria in a ternary system saponin-sunflower oil monoglycerides/water; interactions between aliphatic and alicyclic amphiphiles, *J. Sci. Food Agric.,* 30, 864, 1979.
133. **Heftman, E.,** Biochemistry of steroidal saponins and glycoalkaloids, *Lloydia,* 30, 209, 1967.
134. **Ruzicka, L.,** The isoprene rule and the biogenesis of terpenic compounds, *Experientia,* 9, 253, 1953.
135. **Corey, E. J. and Ortiz de Montellano, P. R.,** Enzymic synthesis of β-amyrin from 2,3 oxidosqualene, *J. Am. Chem. Soc.,* 89, 3362, 1967.
136. **Nowacki, E., Jurzysta, M., and Diettrych-Szostak, D.,** Zur biosynthese der medicagensaure in keimenden Luzernesamen, *Biochem. Physiol. Pflanz.,* 169, 183, 1976.
137. **Peri, I., Mor, U., Heftman, E., Bondi, A., and Tencer, Y.,** Biosynthesis of triterpenoid sapogenols in soybean and alfalfa seedlings, *Phytochemistry,* 18, 1671, 1979.
138. **Fassler, L., Lichtenthaler, H. K., Guriz, K., and Biacs, P. A.,** Accumulation of saponins and sterols in *Avena* seedlings under high-light and low-light growth conditions, *Dev. Plant Biol.,* 9, 225, 1984.
139. **Fuggersburger-Heinz, R. and Franz, G.,** Formation of glycyrrhizinic acid in *Glycyrrhiza glabra* var. *typica, Planta Med.,* 50, 409, 1984.
140. **Cheeke, P. R.,** Biological properties and nutritional significance of legume saponins, in *Leaf Protein Concentrate,* Telek, L. and Graham, H., Eds., AVI Publishing, Westport, CT, 1983, chap. 13.
141. **Birk, Y., Bondi, A., Gestetner, B., and Ishaaya, I.,** A thermostable haemolytic factor in soybeans, *Nature (London),* 197, 1089, 1963.
142. **Lower, E. S.,** Activity of the saponins, *East. Pharm.,* 28, 55, 1985.
143. **Ewart, A. J.,** The poisonous action of ingested saponins, *CSIRO Bull.,* p. 50, 1931.
144. **Segal, R., Mansour, M., and Zaitschek, D. V.,** Effect of ester groups on the haemolytic action of some saponins and sapogenins, *Biochem. Pharmacol.,* 15, 1411, 1966.
145. **Segal, R., Milo-Goldzweig, I., Schupper, H., and Zaitschek, D. V.,** Effect of ester groups on the haemolytic action of sapogenins. II. Esterification with bifunctional acids, *Biochem. Pharmacol.,* 19, 2501, 1970.
146. **Tschesche, R. and Wulff, G.,** Chemistry and biology of saponins, *Fortschr. Chem. Org. Naturst.,* 30, 461, 1973.
147. **Namba, T., Yoshizaki, M., Tomimori, T., Kobashi, K., Mitsui, K., and Hase, J.,** Hemolytic and its protective activity of ginseng saponins, *Chem. Pharm. Bull.,* 21, 459, 1973.
148. **Seeman, P.,** Ultrastructure of membrane lesions in immune lysis, osmotic lysis and drug induced lysis, *Fed. Proc.,* 33, 2116, 1974.
149. **Nishikawa, M., Nojima, S., Akiyama, T., Sankawa, U., and Inoue, K.,** Interaction of digitonin and its analogues with membrane cholesterol, *J. Biochem.,* 96, 1231, 1984.
150. **Segal, L. and Milo-Goldzwerg, I.,** The susceptibility of cholesterol depleted erythrocytes to saponin and sapogenin hemolysis, *Biochim. Biophys. Acta,* 512, 223, 1978.

151. **Akiyama, T., Takagi, S., Sankawa, U., Inari, S., and Saito, H.**, Saponin-cholesterol interactions in multilayers of egg yolk lecithin as studied by deuterium magnetic resonance: digitonin and its analogues, *Biochemistry*, 19, 1904, 1980.

152. **Kinsky, S. C., Luse, S. A., and van Deenen, L. L. M.**, Interaction of polyene antibiotics with natural and artificial membrane systems, *Fed. Proc.*, 25, 1503, 1968.

153. **Dourmashkin, R. R., Dougherty, R. M., and Harris, R. J. C.**, Electron microscope observations on rous sarcoma virus and cell membranes, *Nature (London)*, 194, 1116, 1962.

154. **Brown, J. N. and Harris, J. R.**, The entry of ferritin into haemoglobin-free human erythrocyte ghosts prepared under different conditions, *J. Ultrastruct. Res.*, 32, 405, 1970.

155. **Gogelein, H. and Huby, A.**, Interaction of saponin and digitonin with black lipid membranes and lipid monolayers, *Biochim. Biophys. Acta*, 773, 32, 1984.

156. **Tsang, D., Yeung, H. W., Tso, W. W., and Peck, H.**, Ginseng saponins: influence on neurotransmitter uptake in rat brain synaptosomes, *Planta Med.*, 49, 221, 1985.

157. **Alvarez, J. R. and Torres-Pinedo, R.**, Interactions of soybean lectin, soyasaponins and glycinin with rabbit jejunal mucosa *in vitro*, *Pediatr. Res.*, 16, 728, 1982.

158. **Fukuda, K., Utsumi, H., Shoji, J., and Hamada, A.**, Saponins can cause the agglutination of phospholipid vesicles, *Biochim. Biophys. Acta*, 820, 199, 1985.

159. **Jackson, H. D. and Shaw, R. A.**, Chemical and biological properties of a respiratory inhibitor from alfalfa saponin, *Arch. Biochem. Biophys.*, 84, 411, 1954.

160. **Ishaaya, I. and Birk, Y.**, Soybean saponins: the effect of proteins on the inhibitory activity of soybean saponins on certain enzymes, *J. Food Sci.*, 30, 118, 1965.

161. **Lee, S. W., Lee, J. S., Kim, Y. H., and Jim, K. D.**, Effect of ginseng saponin on the Na^+, K^+-ATPase of dog cardiac sarcolemma, *Arch. Pharm. Res.*, 9, 29, 1986.

162. **Shibata, S.**, Saponins with biological and pharmacological activity, in *New Natural Products and Plant Drugs with Pharmacological, Biological and Therapeutical Activity*, Wagner, H., and Wolff, P., Eds., Springer-Verlag, Berlin, 1977, 177.

163. **Avakadian, E. V., Sugimoto, R. B., Taguchi, S., and Horvath, S. M.**, Effect of *Panax ginseng* extract on energy metabolism during exercise in rats, *Planta Med.*, 50, 151, 1984.

164. **Hiai, S., Oura, H., Tsuhada, K., and Hirai, Y.**, Stimulating effect of *Panax ginseng* extract on RNA-polymerase activity in rat liver nuclei, *Chem. Pharm. Bull.*, 19, 1656, 1971.

165. **Oura, H., Hiai, S., and Seno, H.**, Synthesis and characterization of nuclear RNA induced by *Radix ginseng* extract in rat liver, *Chem. Pharm. Bull.*, 19, 1598, 1971.

166. **Yamamoto, M., Takeuchi, N., Kumagai, A., and Yamamura, Y.**, Stimulatory effect of *Panax ginseng* principles on DNA, RNA, protein and lipid synthesis in rat bone marrow, *Arzneim. Forsch.*, 27, 1169, 1977.

167. **Anisimov, M. M., Prokofieva, N. G., Strigma, L. T., Chetyrina, N. S., Aladjina, N. G., and Elyakov, G. B.**, Comparative study of cytotoxic activity of triterpene glycosides from marine organisms, *Biochem. Pharmacol.*, 26, 2113, 1977.

168. **Cucu, V. and Grecu, L.**, Antimicrobial action of saponins, *Farmacia (Bucharest)*, 19, 641, 1971.

169. **Wolters, B.**, The antibiotic action of saponins. III. Saponins as plant fungistatic compounds, *Planta*, 79, 77, 1968.

170. **Gestetner, B., Assa, Y., Henis, Y., Birk, Y., and Bondi, A.**, Lucerne saponins. IV. Relationship between their chemical constitution and haemolytic and antifungal activities, *J. Sci. Food Agric.*, 22, 168, 1971.

171. **Assa, Y., Gestetner, B., Chet, I., and Henis, Y.**, Fungistatic activity of lucerne saponins and digitonin as related to sterols, *Life Sci.*, 11, 637, 1972.

172. **Lazurevskii, G. V., Zhuchenko, A. A., Kintya, P. K., Balashova, N. N., Moschenko, N. E., Bobieko, V. A., Para, S. P., Andryushchenko, V. K., and Goryanu, G. H.**, Effect of steroid glycosides on *Phytophthora infestans*, *Dokl. Akad. Nauk SSSR*, 243, 1076, 1978.

173. **Hamaya, E., Tsushida, T., Nagata, T., Nishino, C., Eknol, N., and Manobe, S.**, Antifungal components of Camellia plant, *Nippon Shokubutsu Byori Gakkaiho*, 50, 628, 1984.

174. **Betz, H. and Schloesser, E.**, On the use of fungicidal saponins in crop protection, *Tagungsber. Akad. Landwirtschaftswiss. D.D.R.*, 222, 179, 1984.

175. **Anisimov, M. M., Shcheglov, V. V., Strigma, L. I., Chetrina, N. S., Uvarova, N. I., Oshitok, G. I., Aladina, N. G., Vercherko, L. P., and Oshitok, G. I.**, Chemical structure and antifungal activity of the triterpenoids, *Izv. Akad. Nauk SSSR, Ser. Biol.*, 4, 570, 1979; *Chem. Abstr.*, 91, 187216, 1979.

176. **Anisimov, M. M., Shcheglov, V. V., and Dzizenko, S. N.**, Effect of triterpene glycosides on the biosynthesis of sterols and fatty acids in the yeast *Saccharomyces carlsbergensis*, *Prikl. Biokhim. Mikrobiol.*, 14, 573, 1978.

177. **Anisimov, M. M., Shentsova, E. B., Shcheglov, V. V., Strigma, L. I., Uvarova, N. I., Levina, E. V., Oshitok, G. I., and Elyakov, G. B.**, A comparative study of the cytotoxic effect of dammaran triterpenes and betulin in early embryogenesis of the sea urchin, *Toxicon*, 16, 31, 1978.

178. Applebaum, S. W. and Birk, Y., in *Insect and Mite Nutrition*, Rodriguez, J. G., Ed., North-Holland, Amsterdam, 1972, 629.

179. Tschesche, von R., Wulff, G., Weber, A., and Schmidt, H., Damaging effect of saponin on termites *(Isoptera reticulitermes)*, Z. *Naturforsch. Teil B*, 25, 499, 1970.

180. Applebaum, S. W. and Birk, Y., in *Herbivores - Their Interaction with Secondary Plant Metabolites*, Rosenthal, G. A. and Janzen, D. H., Eds., Academic Press, New York, 1979, 539.

181. Krzymanska, J. and Waligoria, D., Significance of saponins in resistance of alfalfa against green pea aphid (*Acyrthosiphon pisum* Harris), *Pr. Nauk. Inst. Ochr. Rosl.*, 24, 153, 1983; *Chem. Abstr.*, 102, 128998, 1985.

182. Krzymanska, J., Rosada, J., Waligoria, D., and Jakubiak, W., Investigation on the role of saponins in resistance of alfalfa *(Medicago sp.)* to the green pea aphid *(Acyrthosiphon pisum* Harris) with the use of a radioisotope method, *Pr. Nauk. Inst. Ochr. Rosl.*, 25, 207, 1983; *Chem. Abstr.*, 102, 129098, 1985.

183. Hosttetmann, K., Saponins and schistosomiasis: isolation and determination of saponin structure, *Schweiz. Apoth. Ztg.*, 123, 223, 1985.

184. Parkhurst, R. M., Thomas, D. W., and Skinner, W. A., Molluscicidal saponins of *Phytolacca dodecanara:* oleanoglycotoxin-A, *Phytochemistry*, 12, 1437, 1973.

185. Gafner, F., Msonthi, J. D., and Hostettman, K., Phytochemistry of African medicinal plants. III. Molluscicidal saponins from *Talinum tenuissimum*, *Helv. Chim. Acta*, 68, 555, 1985.

186. Seeman, P., Ultrastructure of membrane lesions in immune lysis, osmotic lysis and drug induced lysis, *Fed. Proc.*, 33, 2116, 1974.

187. Roy, P. K., Munshi, J. S.-D., and Munshi, J. F., Scanning electron microscopic evaluation of effects of saponins on the gills of the climbing perch (*Anabas testudineus* Bloch) (Anabantidae: Pisces), *Indian J. Exp. Biol.*, 24, 511, 1986.

188. Gestetner, B., Birk, Y., and Tencer, Y., Fate of ingested soy saponins and physiological aspects of their hemolytic activity, *J. Agric. Food Chem.*, 16, 1031, 1968.

189. Chen, S. E., Sawchuk, R. J., and Staba, E. J., American ginseng. III. Pharmacokinetics of ginsengosides in the rabbit, *Eur. J. Drug Metab.*, 5, 161, 1980.

190. Odani, T., Tanizawa, H., and Takino, Y., Studies on the absorption, distribution, excretion and metabolism of ginseng saponins. II. The absorption, distribution and excretion of ginsengoside Rg_1 in the rat, *Chem. Pharm. Bull.*, 31, 292, 1983.

191. Odani, T., Tanizawa, H., and Takino, Y., Studies on the absorption, distribution, excretion and metabolism of ginseng saponins. III. The absorption, distribution and excretion of ginsengoside Rb_1 in the rat, *Chem. Pharm. Bull.*, 31, 1059, 1983.

192. Shibata, S., Saponins with biological and pharmaceutical activity, in *New Natural Products and Plant Drugs with Pharmacological, Biological and Therapeutical Activity*, Wagner, H. and Wulff, P., Eds., Springer-Verlag, Berlin, 1972, 177.

193. Peterson, D. W., Effect of sterols on the growh of chicks fed high alfalfa diets or a diet containing quillaja, *J. Nutr.*, 42, 597, 1950.

194. Cayen, M. N., Effect of dietary tomatine on cholesterol metabolism in the rat, *J. Lipid Res.*, 12, 482, 1971.

195. Morgan, B., Heald, M., Brooks, S. G., Tee, J. L., and Green, J., The interactions between dietary saponins, cholesterol and related sterols in the chick, *Poult. Sci.*, 51, 677, 1972.

196. Malinow, M. R., McLaughlin, P., Papworth, L., Stafford, C., Kohler, G. O., Livingston, A. L., and Cheeke, P. R., Effects of alfalfa saponins on intestinal cholesterol absorption in rats, *Am. J. Clin. Nutr.*, 30, 2061, 1977.

197. Malinow, M. R., McLaughlin, P., and Stafford, C., Prevention of hypercholesterolemia in monkeys *(Macaca fascicularis)* by digitonin, *Am. J. Clin. Nutr.*, 31, 814, 1978.

198. Malinow, M. R., McLaughlin, P., and Stafford, C., Alfalfa seeds: effects on cholesterol metabolism, *Experientia*, 36, 562, 1980.

199. Malinow, M. R., Connor, W. E., McLaughlin, P., Stafford, C., Lin, D. S., Livingston, A. L., Kohler, G. O., and McNulty, W. P., Cholesterol and bile acid balance in *Macaca fascicularis:* effects of alfalfa saponin, *J. Clin. Invest.*, 67, 156, 1987.

200. Malinow, M. R., Effects of synthetic glycosides on cholesterol absorption, *Ann. N.Y. Acad. Sci.*, 454, 23, 1985.

201. Cayen, M. N. and Dvornik, D., Effect of diosgenin on lipid metabolism in rats, *J. Lipid Res.*, 20, 162, 1979.

202. Sidhu, G. S., Upson, B., and Malinow, M. R., Effects of soy saponins and trigogenin cellobioside on intestinal uptake of cholesterol, cholate and glucose, *Nutr. Rep. Int.*, 35, 615, 1987.

203. Sidhu, G. S. and Oakenfull, D. G., A mechanism for the hypocholesterolaemic activity of saponins, *Br. J. Nutr.*, 55, 643, 1986.

204. Oakenfull, D. G., Topping, D. L., Illman, R. J., and Fenwick, D. E., Prevention of dietary hypercholesterolaemia in the rat by soyabean and quillaja saponins, *Nutr. Rep. Int.*, 29, 1039, 1984.

205. **Sidhu, G. S. and Oakenfull, D. G.,** How dietary saponins lower plasma cholesterol, *Proc. Nutr. Soc. Aust.*, 8, 153, 1983.

206. **Cheeke, P. R.,** Nutritional and physiological properties of saponins, *Nutr. Rep. Int.*, 13, 315, 1976.

207. **Jenkins, D. J. A., Wolever, T. M. S., Leeds, A. R., Gassui, M. A., Haisman, P., Dilawari, J., Goff, D. V., Metz, G. L., and Alberti, K. G. M. M.,** Dietary fibres, fibre analogues and glucose tolerance: importance of viscosity, *Br. Med. J.*, 1, 1392, 1978.

208. **Urban, E.,** Gut mucosa, adaptive growth and enteroglucagon, *Gastroenterolology*, 91, 251, 1986.

209. **Hiai, S.,** Chinese medicinal material and secretion of ACTH and corticosteroid, in *Advances in Chinese Medicinal Materials Research*, Chang, H. W., Yeung, H. W., Tso, W. W., and Koo, A. Eds., World Scientific, Singapore, 1985, 49.

210. **Pearce, P. T., Zois, I., Wynne, K. W., and Funder, J. W.,** *Panax ginseng* and *Eleutherococcus senticosus* extracts - *in vitro* studies on binidng to steroid receptors, *Endocrinol. Jpn.*, 29, 567, 1982.

211. **Petkov, V.,** Effect of ginseng on the brain biogenic monoamines and 3' -5' -AMP system. Experiments on rats, *Arzneim. Forsch.*, 28, 388, 1978; *Chem. Abstr.*, 88, 182983, 1978.

212. **Zhang, Y., Shou, J., Song, J., Wang, Y., Shao, Y., Li, C., Zhou, S., Li, Y., and Li, D.,** Effects of *Astragalus* saponin 1 on cAMP and cGMP levels in plasma and DNA synthesis in regenerating liver, *Yao Hsueh Hsueh Pao*, 19, 619, 1984; *Chem. Abstr.*, 101, 204288, 1985.

213. **Zong, R., Zheng, S., and Liu, J.,** Adrenocorticotrophic hormone-like effects of ginseng saponin, *Baiqiuen Yike Daxue Xuebao*, 11, 254, 1985; *Chem. Abstr.*, 105, 218698, 1986.

214. **Williams, M. C., Rodewijk, J. C. M., and Olsen, J. D.,** Intoxication in cattle, chicks and hamsters from seed of the palo santo tree *(Bulnesia sarmientii)*, *Vet. Rec.*, p. 646, 1984.

215. **Ishaaya, I., Birk, Y., Bondi, A., and Tencer, Y.,** Soy bean saponins. IX. Studies of their effect on birds, mammals and cold blooded organisms, *J. Sci. Food Agric.*, 20, 433, 1969.

216. **Malinow, M. R., McNulty, A. L., and Kohler, G. O.,** The toxicity of alfalfa saponin in rats, *Food Cosmet. Toxicol.*, 19, 443, 1981.

217. **Phillips, J. C., Butterworth, K. R., Gaunt, I. F., Evans, J. G., and Grasso, P.,** Long term toxicity studies of quillaia extract in mice, *Food Cosmet. Toxicol.*, 17, 23, 1979.

218. **Oser, B. L.,** An evaluation of *Yucca mohavensis* as a source of food grade saponin, *Food Cosmet. Toxicol.*, 4, 57, 1966.

219. **Ominami, H., Kimura, Y., Okuda, H., Arichi, S., Yoshikawa, M., and Kitagawa, I.,** Effects of saponins on liver injury induced by highly peroxidized fat in rats, *Planta Med.*, 50, 440, 1984.

220. **Freeland, W. J., Calcott, P. H., and Anderson, L. R.,** Tannins and saponin: interaction in herbivore diets, *Biochem. Syst. Ecol.*, 13, 189, 1985.

221. **Cheeke, P. R., Kinzell, J. H., and Pederson, M. W.,** Influence of saponins on alfalfa utilization by rats, rabbits and swine, *J. Anim. Sci.*, 45, 476, 1977.

222. **Cooney, W. J., Butts, J. S., and Bacon, L. E.,** Alfalfa in chick rations, *Poult. Sci.*, 27, 828, 1948.

223. **Livingston, A. L., Knuckles, B. E., Edwards, R. H., de Fremery, D., Miller, R. E., and Kohler, G. O.,** Distribution of saponin in alfalfa protein recovery systems, *J. Agric. Food Chem.*, 27, 362, 1979.

224. **Cheeke, P. R., Pedersen, M. W., and England, D. C.,** Responses of rats and swine to alfalfa saponins, *Can. J. Anim. Sci.*, 58, 783, 1978.

225. **Oakenfull, D. G., and Topping, D. L.,** Saponins and plasma cholesterol, *Atherosclerosis*, 48, 301, 1983.

226. **Newman, H. A., Kummerow, F. A., and Scott, H. H.,** Dietary saponin, a factor which may reduce liver and serum cholesterol levels, *Poult. Sci.*, 37, 42, 1957.

227. **Griminger, P. and Fisher, H.,** Dietary saponin and plasma cholesterol in the chicken, *Proc. Soc. Exp. Biol. Med.*, 99, 424, 1958.

228. **Qureshi, A. A., Auirmeileh, N., Din, Z. Z., Ahmad, Y., Burger, W. C., and Elson, C. E.,** Suppression of cholesterolgenesis and reduction of LDL cholesterol by dietary ginseng and its fractions in chicken liver, *Atherosclerosis*, 48, 81, 1983.

229. **Reshef, G., Gestetner, B., Birk, Y., and Bondi, A.,** Effect of alfalfa saponins on the growth and some aspects of lipid metabolism of mice and quails, *J. Sci. Food Agric.*, 27, 63, 1976.

230. **McNaughton, J. L.,** Effect of dietary fiber on egg yolk, liver and plasma cholesterol concentrations in the laying hen, *J. Nutr.*, 108, 1842, 1978.

231. **Nakaue, H. S., Lowry, R. R., Cheeke, P. R., and Arscott, G. H.,** The effect of dietary alfalfa of varying saponin content on egg cholesterol and laying performance, *Poult. Sci.*, 59, 2744, 1980.

232. **Sims, J. S., Kitts, W. D., and Bragg, D. B.,** Effect of dietary saponin on egg cholesterol level and laying hen performance, *Can. J. Anim. Sci.*, 64, 977, 1984.

233. **Kritchevsky, D., Tepper, S. A., and Story, J. A.,** Isocaloric, isogravic diets in rats. III. Effect of nonnutritive fiber (alfalfa or cellulose) on cholesterol metabolism, *Nutr. Rep. Int.*, 9, 301, 1975.

234. **Malinow, M. R., McLaughlin, P., Stafford, C., Livingston, A. L., and Kohler, G. O.,** Alfalfa saponins and alfalfa seeds: dietary effects on cholesterol fed rabbits, *Atherosclerosis*, 37, 433, 1980.

235. **Oakenfull, D. G., Fenwick, D. E., Hood, R. L., Topping, D. L., Illman, R. J., and Storer, G. B.,** Effects of saponins on bile acids and plasma lipids in the rat, *Br. J. Nutr.*, 42, 209, 1979.

236. **Oakenfull, D. G., and Sidhu, G. S.,** Prevention of dietary hypercholesterolaemia by chickpea saponin and navy beans, *Proc. Nutr. Soc. Aust.,* 9, 104, 1984.

237. **Sautier, C., Doucet, C., Flamant, C., and Lemonnier, D.,** Effect of soyprotein and saponins on serum, tissue and feces steroids in rat, *Atherosclerosis,* 34, 233, 1979.

238. **Gibney, M. J., Pathirana, C., and Smith, L.,** Saponins and fiber — lack of interactive effects on serum and liver cholesterol in rats and hamsters, *Atherosclerosis,* 45, 365, 1982.

239. **Bingham, R., Harris, D. H., and Laga, T.,** Yucca plant saponin in the treatment of hypertension and hypercholesterolemia, *J. Appl. Nutr.,* 30, 127, 1978.

240. **Mathur, K. S.,** Hypocholesterolaemic effect of Bengal gram: a long-term study in man, *Br. Med. J.,* 1, 30, 1968.

241. **Anon.,** Effect of legume seeds on serum cholesterol, *Nutr. Rev.,* 38, 159, 1980.

242. **Potter, J. D., Topping, D. L., and Oakenfull, D. G.,** Soya products, saponins and plasma cholesterol, *Lancet,* 1, 223, 1979.

243. **Heaton, K. W.,** *Bile Salts in Health and Disease,* Churchill-Livingstone, Edinburgh, 1972.

244. **Topping, D. L., Storer, G. B., Calvert, G. D., Illman, R. J., Oakenfull, D. G., and Weller, R. A.,** Effects of dietary saponins on fecal bile acids and neutral sterols, plasma lipids and lipid turnover in the pig, *Am. J. Clin Nutr.,* 33, 783, 1980.

245. **Potter, J. D., Illman, R. J., Calvert, G. D., Oakenfull, D. G., and Topping, D. L.,** Soya saponins, plasma lipids, lipoproteins and fecal bile acids: a double blind cross-over study, *Nutr. Rep. Int.,* 22, 521, 1980.

246. **Takashi, I., Keiichi, U., and Hisayuki, K.,** Gourd saponins as antioxidants in oils in food, cosmetics and pharmaceuticals, Japan Kokai Tokkyo Koho 61,130,390; *Chem. Abstr.,* 105, 151815, 1986.

247. **Jai Ho, C. and Ki, O. S.,** Studies on the anti-aging action of Korean ginseng, *Han guk Sikp um Kwahakhoe Chi,* 17, 566, 1985; *Chem. Abstr.,* 104, 123111, 1986.

248. **Kunihiko, S.,** Current health foods, *New Food Ind.,* 72, 61, 1985; *Chem. Abstr.,* 105, 151815, 1986.

249. **Asaki, D. and Kogyo, K.,** Preparation of powder or granules containing vitamin E, *Jpn. Kokai Tokkyo Koho,* 60, 64, 919; *Chem. Abstr.,* 103, 76253, 1985.

250. **Asaki, D. and Kogyo, K.,** Aqueous vitamin E solution, *Jpn. Kokai Tokkyo Koho,* 60, 51, 104; *Chem. Abstr.,* 103, 86796, 1985.

251. *Approved Food Standards and Approved Food Additives,* National Health and Medical Research Council, Canberra, 1979.

252. **Essman, E. J.,** The medicinal uses of herbs, *Fitoterapia,* 55, 279, 1984.

253. **Hiromichi, H. and Nakao, I.,** The immune response of splenic lymphocytes after cimicifugoside treatment *in vitro* and pretreatment *in vivo, J. Pharmacobiol. Dyn.,* 3, 643, 1980.

254. **Hikokichi, O., Shoichi, N., Akita, K., and Masaru, T.,** Immunosuppressive effect of licorice root extracts, *Taisha,* 10, 651, 1973; *Chem. Abstr.,* 81, 76115, 1974.

255. **Nozaki, H., Suzuki, H., Hirayama, T. Kasai, R., Wa, R. Y., and Lee, K. H.,** Antitumor triterpenes of *Maytenus diversifolia, Phytochemistry,* 25, 479, 1986.

256. **Gupta, S. S., Bhagwat, A. W., and Ram, A. K.,** Cardiac stimulant activity of the saponins of *Achyranthes aspera, Indian J. Med. Res.,* 60, 462, 1972.

257. **Gupta, S. S., Ram, A. K., and Tripathi, R. H.,** Inhibition of slow reacting substance of anaphylaxis (SRS-A) by a saponin of *Gardenia turgida, Curr. Sci.,* 41, 614, 1972.

258. **Johri, R. K., Zutshi, H., Kameshwaran, L., and Atal, C. K.,** Effect of queticin and albizzia saponins on rat mast cells, *Indian J. Physiol. Pharmacol.,* 29, 43, 1985.

259. **Kubo, M., Matsuda, H., Tani, T., Namba, K., Arichi, G., and Kitagawa, I.,** Effects of soya saponin on experimental disseminated intravascular coagulation. I, *Chem. Pharm. Bull.,* 32, 1467, 1984.

260. **Kereselidze, E. V., Pkheidze, T. A., Kemertelidize, E. P., Khardziani, S. D., Dzhaparidze, T. N., and Makharadze, Sh. K.,** Fibrinolytic activity of *Ruscus ponticus* and *Ruscus hyphophyllum* saponins, *Soobshch. Akad. Nauk Gruz. S.S.R.,* 78, 485, 1975.

261. **Osaka Yakuhin Kenkyusho, K. K.,** Cosmetics containing soybean saponins, *Jpn. Kokai Tokkyo Koho,* 59, 106, 410; *Chem. Abstr.,* 101, 116595P, 1984.

262. **Tomohisa, A. and Shintara, A.,** Topical formulations containing saponins and hydroquinone glycosides for cosmetics and pharmaceuticals, *Jpn. Kokai Tokkyo Koho,* 61, 210, 004; *Chem. Abstr.,* 106, 72705P, 1987.

263. **Susumo, O. and Toshihiro, Y.,** Surfactants, *Jpn. Kokai Tokkyo Kohc,* 60, 190, 224; *Chem. Abstr.,* 104, 151295S, 1986.

264. **Lion Corp.,** Shampoos containing sugar alcohol esters and saponins, *Jpn. Kokai Tokkyo Koho,* 59, 184, 299; *Chem. Abstr.,* 102, 1191122, 1985.

265. **Lion Corp.,** Shampoos and cleaning solutions containing acylamino acid salts and saponins, *Jpn. Kokai Tokkyo Koho,* 59, 179, 597; *Chem. Abstr.,* 102, 119421, 1985.

266. **Lion Corp.,** Hair preparations containing surfactants, alcohols, fatty acid esters and saponins, *Jpn. Kokai Tokkyo Koho* 60, 194, 710; *Chem. Abstr.,* 102, 209145, 1985.

267. **Ricoh, Co. Ltd.,** Diazo coyping process and material, *Jpn. Kokai Tokkyo Koho,* 59, 55, 435; *Chem. Abstr.,* 102, 54012, 1985.
268. **Ricoh Co. Ltd.,** Diazo copying material, Japan Kokai Tokkyo Koho 59, 131, 928; *Chem. Abstr.,* 102, 54014, 1985.
269. **Kanie, T.,** Application of saponin to biological treatment, *PPM,* 17, 28, 1986; *Chem. Abstr.,* 106, 55238W, 1987.
270. **Pavlista, B. P. R.,** Impregnating agent for filter systems of tobacco products, German Patent 3,411,057; *Chem. Abstr.,* 104, 3504, 1986.

Chapter 5

ALIPHATIC NITROCOMPOUNDS

Walter Majak and Michael A. Pass

TABLE OF CONTENTS

I. INTRODUCTION

This review is an attempt to comprehensively cover the chemistry, biochemistry, and biological effects of aliphatic nitrocompounds from higher plants, particularly those of the genus *Astragalus* (Leguminosae). The impact of these natural toxins on livestock production is also emphasized. For reviews on the enzymology and pharmacology of synthetic nitrocompounds and those of microbial and invertebrate origin, the reader is referred to a review by Venulet and Van Etten[1] and to a recent publication by Alston et al.[2] The genus *Astragalus* also contains species that synthesize other toxins. Studies on the selenium accumulators and locoweeds that contain the indolizidine alkaloid, swainsonine, have been reviewed by James.[3] Briefer accounts on the mode of action and toxicological effects of 3-nitropropanol and 3-nitropropionic acid are also available,[3-6] but more recent developments are reviewed here. Aromatic nitrocompounds also occur in higher plants. The aristolochic acids, which are derivatives of phenanthrene, occur widely in the family Aristolochiaceae,[7,8] but nitrobenzene substitution has only been detected in a few instances.[9] Aristolochic acid has neoplastic and nephrotoxic properties,[9a] but the biological activity of the nitrobenzenes in higher plants has not been ascertained.

II. CHEMISTRY, ISOLATION, AND IDENTIFICATION

Primary nitrocompounds have the general formula RCH_2NO_2 in which the nitro (NO_2) group is linked through nitrogen to the α-carbon. Nitroalkanes are practically insoluble in water, but they dissolve in aqueous alkali because the hydrogen on the α-carbon is acidic, reacting with base to form the nitronate anion. When the anion is acidified, the proton adds more readily to the oxygen atoms producing the unstable aci isomer, which tautomerizes to the nitro form (Scheme 1).[10]

$$RCH_2NO_2 \xrightarrow{^-OH} RCH=\overset{+}{N}\overset{O^-}{\underset{O^-}{\diagup}} \xrightarrow{H^+} RCH=\overset{+}{N}\overset{OH}{\underset{O^-}{\diagup}} \rightarrow RCH_2NO_2 \qquad (1)$$

In higher plants, the primary nitro group occurs in four aliphatic systems: most commonly as 3-nitro-1-propanol (NPOH) and 3-nitro-1-propionic acid (NPA) and less abundantly as 1-phenyl-2-nitroethane and 1-(4'-hydroxyphenyl)-2-nitroethane. The alcohols occur as ether glycosides (acetals) and should be distinguished from the NPA conjugates which are glucose esters of NPA. The ten NPA esters which are known to occur in nature are listed in Table 1. The most common glycoside of NPOH is miserotoxin (3-nitro-1-propyl-β-D-glucopyranoside). The hydroxyphenyl derivative has also been detected as the β-D-glucoside and as the β-D-primeveroside (see next section). The difference in structure between miserotoxin and the β-form of the 6-monoester is illustrated in (Scheme 2). Recently, the gentiobioside of NPOH was isolated.[10a]

miserotoxin, $R_1 = CH_2CH_2CH_2NO_2$, $R_2 = H$

6 monoester, $R_2 = COCH_2CH_2NO_2$, $R_1 = H$ $\qquad (2)$

Nitropropanol is a liquid with a high boiling point (85°C at 2 mmHg), and NPA is a crystalline solid (mp 65 to 66°). The nitroalcohol can be liberated from miserotoxin by mild acid hydrolysis or through the enzymatic action of β-glucosidase. Partial acid hydrolysis can be used to degrade complex NPA conjugates into simpler components, such as the 6-

Table 1
SUBSTITUTION PATTERNS
FOR NPA GLUCOSE ESTERS

Common name	Position of NPA and anomeric H	Ref.
6-Monoester	6-α,β	35
Cibarian	1,6-β	11
Coronarian	2,6-α	36
	4,6-α	11
	4,6-β	11
Karakin	1,2,6-β	40
Coronillin	1,2,6-α	36
Corynocarpin	1,4,6-β	37
Corollin	2,3,6-α	36
Hiptagin	1,2,4,6-β	38

monoester, NPA, and glucose. Extraction with acetone and purification on silca gel columns is a preferred method for isolating NPA conjugates.[11,12] Miserotoxin was originally isolated from ethanol extracts subjected to countercurrent distribution.[13] Simpler methods of isolation were subsequently devised using paper chromatography and preparative thin-layer chromatography (TLC),[14] or deactivated coconut charcoal and silica gel columns.[15,16] Visualization of aliphatic nitrocompounds on TLC is best accomplished with the diazotized *p*-nitroaniline or sulfanilic acid reagents.[14] The method is based on the coupling of the aromatic diazonium salts to the nitronate anion, and it can be used quantitatively as a spectrophotometric procedure.[14] Nitrocompounds can also be determined indirectly by measuring the nitrite ion after alkaline displacement of the nitro group.[17,18] Applications of these colorimetric methods for the determination of NPA or NPOH in biological samples as well as the newer methods of analysis utilizing gas chromatography (GC) and high-performance liquid chromatography (HPLC) procedures were recently reviewed.[6] HPLC, which can resolve NPA, NPOH, and miserotoxin, employs a reverse phase octadecylsilane column eluted isocractically with 0.1% orthophosphoric acid.[19] The three compounds can also be resolved on silca gel TLC.[12,16] Proton NMR and ^{13}C-NMR appear to be the most valuable spectroscopic methods for determining the structures of naturally occurring NPA and NPOH conjugates isolated from higher plants.[10a,12,20,21] Near infrared reflectance spectroscopy (NIRS) was recently evaluated as a screening method for determining NPA in *Astragalus* species.[21a] The NIRS equation predicted NPA concentrations with a correlation coefficient of 0.77.

III. DISTRIBUTION

Miserotoxin, the glucoside of NPOH, was first isolated from *Astragalus miser* var. *oblongifolius* by Stermitz et al.[13] Subsequently, it was detected in two other varieties of timber milkvetch (*A. miser* var. *serotinus*, *A. miser* var. *hylophilus*) and in nine additional species of *Astragalus*.[12,22] The identification of this new poisonous principle in timber milk vetch prompted an immediate search for other toxic nitrocompounds. Williams and co-workers[22,23] screened many species of *Astraglus* to determine inorganic nitrite formation from the organonitro group using the method of Cooke.[17] Initial tests, using leaflet or stem samples from herbarium specimens, indicated that 14 to 22% of *Astragalus* species in the western U.S. contained nitrocompounds and that the nitro content could exceed 30 mg NO_2/per gram dry plant material.[13,23] Surveys for nitrocompounds in the genus *Astragalus* were expanded to include species and varieties distributed throughout North and South America, Europe, and Asia. In these studies, Williams and Barneby detected nitrocompounds in 296 species and varieties of *Astragalus*.[24,25] They concluded that the detection of a nitro-synthesizing species often indicated that other species within the taxonomic section could also contain nitrotoxins, and this feature could be used to predict *Astragalus* toxicity. Further examination of *Astragalus* leaflets from European herbaria yielded an additional 187 species that gave a positive test for nitrocompounds.[26] Hydrolysis products were then examined by chromatography to determine the unconjugated forms of the nitrocompounds. More than 100 species were screened by TLC, and this showed that either NPA or NPOH was present, but both compounds were not detected simultaneously in any one plant.[27,28] Species within a section of *Astragalus* were chemotaxonomically related since they synthesized the same type of nitrocompound. Some references have been made to the presence of both types of nitro compounds in single species,[2] but these references do not appear to be correct.[12] Recently, an exception to this rule was found when both NPA and NPOH were detected in *A. palenae* which is toxic to livestock in Argentina.[29] In Europe, nitrocompounds were recently detected in *A. lusitanicus* which is toxic to sheep.[29a]

NPA was probably the first nitrocompound to be detected in nature, and the history on the elucidation of its structure, including the first reports on the identification of glucose esters of NPA, has been summarized by Wilson.[30] Derivatives of NPA occur in at least nine genera of higher plants. In addition to *Astragalus*, NPA conjugates are found in three other legume genera (*Coronilla varia*, *Indigofera* spp., and *Lotus* spp.), in members of the Malpighiaceae (*Hiptage benghalensis* and *Heteropteris angustifolia*), the Corynocarpaceae (*Corynocarpus laevigatus* and *C. similis*), and the Violaceae (*Viola odorata*). NPA conjugates have not been characterized in the violet species, in most of the lotus species, where at least 17 members can synthesize NPA,[31] nor in most of the indigo species, where 65 members gave a positive test for nitrocompounds.[32,32a] However, the NPA esters of *Lotus pedunculatus* have been identified.[33] Recently, nitrocompounds were also detected in 13 additional species of *Heteropteris* and in two species of *Janusia* (Malpighiaceae).[34] Stermitz and Yost[12] have reviewed the distribution of glucose esters of NPA in five species of *Astragalus*. The NPA esters in *Coronilla varia*[35,36] and *Corynocarpus laevigatus*[37] have also been identified. Karakin was recently isolated from *C. similis*.[37a] Hiptagin, identified originally in *H. benghalensis* (*H. mandoblata*)[38] is also the principle NPA ester in *H. angustifolia*.[39] In addition to karakin, *Indigofera spicata* (*I. endecaphylla*) contains other triesters of NPA as well as diesters.[40] Species of *Indigofera* may also contain a second toxicant, the hepatotoxic amino acid indospicine, which was first isolated from *I. spicata*.[41] The distribution of NPA esters as of 1977 and their structural data were summarized by Moyer et al.[36]

1-Phenyl-2-nitroethane has been detected in at least five species of higher plants including *Citrus unshiu*,[42] *Tropaeolum majus*,[43] *Dennettia tripetala*, *Aniba canelilla*, and *Ocotea pretiosa*.[44] The glucoside of 1-(4-hydroxyphenyl)-2-nitroethane occurs in *Thalictrum aquilegifolium*[45] and in cell suspension cultures of *Eschscholtzia californica*,[46] and the primeveroside occurs in *Annona squamosa*.[47]

IV. BIOSYNTHESIS

Biosynthetic studies were originally conducted with the ascomycete fungus *Penicillium atrovenetum* which synthesizes NPA. Tracer studies were conducted at various laboratories, and these have been reviewed by Venulet and Van Etten[1] and by Wilson.[48] In essence, it was concluded that carbons 2, 3, and 4 of L-aspartic acid were incorporated intact into carbons 3, 2 and 1 of NPA.[49,50] It has been shown[50] that the amino group of aspartic acid is utilized in preference to ammonium ions in the formation of the nitro group, but $^{15}NH_4^+$ was more preferred than $^{15}NO_3^-$. These results conflicted with earlier ones which showed that NPA was not synthesized in cultures of *P. atrovenetum* when aspartic acid was the sole source of nitrogen. However, ammonium ions appeared to stimulate NPA synthesis in the presence of aspartic acid.[51] The intact incorporation of the nitrogen of aspartic acid into

NPA in *P. atrovenetum* and the oxidation of the alkyl amino precursor was recently demonstrated with stable isotopes.[51a,51b] A novel intermediate was also proposed for the biosynthesis of NPA. Extracts of the fungus catalyzed the specific reduction of β-nitroacrylic acid to NPA in the presence of NADPH.[52] However, the natural occurrence of this synthetic intermediate has not been demonstrated.

In higher plants, biosynthesis of NPA was examined in *Indigofera spicata*, but activity was not incorporated from ^{14}C-aspartate to NPA.[53] Label was found in NPA when malonate or malonyl monohydroxamate was administered, but the percent incorporation was very low (<0.1%). With one exception,[29] the co-occurrence of NPA and NPOH has not been demonstrated, but the nitroalcohol appears to be a good precursor for the nitroacid. When [15]NPOH was fed to *Astragalus falcatus* (using the stem wick method), the incorporation into NPA was 1.24%.[20] This is not surprising in view of the rapid conversion of NPOH to NPA in mammalian tissue where the oxidation is effected by alcohol dehydrogenase,[54] an enzyme which is also widely distributed among plants. The possibility that NPA-containing plants have a very small, but active pool of NPOH was considered by Stermitz and Yost, but they concluded that each nitrocompound was probably synthesized by an independent pathway.[12] They suggested homoserine as a possible intermediate in the biosynthesis of NPA or NPOH from aspartic acid. To test this hypothesis, Majak and Towers[55] administered ^{14}C-aspartate and ^{14}C-homoserine to cuttings of *A. miser* var. *serotinus*, *A. collinus*, and *Coronilla varia*. The plants were also given ^{14}C-acetate and ^{14}C-β-alanine. However, when labeled NPA or NPOH was isolated, dilution factors were usually very high (>100,000). It was concluded that either the labeled compounds were poor precursors or that the method of isotope administration did not result in adequate absorption. Yost, for example, found the stem wick method to be superior to the hydroponic route for the administration of [15]NPOH to *A. falcatus*.[20]

The biological oxidation of an oxime to an aliphatic nitrocompound was first detected in *Tropaeolum majus* where phenylacetaldoxime was shown to be a precursor for 1-nitro-2-phenylethane.[43] Recently, the biogenesis of the nitro group of 1-(4'-hydroxyphenyl)-2-nitroethane (HPNE) was elucidated.[46] Microsomal fractions from cell suspension cultures of *Eschscholtzia californica* catalyzed the synthesis of HPNE from tyrosine, and 4-hydroxyphenylacetaldoxime was identified as an intermediate. These studies provide further insight on the biological oxidation of organic nitrogen, but the role of oximes in the biosynthesis of NPA or NPOH is unknown.

Studies were recently conducted by Pishak and Gustine on the incorporation of uridine diphosphate-^{14}C-glucose (UDPG) into the NPA esters of *C. varia*.[56] The authors concluded that UDPG was not the glucose donor in the formation of NPA glucose esters. They suggested that a carboxy-activated NPA derivative such as NPA-CoA could serve as the starting point for esterification of NPA to glucose.[57] In contrast, a glucosylating enzyme for HPNE utilizing UDPG was detected in cell suspension cultures of *E. californica*.[46] In suspension cultures of *C. varia*, exogenous NPA was readily incorporated into 3-nitropropanoyl-D-glucopyranoses.[57a]

V. METABOLISM IN ANIMALS

Conjugates of NPA and NPOH are rapidly hydrolyzed by rumen microbial enzymes. Esterase activity in rumen fluid liberates NPA, and β-glucosidase, which is widely distributed among species of rumen bacteria,[58] releases NPOH. In monogastric animals such as rats, the hydrolysis of miserotoxin proceeds at a much slower rate owing to a lower level of microbial β-glucosidase in the gastrointestinal tract, and the glycoside can be absorbed intact.[16] However, esters of NPA should be readily hydrolyzed by mammalian esterases which occur abundantly in the digestive secretions of the small intestine. Interconversion of

NPOH to NPA was not detected in the gastrointestinal system of rats[16] nor in the reticulorumen of cattle.[59] Miserotoxin is not converted to NPA in the digestive tract of nonruminants,[16] although this misconception has been stated on a number of occasions.[5,60] James[3] proposed that in the rumen, miserotoxin is catabolized to inorganic nitrite, but he offered no data to support this biotransformation. In the absence of microbial activity, NPA and NPOH remain intact in the rumen and are readily absorbed from the reticulorumen, which is the major site for their absorption.[59,61] When NPA and NPOH were given intraruminally to sheep, the nitroalcohol was absorbed more rapidly than the nitroacid.[61] This agreed with earlier studies on the rate of methemoglobin formation after oral administration of each compound to sheep.[62] These differential rates of absorption could explain, in part, the lesser toxicity of NPA to ruminants.

After absorption from the digestive tract, NPOH is rapidly converted to NPA, and the reaction is not reversible. This has been demonstrated in sheep, cattle, and rats and the oxidation probably occurs in the liver.[54,59,61,63] Recent *in vitro* studies with equine hepatic alcohol dehydrogenase have shown that NPOH is a substrate for the enzyme which can metabolize NPOH to NPA.[64] In rats, when alcohol dehydrogenase was inhibited with ethanol or 4-methylpyrazole, the oxidation of NPOH to NPA was suppressed, and clinical signs of poisoning did not develop.[54] It was demonstrated that NPOH and NPA are equally toxic when given intraperitoneally to rats and that the toxicity of NPOH is due to its rapid conversion to NPA.[54]

Higher levels of plasma nitrite are usually observed when NPOH is given intraruminally than when NPA is given,[61] and this is reflected in the higher level of methemoglobin after administration of NPOH.[62] The higher levels of nitrite after administration of the nitroalcohol are likely due to differences in rates of absorption or ruminal detoxification, but could also be attributed to the action of hepatic alcohol dehydrogenase. *In vitro*, this enzyme may effect the release of nitrite from NPOH but not from NPA.[64] Alston et al.[65] also detected nitrite as a product in the *in vitro* oxidation of NPOH by equine alcohol dehydrogenase. However, in experiments with sheep, intravenous doses of NPOH and NPA (20 mg/kg) yielded similar plasma nitrite levels.[66] In addition, NPA and NPOH were equally toxic when given to rats intraperitoneally.[54] These experiments indicate an efficient conversion of NPOH to NPA by alcohol dehydrogenase *in vivo* with little or no production of nitrite by the enzyme. If some NPOH is degraded to nitrite rather than oxidized to NPA as observed *in vitro*, then, presumably, the nitroalcohol would yield higher levels of plasma nitrite when given intravenously.

The production of nitrite from NPA may occur by a metabolic pathway similar to that described for other nitrocompounds. Primary nitroalkanes are substrates for amino acid oxidase and glucose oxidase. These enzymes can reduce O_2 to H_2O_2 and oxidize nitroethane to nitrite and acetaldehyde.[67] A similar reaction occurs with 1- and 2-nitropropane in the presence of hepatic microsomal monooxygenase.[68] It has also been proposed that cytotoxic acrolein is a product in the oxidation of NPOH by alcohol dehydrogenase. Alston et al.[65] proposed the enzymatic formation of 3-nitropropionaldehyde which then spontaneously decomposed to nitrite and acrolein. Acrolein was determined by indirect methods, so unequivocal proof of its formation remains to be shown. This may be difficult to achieve quantitatively because the compound is highly reactive, binding very rapidly to macromolecules.

Nitrite, generated during the metabolism of NPA and NPOH, oxidizes hemoglobin to methemoglobin, but the level of oxidation is usually <30%, even in fatal poisoning.[62,69,70] Acute methemoglobinemia is associated with the formation of >80% methemoglobin, and the syndrome can be reversed with an intravenous injection of methylene blue. However, this is not an effective treatment for NPA or NPOH poisoning in ruminants.[71] Thus, NPA or a metabolite of NPA is the primary toxic agent, but not nitrite. In recent years, the toxic mode of action of NPA has been examined by two groups of investigators who both agree that NPA inactivates succinate dehydrogenase, a mitochondrial enzyme essential for respiration. Alston et al.[72] proposed a mechanism whereby the NPA nitronate binds irreversibly as a suicide inhibitor to the flavin component of the enzyme. Coles et al.[73] suggested a different mechanism whereby the flavin oxidizes NPA to 3-nitroacrylate, and this is concomitantly attacked by the sulfhydryl group of the enzyme, causing inactivation. Two additional enzymes, fumarase, which is associated with Krebs cycle, and isocitrate lyase, which is involved in the glyoxylate cycle, are also inhibited by NPA.[74,75] The studies on the mechanisms of NPA inhibition have been reviewed.[76,77] Recently, the inhibition of rat brain acetylcholinesterase by NPA was reported.[77a]

VI. TOXICOSES IN ANIMALS AND INSECTS

Two clinical syndromes, acute and chronic, have been described in livestock intoxicated by nitrocompounds.[69,71,78-80] The acute syndrome has a rapid onset, and death occurs from a few hours to 1 day after ingestion of the toxin. The major clinical signs include incoordination, distress, labored breathing, cyanosis, muscular weakness, collapse, and sometimes excitability if stimulated. Sheep sometimes collapse and die with few clinical signs.[3] In chronic poisoning, the animals lose weight and develop respiratory distress, a poor hair coat, hind limb paresis, poor exercise tolerance, nasal discharge, and a roaring sound. Knuckling of the fetlocks is common. Animals may die if they become excited or stressed. Animals can recover, but in some ill health may persist for many months. In sheep the respiratory signs of chronic poisoning are more prominent than the nervous system signs.[3] The clinical syndrome of poisoning in horses tends to be more acute than in ruminants.[81] Lactating cattle and sheep appear to be more susceptible to intoxication than nonlactating ones, but the reason is not known.[80] The acute and chronic syndromes can also be produced in experimental animals including rats,[16,54,70] pigs,[82] chicks,[71,82,83] pigeons,[84] rabbits,[71] mice,[85] and meadow voles.[82] In those species, the development of the acute or chronic disease depends on the dose and duration of exposure to the toxin.

Cattle and sheep can be intoxicated by feeding NPOH or NPA at 20 to 60 mg/kg body weight.[78] Given intravenously, NPOH was lethal to cattle at 30 to 35 mg/kg.[69] The oral acute LD_{50} of NPOH in rats was estimated as 77 mg/kg, and as 61 mg/kg when given intraperitoneally.[16,54] The LD_{50} of NPA was 67 mg/kg when given intraperitoneally to rats[54] and 140 mg/kg in mice.[85] Chronic toxicity occurs in rats given NPOH or NPA intraperitoneally at 25 mg/kg twice daily for several days.[86] The oral lethal dose of NPA is about 75 mg/kg in pigeons.[84]

The literature on the susceptibility of animals to intoxication by nitrocompounds is somewhat confusing.[87] The current understanding of the pathogenesis of miserotoxin poisoning of ruminants is summarized in (Scheme 3).[54,59,61,63]

Miserotoxin and NPOH themselves do not appear to be toxic, but must be converted to NPA for toxicity to occur.[16,54] In rats, miserotoxin is not hydrolyzed after absorption from the gastrointestinal tract, and little hydrolysis occurs in the stomach or duodenum.[16] In addition, studies indicate that in rats, the glycoside can be excreted intact in urine.[66] Some monogastric animals such as horses have microbes in the large intestine which are capable of hydrolyzing miserotoxin. However, studies in rats with purified miserotoxin indicated that the glycoside is readily absorbed from the stomach and small intestine, and little, if any, reaches the large intestine.[16] It has been reported that horses have been poisoned by plants containing miserotoxin.[3,81] A possible explanation is that in some monogastric animals, the glycoside is toxic by some mechanism other than conversion to NPA. In rabbits, for instance, oral administration of timber milkvetch extracts produced fatal methemoglobinemia, but clinical signs did not develop when lethal doses were followed by an intravenous

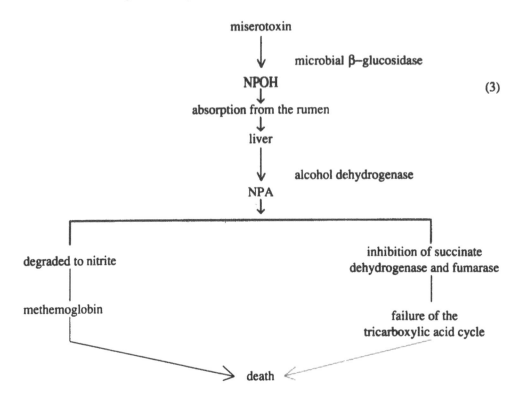

injection of methylene blue.[71] Williams et al.[71] concluded that in rabbits, miserotoxin was not converted to NPOH before absorption and that clinical signs were induced by nitrite poisoning.

Mention should be made of some preliminary experiments which we have recently conducted with rabbits.[66] When these animals were given miserotoxin intraperitoneally at 100 mg/kg, the glycoside was rapidly absorbed as shown by the high levels of miserotoxin (up to 0.32 μM/ml) in plasma. However, when the glycoside was given orally at the same dose, it was not detected in the plasma. This experiment was repeated four times, and the results were similar. The oral route of administration produced high levels of plasma nitrite (up to 0.13 μM/ml) compared to the intraperitoneal route which yielded lower plasma nitrite levels (0.05 μM/ml). Other possible metabolites of miserotoxin, such as NPA or NPOH, were not detected in samples of rabbit plasma analyzed by HPLC. The results suggest that in rabbits the glycoside is rapidly metabolized to release nitrite, but this metabolism does not involve hydrolysis of the glycosidic bond. These results are in general agreement with those reported earlier for rabbits.[71]

The early investigations of the pathology of nitrotoxin poisoning in cattle and sheep revealed cardiac failure together with some degeneration in the central nervous system.[81] Another study reported ulceration of the small intestine in sheep poisoned with timber milk vetch,[79] although this was disputed.[88] Recently, a more extensive study of the pathology associated with these intoxications has been reported. The major findings in chronically affected animals were congestion of the liver, ulceration of the cardiac region of the abomasum, emphysema, bronchoconstriction, Wallerian degeneration of the spinal cord and peripheral nerves, and focal hemorrhages in the brain.[78] Secondary bronchopneumonia has been reported in some poisoned animals.[78,81]

The central nervous system lesions of NPA toxicity have been studied in mice and rats. The major changes seen in rats are shrinkage of neurons, swelling of glial cells, vacuolation of white matter, and bilateral lesions in the caudate putamen, hippocampus, dorsolateral thalamus, and cerebellum.[89] Three patterns of lesions have been recognized in the mouse: swelling and pyknosis of neurons in the lateral caudate putamen; swelling of dendrites in the globus pallidus, entopeduncular nucleus, and anterior substantia nigra pars reticularis; and adaxonal, intramyelinic cleft formation in the midbrain, medulla, and spinal cord.[85] Histochemical staining has demonstrated a decrease in succinate dehydrogenase activity in the brain of poisoned rats and mice.[89,90] The neurological changes have been shown not to be a consequence of methemoglobinemia and are probably due to histotoxic anoxia as a result of the inhibition of succinate dehydrogenase activity.[85]

No specific antidote is available for the treatment of animals poisoned with nitrotoxins. Administration of methylene blue reduced the concentration of methemoglobin, but, with the exception of rabbits, poisoned animals still died.[71] Thiamine (400 mg/cow or 100 mg/ sheep intramuscularly) has been recommended as a treatment,[91] but some ranchers in Canada have questioned its efficacy. Thiamine (10 mg/kg intraperitoneally) given to rats at approximately ten times the dose recommended for cattle was ineffective in preventing acute or chronic NPA toxicity.[86] The value of thiamine in treating affected livestock requires further evaluation.

At present, the best way to prevent poisoning is by restricting the animals' access to the plant and by proper grazing management. Poisoning is often associated with conditions under which forage plants other than poisonous ones are in short supply, such as on overgrazed or early spring pastures.[80] Lactating cows and ewes are most susceptible, and efforts should be made to prevent their exposure to toxic plants.

Nitrocompounds for which structures are based upon NPA or NPOH are also toxic to insects. Honeybees (*Apis mellifera*) have been poisoned by foraging on timber milkvetch, the nectar of which contained miserotoxin.[92] Feeding trials with caged bees also demonstrated the acute toxicity of NPA and NPOH to bees.[92] Under field conditions, signs of poisoning included loss of coordination, weakness, inability to fly, and, for the colony population, an average daily mortality rate of 0.1 to 0.8%. Nectar from the karaka tree (*Corynocarpus laevigatus*) was also shown to be toxic to the honeybee, but whether this can be attributed to the natural abundance of NPA or karakin is not known.[93] Glucose esters of NPA isolated from crown vetch (*Coronilla varia*) as well as free NPA were shown to be toxic to the cabbage looper (*Trichoplusia ni*).[94] The nitrocompounds increased the mortality of the first instar larvae and decreased pupal weights. The authors concluded that the compounds may act as feed deterrents to insects that attempt to forage on crown vetch.[94] NPA was also toxic to the larvae of the seed-eating beetle *Callosobruchus maculatus*.[95] The leaf beetle *Chrysomela tremulae*, on the other hand, has incorporated NPA into its defensive secretion where the major component is 2-[6'-(3''-nitropropanoyl)-β-D-glucopyranosyl]-3-isoxazolin-5-one.[96,96a] Glucose esters of NPA, derived from *Lotus pedunculatus* and *Coronilla varia*, were also toxic to third instar larvae of the grass grub *Costelytra zealandica*.[33,97] When karakin was administered to the grub by oral injection, the ester was rapidly hydrolyzed as shown by the presence of NPA in visceral extracts.[98] Hydrolysis of miserotoxin in the honeybee was also reported.[92] These results indicate that insect organs contain the hydrolytic enzymes that can liberate NPA or NPOH. Although NPA and NPOH appeared to be similar in toxicity to the honeybee,[92] it is not known whether the toxic effect of NPOH was due to its conversion to NPA.

VII. DETOXIFICATION

The ability of ruminal microorganisms to modify dietary substances that are potentially toxic is well established.[60,99] Toxic metabolites can be formed when glycosides are hydrolyzed by microbial enzymes, but the glycosides themselves may be innocuous. The cyanogenic glycoside amygdalin, when given orally to germ-free rats at 600 mg/kg (equivalent to approximately nine times the lethal dose of cyanide), did not produce visible signs of poisoning.[100] Compared to NPOH (LD$_{50}$ 77 mg/kg), miserotoxin (LD$_{50}$ >2.5 g/kg) was relatively innocuous when given orally to rats, presumably because of the low level of β-glucosidase activity associated with the enteral microflora.[16] Rabbits were also less susceptible to miserotoxin poisoning than cattle.[71] In monogastric mammals, glycosides can be absorbed from the gastrointestinal tract and eliminated intact, but microbial hydrolysis in the rumen precludes this route of detoxification for ruminants. Inhibitors of β-glucosidase are known, however, and it is conceivable that rates of hydrolysis could be significantly altered to promote more glycoside absorption and less production of the toxic aglycone in ruminants. The hydrolysis of miserotoxin was delayed for approximately an hour, for example, when the glycoside was incubated with rumen inocula containing 0.1% glucono-1:5-lactone.[66] However, more potent inhibitors of β-glucosidase activity have been identified in nature.[101,102] In contrast, the inhibition of esterase activity in the rumen would be of little consequence because these hydrolytic enzymes are widely distributed in mammalian organs beyond the forestomach, hence, the acute toxicity of NPA conjugates to monogastric animals.[82]

Detoxification of aliphatic nitrocompounds can also be achieved through microbial reduction of the nitro group. Extensive studies were conducted in the last decade on the degradation of NPA and NPOH by pure and mixed cultures of rumen bacteria, and these have been summarized.[6] The anaerobic degradation may occur through a reductive cleavage of the nitro group to yield inorganic nitrite which was detected at low levels in cultures of bacteria incubated with NPA and NPOH.[103] The nitrite is rapidly reduced to ammonia, thereby resulting in its detoxification. The other carbon-containing degradation products of NPA or NPOH have not been detected. However, nitrite, acetone, and acetaldehyde have been identified as products in the aerobic degradation of 2-nitropropane and nitroethane by fungi.[104]

Recently, nitrocompounds that are less toxic than NPOH were tested as potential inducers of NPOH detoxification in the bovine rumen. The best enhancement of the *in vitro* rate of NPOH degradation was achieved with supplements of nitroethane given intraruminally at 20 mg/kg/d.[105] These *in vitro* rates of metabolism compared favorably with those reported for NPA in sheep which were not affected by large intraruminal doses of NPA.[106] The lower toxicity of NPA than NPOH to ruminants could be partly due to the more rapid rate of NPA detoxification by rumen microorganisms. The use of nitroethane (stabilized as the sodium salt) under field conditions may offer a means of protection for animals exposed to toxic, NPOH-containing plants. Preliminary tests suggest that at 20 mg/kg, nitroethane is innocuous as shown by the normal methemoglobin levels in treated animals[105] and by long-term feeding trials which did not debilitate treated cattle.[66] A more recent development indicates that the induced capacity to detoxify can be transmitted to other animals. The enhanced capacity of rumen microbes to degrade nitrate and nitrite in animals adapted to nitrate (0.1 g/kg/d) was transferred to noninduced (untreated) animals housed in adjacent pens.[107] A similar phenomenon has been observed in Australia and North America with rumen microbes capable of mimosine degradation.[108,108a] In short, induction may provide not only protection for treated animals, but this protection may be transferable to other animals at risk.

The role of intestinal microflora in the reduction of the nitro group has been extensively studied for aromatic substituents,[109] but not for aliphatic ones. The two types of compounds were compared in one study where *p*-nitrobenzoate was reduced 5 to 12 times faster than 2-nitroethanol by pure cultures of *Clostridium* species.[110]

The detoxification of NPOH can also be achieved with inhibitors of alcohol dehydrogenase which are used in the treatment of other alcohol poisonings.[111] The conversion of NPOH to NPA was suppressed when rats were pretreated with ethanol or 4-methylpyrazole, and clinical signs of poisoning did not develop at dose levels of 100 mg NPOH per kilogram, a dose that killed untreated animals.[54] Enzyme inhibition on a partial basis might be an effective method for the prevention of timber milkvetch poisoning. In experimental acute NPOH poisoning in rats, however, the inhibitors were not an effective treatment if they were given after the NPOH dose.[54] This suggests that the conversion of NPOH to NPA was extremely rapid and irreversible.

VIII. AGRONOMIC ASPECTS

Three legumes that contain NPA esters in their foliage have been cultivated for forage production. In grazing trials with dairy cattle in Puerto Rico, creeping indigo (*I. spicata*) was highly preferred over other legumes tested, and it led in forage production.[112] In Hawaii, however, it was demonstrated that creeping indigo was hepatotoxic to ruminants and rabbits,[113] and this was subsequently attributed to the toxic effects of indospicine.[41,114] The genus of *Indigofera* appears to contain relatively low levels of NPA,[32] and this was confirmed recently when *Indigofera* species with agronomic potential were screened for both NPA and indospicine.[114a]

Cultivars of *Coronilla varia* (crown vetch) and *Lotus pedunculatus* also contain NPA.[4,31,33] Both of these species have been used in pastures for grazing ruminants, but poisoning under field conditions has not been reported.[31,115] If improved varieties are to be developed, as suggested for crown vetch,[116] selections should be screened for NPA to minimize its content, and, in the case of *Lotus*, selections should also be screened for their potential to produce hydrogen cyanide.[117] However, these constituents may function as insect toxins or deterrents, in which case their complete removal may generate new problems related to insect herbivory. Crown vetch is widely used for erosion control of disturbed soils in the eastern U.S. In view of its toxicity to monogastric animals,[82] the advisability of widespread seeding of crown vetch is questionable. Birds of prey could be adversely affected, for example, if the rodent population declined due to the toxic effects of crown vetch. In general, new plant introductions, like crown vetch, should be initially screened for toxic properties and intercepted, if necessary, to prevent the potential spread of poisonous plants which occasionally become weedy infestations.[118]

On native rangelands poisonous plants can be controlled with herbicides. The phenoxy herbicides 2,4,5-T and Silvex® (Fenoprop) were shown to be effective in the control of timber milkvetch (*A. miser* var. *serotinus* and *A. miser* var. *oblongifolius*) by reducing the content of chlorophyll (as in senescence) and by decreasing the level of miserotoxin.[119,120] However, the availability of these herbicides is very limited, and their use is restricted in many areas. Triazine and urea herbicides did not significantly reduce miserotoxin levels in timber milkvetch.[119] Mecoprop, 2,4,5-T, and 2,4-D were effective in the control of Kelsey milkvetch (*A. atropubescens*), which also contains miserotoxin, but not to the same degree as Silvex® which eradicated the plant at applications of 2.2 kg/ha.[121] More recently, triclopyr was shown to be very effective in the control of *A. miser* var. *oblongifolius*.[121a]

Forage productivity on native rangeland can be improved with fertilizers, especially nitrogen, a nutrient which is frequently deficient in soils. Nitrogen is a significant component of the miserotoxin molecule, but nitrogen, as KNO_3, $(NH_4)_2SO_4$, or NH_4NO_3, applied at 56 or 112 kg N per hectare did not affect the growth or toxicity of timber milkvetch.[119,122] Similar results were obtained in a recent study in Canada where urea was applied aerially at 100 and 200 kg N per hectare.[123] Levels of miserotoxin were not elevated in the first year after fertilizer application, but in the second year, a significant increase was detected at a clear-cut site treated with 200 kg N per hectare.[123] Favorable moisture conditions, residual N, and reduced interspecific competition were cited as possible reasons for increased toxicity in 1983 (4.69% miserotoxin on a dry weight basis) compared to 1982 (2.95%).

The use of herbicides to eradicate poisonous plants or N fertilizers to enhance the growth of grass is not economical or feasible in many situations. The widespread application of herbicides, for example, could have a negative impact on the food chain of wildlife. An alternate method for the prevention of livestock poisoning involves understanding the dynamics of the toxin in the living plant and determining where and when the animals are most likely to be at risk. To this end, the dynamic changes in miserotoxin concentration of timber milk vetch were extensively studied in British Columbia,[124-126] and guidelines were formulated that predict the toxicity of the plant.[127,128] These guidelines can be used as range-management tools to prevent livestock poisoning. For example, rainfall can cause sudden increases in toxicity, especially when prevailing dry conditions are interrupted by heavy rain storms. The plant is most toxic before flowering, and if dry conditions persist, the toxin levels will approach zero with advancing stages of growth, in agreement with the changes in miserotoxin concentrations occurring in other varieties of timber milk vetch.[119] The field studies also showed that different plant communities show different levels of the toxin. Grassland sites were more toxic than forest range, which could be further divided into risk situations depending on the availability of light to the understory and the density and type of tree cover. Studies were recently conducted to determine the effect of clipping on the growth and toxicity of timber milk vetch. The biomass and miserotoxin content of *A. miser* var. *serotinus* were substantially reduced in response to early clipping in spring. The results indicated that early grazing may reduce the potential hazard of timber milkvetch to livestock.[128a]

Higher plant species containing nitrotoxins are not generally consumed by humans, but a potential public health problem may exist with fermented foods which may contain NPA produced by strains of molds from *Aspergillus oryzae*[129] or *A. soyae*.[130]

REFERENCES

1. **Venulet, J. and Van Etten, R. L.,** Biochemistry and pharmacology of the nitro and nitroso group, in *The Chemistry of the Nitro and Nitroso Groups,* (Part 2), Feuer, H., Ed., Interscience, New York, 1970, chap. 4.
2. **Alston, T. A., Porter, D. J. T., and Bright, H. J.,** The bioorganic chemistry of the nitroalkyl group, *Bioorg. Chem.,* 13, 375, 1985.
3. **James, L. F.,** Neurotoxins and other toxins from *Astragalus* and related genera, in *Handbook of Natural Toxins,* Vol. 1, Keeler, R. F. and Tu, A. T., Eds., Marcel Dekker, New York, 1983, chap. 13.
4. **Gustine, D. L.,** Aliphatic nitro compounds in crownvetch: a review, *Crop Sci.,* 19, 197, 1979.
5. **Cheeke, P. R. and Shull, L. R.,** *Natural Toxicants in Feeds and Poisonous Plants,* AVI Publishing, Westport, CT. 1985, 199.
6. **Majak, W., Cheng, K.-J., Muir, A. D., and Pass, M. A.,** Analysis and metabolism of nitrotoxins in cattle and sheep, in *Plant Toxicology,* Seawright, A. A., Hegarty M. P., James, L. F., and Keeler, R. F., Eds., Queensland Poisonous Plants Committee Animal Research Institute, Yeerongpilly, 1985, 446.
7. **Priestap, H. A.,** Minor aristolochic acids from *Aristolochia argentina* and mass spectral analysis of aristolochic acids, *Phytochemistry,* 26, 519, 1987.
8. **Shamma, M. and Moniot, J. L.,** The aristolochic acids and aristolactams, in *Isoquinoline Alkaloid Research 1972-1977,* Plenum Press, New York, 1978, chap. 17.
9. **Neme, G., Nieto, M., D'Arcangelo, A. T., and Gros, E. G.,** 3-Nitro-4-hydroxy-phenethylamine from *Cereus validus, Phytochemistry,* 16, 277, 1977.
9a. **Mengs, U.,** On the histopathogenesis of rat forestomach carcinoma caused by aristolochic acid, *Arch. Toxicol.,* 52, 209, 1983.
10. **Noller, C. R.,** *Chemistry of Organic Compounds,* W. B. Saunders, Philadelphia, 1965, 279.
10a. **Majak, W. and Benn, M. H.,** 3-Nitro-1-propyl-β-D-gentiobioside from *Astragalus miser* var. *serotinus, Phytochemistry,* 27, 1089, 1988.
11. **Stermitz, F. R., Lowry, W. T., Ubben, E., and Shariff, I.,** 1,6-Di-3-nitropropanoyl-β-D-glucopyranoside from *Astragalus cibarius, Phytochemistry,* 11, 3525, 1972.

12. **Stermitz, F. R. and Yost, G. S.**, Analysis and characterization of nitro compounds from *Astragalus* species, in *Effects of Poisonous Plants on Livestock*, Keeler, R. F., van Kampen, K. R., and James, L. F., Eds., Academic Press, New York, 1978, 371.

13. **Stermitz, F. R., Lowry, W. T., Norris, F. A., Buckeridge, F. A., and Williams, M. C.**, Aliphatic nitro compounds from *Astragalus* species, *Phytochemistry*, 11, 1117, 1972.

14. **Majak, W. and Bose, R. J.**, Chromatographic methods for the isolation of miserotoxin and detection of aliphatic nitro compounds, *Phytochemistry*, 13, 1005, 1974.

15. **Majak, W., Lindquist, S. G., and McDiarmid, R. E.**, A simple method for purification and determination of miserotoxin, *J. Agric. Food Chem.*, 31, 650, 1983.

16. **Majak, W., Pass, M. A., and Madryga, F. J.**, Toxicity of miserotoxin and its aglycone (3-nitropropanol) to rats, *Toxicol. Lett.*, 19, 171, 1983.

17. **Cooke, A. R.**, The toxic constituent of *Indigofera endecaphylla*, *Arch. Biochem. Biophys.*, 55, 114, 1955.

18. **Matsumoto, H., Unrau, A. M., Hylin, J. W., and Temple, B.**, Spectrophotometric determination of 3-nitropropanoic acid in biological extracts, *Anal. Chem.*, 33, 1442, 1961.

19. **Muir, A. D., and Majak, W.**, Quantitative determination of 3-nitropropionic acid and 3-nitropropanol in plasma by HPLC, *Toxicol. Lett.*, 20, 133, 1984.

20. **Yost, G. S.**, Toxic Secondary Metabolites of Plants, Ph.D. thesis, Colorado State University, Fort Collins, 1977.

21. **Pfeffer, P. E. and Valentine, K. M.**, Assessment of ^{13}C-shift parameters in di- and tri-*O*-(3-nitropropanoyl)-D-glucopyranoses, *Carbohydr. Res.*, 73, 1, 1979.

21a. **Rumbaugh, M. D., Clark, D. H., and Lamb, R. C.**, Near infrared reflectance spectroscopy screening of *Astragalus* species for nitro-toxins, in *Proc. 15th Int. Grassland Congr.*, National Grassland Research Institute, Nishi-nasuno, Tochigi-ken, 329-27, Japan, 1985, 315.

22. **Williams, M. C. and Norris, F. A.**, Distribution of miserotoxin in varieties of *Astragalus miser* Dougl. ex Hook, *Weed Sci.*, 17, 236, 1969.

23. **Williams, M. C., and Parker, R.**, Distribution of organic nitrites in *Astragalus*, *Weed Sci.*, 22, 259, 1974.

24. **Williams, M. C. and Barneby, R. C.**, The occurrence of nitro-toxins in North American *Astragalus* (Fabaceae), *Brittonia*, 29, 310, 1977.

25. **Williams, M. C. and Barneby, R. C.**, The occurrence of nitro-toxins in Old World and South American *Astragalus* (Fabaceae), *Brittonia*, 29, 327, 1977.

26. **Williams, M. C.**, Nitro compounds in foreign species of *Astragalus*, *Weed Sci.*, 29, 261, 1981.

27. **Williams, M. C.**, 3-Nitropropionic acid and 3-nitro-1-propanol in species of *Astragalus*, *Can. J. Bot.*, 60, 1956, 1982.

28. **Williams, M. C. and Davis, A. M.**, Nitro compounds in introduced *Astragalus* species, *J. Range Manage.*, 35, 113, 1982.

29. **Williams, M. C. and Gomez-Sosa, E.**, Toxic nitro compounds in species of *Astragalus* (Fabaceae) in Argentina, *J. Range Manage.*, 39, 341, 1986.

29a. **Tarazona, J. V. and Sanz, F.**, Aliphatic nitro compounds in *Astragalus lusitanicus* Lam, *Vet. Hum. Toxicol.*, 29, 437, 1987.

30. **Wilson, B. J.**, Miscellaneous *Aspergillus* toxins, in *Microbial Toxins*, Vol. 6, Ciegler, A., Kadis, S., and Ajl, S. J., Eds., Academic Press, New York, 1971, chap. 3.

31. **Williams, M. C.**, Toxic nitro compounds in *Lotus*, *Agron. J.*, 75, 520, 1983.

32. **Williams, M. C.**, Nitro compounds in *Indigofera* species, *Agron. J.*, 73, 434, 1981.

32a. **Murray, L. R., Moore, T., and Sharman, I. M.**, The toxicity of *Indigofera enneaphylla* in rats, *Aust. J. Agric. Res.*, 16, 713, 1965.

33. **Gnanasunderam, C. and Sutherland, O. R. W.**, Hiptagin and other aliphatic nitro esters in *Lotus pedunculatus*, *Phytochemistry*, 25, 409, 1986.

34. **Williams, M. C.**, Poisonous Plant Newsletter, 4, 1, Agricultural Research Service, U.S. Department of Agriculture, Logan, Utah, 1981.

35. **Majak, W. and Bose, R. J.**, Nitropropanoylglucopyranoses in *Coronilla varia*, *Phytochemistry*, 15, 415, 1976.

36. **Moyer, B. G., Pfeffer, P. E., Moniot, J. L., Shamma, M., and Gustine, D. L.**, Corollin, coronillin and coronarian: three new 3-nitropropanoyl-D-glucopyranoses from *Coronilla varia*, *Phytochemistry*, 16, 375, 1977.

37. **Moyer, B. G., Pfeffer, P. E., Valentine, K. M., and Gustine, D. L.**, 3-Nitropropanoyl-D-glucopyranoses of *Corynocarpus laevigatus*, *Phytochemistry*, 18, 111, 1979.

37a. **Cabalion, P. and Poisson, J.**, *Corynocarpus similis* Hemsley, plante alimentaire et toxique de Vanuatu (ex-Nouvelles-Hebrides), *J. Ethnopharmacol.*, 21, 189, 1987.

38. **Finnegan, R. A. and Stephani, R. A.**, Structure of hiptagin as 1,2,4,6-tetra-*O*-(3-nitropropanoyl)-β-D-glucopyranoside, its identity with endecaphyllin X, and the synthesis of its methyl ether, *J. Pharm. Sci.*, 57, 353, 1968.

39. Stermitz, F. R., Hnatyszyn, O., Bandoni, A. L., Rondina, R. V. D., and Coussio, J. D., Screening of Argentine plants for aliphatic nitro compounds: hiptagin from *Heteropteris angustifolia*, *Phytochemistry*, 14, 1341, 1975.

40. Finnegan, R. A. and Stephani, R. A., The structure of karakin, *Lloydia*, 33, 492, 1968.

41. Hegarty, M. P. and Pound, A. W., Indospicine, a hepatotoxic amino acid from *Indigofera spicata*: isolation, structure and biological studies, *Aust. J. Biol. Sci.*, 23, 831, 1970.

42. Sakurai, K., Toyoda, T., Muraki, S., and Yoshida, T., Odorous constituents of the absolute from flower of *Citrus unshiu* Marcovitch, *Agric. Biol., Chem.*, 43, 195, 1979.

43. Matsuo, M., Kirkland, D. F., and Underhill, E. W., 1-Nitro-2-phenylethane, a possible intermediate in the biosynthesis of benzylglucosinolate, *Phytochemistry*, 11, 697, 1972.

44. Leboeuf, M., Cave, A., Bhaumik, P. K., Mukherjee, B., and Mukherjee, R., The phytochemistry of the Annonaceae, *Phytochemistry*, 21, 2783, 1982.

45. Iida, H., Kikuchi, T., Kobayashi, K., and Ina, H., Structure of thalictoside, *Tetrahedron Lett.*, 21, 759, 1980.

46. Hösel, W., Berlin, J., Hanzlik, T. N., and Conn, E. E., In-vitro biosynthesis of 1-(4'-hydroxyphenyl)-2-nitroethane and production of cyanogenic compounds in osmotically stressed cell suspension cultures of *Eschscholtzia californica*, *Planta*, 166, 176, 1985.

47. Forgacs, P., Desconclois, J. F., Provost, J., Tiberghien, R., and Touché, A., Un novel heteroside nitre extrait d' *Annona squamosa*, *Phytochemistry*, 19, 1251, 1980.

48. Wilson, B. J., Miscellaneous *Penicillium* toxins, in *Microbial Toxins*, Vol. 6, Ciegler, A., Kadis, S., and Ajl, S. J., Eds., Academic Press, New York, 1971, chap. 8.

49. Birkinshaw, J. H. and Dryland, A. M. L., Studies on the biochemistry of microorganisms. 116. Biosynthesis of β-nitropropionic acid by the mould *Penicillium atrovenetum* G. Smith, *Biochem. J.*, 93, 478, 1964.

50. Shaw, P. D. and McCloskey, J. A., Biosynthesis of nitro compounds. II. Studies on potential precursors for the nitro group of β-nitropropionic acid, *Biochemistry*, 6, 2247, 967.

51. Shaw, P. D. and Wang, N., Biosynthesis of nitro compounds. I. Nitrogen and carbon requirements for the biosynthesis of β-nitropropionic acid by *Penicillium atrovenetum*, *J. Bacteriol.*, 88, 1629, 1964.

51a. Baxter, R. L., Abbott, E. M., Greenwood, S. L., and McFarlane, I. J., Conservation of the carbon-nitrogen bond of aspartic acid in the biosynthesis of 3-nitropropionic acid, *J. Chem. Soc. D.*, 564, 1985.

51b. Baxter, R. L. and Greenwood, S. L., Application of the ^{18}O isotope shift in ^{15}N N.M.R. spectra to a biosynthetic problem: experimental evidence for the origin of the nitro group oxygen atoms of 3-nitropropionic acid, *J. Chem. Soc. D.*, 175, 1986.

52. Shaw, P. D., Biosynthesis of nitro compounds. III. The enzymatic reduction of β-nitroacrylic acid to β-nitropropionic acid, *Biochemistry*, 6, 2253, 1967.

53. Candlish, E., La Croix, L. J., and Unrau, A. M., The biosynthesis of 3-nitropropionic acid in creeping indigo (*Indigofera spicata*), *Biochemistry*, 8, 182, 1969.

54. Pass, M. A., Muir, A. D., Majak, W., and Yost, G. S., Effect of alcohol and aldehyde dehydrogenase inhibitors on the toxicity of 3-nitropropanol in rats, *Toxicol. Appl. Pharmacol.*, 78, 310, 1985.

55. Majak, W. and Towers, G. H. N., unpublished data, 1975.

56. Pishak, M. R. and Gustine, D. L., Formation of 3-nitropropanoyl-D-glucopyranoses in *Coronilla varia* L., Abstr., *Plant Physiol.*, 72, (Suppl.), 154, 1983.

57. Pishak, M. R. and Gustine, D. L., Mechanisms for synthesis of 3-nitropropanoyl-D-glucopyranoses in *Coronilla varia* L., Abstr., *Phytochem. Soc. N. Am. Newsl.*, 24(2), 20, 1984.

57a. Moyer, G. B. and Gustine, D. L., Esterification of 3-nitropropanoic acid to glucose by suspension cultures of *Coronilla varia*, *Phytochemistry*, 26, 139, 1987.

58. Majak, W. and Cheng, K.-J., Cyanogensis in bovine rumen fluid and pure cultures of rumen bacteria, *J. Anim. Sci.*, 59, 784, 1984.

59. Majak, W., Pass, M. A., Muir, A. D., and Rode, L. M., Absorption of 3-nitropropanol (miserotoxin aglycone) from the compound stomach of cattle, *Toxicol. Lett.*, 23, 9, 1984.

60. Allison, M. J., The role of ruminal microbes in the metabolism of toxic constituents from plants, in *The Effects of Poisonous Plants on Livestock*, Keeler, R. F., Van Kampen, K. R., and James, L. F., Eds., Academic Press, New York, 1978, 101.

61. Pass, M. A., Majak, W., Muir, A. D., and Yost, G. S., Absorption of 3-nitropropanol and 3-nitropropionic acid from the digestive system of sheep, *Toxicol. Lett.*, 23, 1, 1984.

62. Williams, M. C. and James, L. F., Toxicity of nitro-containing *Astragalus* to sheep and chicks, *J. Range Manage.*, 28, 260, 1975.

63. Muir, A. D., Majak, W., Pass, M. A., and Yost, G. S., Conversion of 3-nitropropanol (miserotoxin aglycone) to 3-nitropropionic acid in cattle and sheep, *Toxicol. Lett.*, 20, 137, 1984.

64. McDiarmid, R. E., Majak, W., and Yost, G. S., Conversion of 3-nitropropanol to 3-nitropropionic acid by equine alcohol dehydrogenase, *J. Toxicol. - Toxin Reviews*, 5(Abstr.), 253, 1986.

65. **Alston, T. A., Seitz, S. P., and Bright, H. J.,** Conversion of 3-nitro-1-propanol (miserotoxin aglycone) to cytotoxic acrolein by alcohol dehydrogenase, *Biochem. Pharmacol.,* 30, 2719, 1981.

66. **Majak, W.,** unpublished data.

67. **Porter, D. J. T., Voet, J. G., and Bright, H. J.,** Nitroalkanes as reductive substrates for flavoprotein oxidases, *Z. Naturforsch. Teil B,* 27, 1052, 1972.

68. **Ullrich, V., Hermann, G., and Weber, P.,** Nitrite formation from 2-nitropropane by microsomal monooxygenases, *Biochem. Pharmacol.,* 27, 2301, 1978.

69. **Majak, W., Udenberg, T., McDiarmid, R. E., and Douwes, H.,** Toxicity and metabolic effects of intravenously administered 3-nitropropanol in cattle, *Can. J. Anim. Sci.,* 61, 639, 1981.

70. **Matsumoto, H., Hylin, J. W., and Miyahara, A.,** Methemoglobinemia in rats injected with 3-nitropropanoic acid, sodium nitrite and nitroethane, *Toxicol. Appl. Pharmacol.,* 3, 493, 1961.

71. **Williams, M. C., Van Kampen, K. R., and Norris, F. A.,** Timber milkvetch poisoning in chickens, rabbits and cattle, *Am. J. Vet. Res.,* 30, 2185, 1969.

72. **Alston, T. A., Mela, L., and Bright, H. J.,** 3-Nitropropionate, the toxic substance of *Indigofera,* is a suicide inactivator of succinate dehydrogenase, *Proc. Natl. Acad. Sci. U.S.A.,* 74, 3767, 1977.

73. **Coles, C. J., Edmondson, D. E., and Singer, T. P.,** Inactivation of succinate dehydrogenase by 3-nitropropionate, *J. Biol. Chem.,* 254, 5161, 1979.

74. **Porter, J. T. and Bright, H. J.,** 3-Carbanionic substrate analogues bind very tightly to fumarase and aspartase, *J. Biol. Chem.,* 255, 4772, 1980.

75. **Schloss, J. V. and Cleland, W. W.,** Inhibition of isocitrate lyase by 3-nitropropionate, a reaction-intermediate analogue, *Biochemistry,* 21, 4420, 1982.

76. **Alston, T. A., Porter, D. J. T., and Bright, H. J.,** Enzyme inhibition by nitro and nitroso compounds, *Acc. Chem. Res.,* 16, 418, 1983.

77. **Gustine, D. L. and Moyer, B. G.,** Review of mechanisms of toxicity of 3-nitropropionic acid in nonruminant animals, in *Proc. 14th Int. Grassland Cong.,* Smith, J. A. and Hays, V. W., Eds., Westview Press, Boulder, CO, 1983, 736.

77a. **Osman, M. Y.,** Effect of β-nitropropionic acid on rat brain acetylcholineesterase, *Biochem. Pharmacol.,* 31, 4067, 1982.

78. **James, L. F., Hartley, W. J., Williams, M. C., and Van Kampen, K. R.,** Field and experimental studies in cattle and sheep poisoned by nitro-bearing *Astragalus* or their toxins, *Am. J. Vet. Res.,* 41, 377, 1980.

79. **Mosher, G. A., Krishnamurti, C. R., and Kitts, W. D.,** Physiological effects of timber milkvetch, *Astragalus miser* var. *serotinus,* on sheep, *Can. J. Anim. Sci.,* 51, 465, 1971.

80. **MacDonald, M. A.,** Timber milkvetch poisoning on British Columbia ranges, *J. Range Manage.,* 5, 16, 1952.

81. **Bruce, E. A.,** *Astragalus camprestris* and other stock poisoning plants of British Columbia, *Can. Dep. Agric. Bull.,* 88, 1927.

82. **Shenk, J. S., Wangsness, P. J., Leach, R. M., Gustine, D. L., Gobble, J. L., and Barnes, R. F.,** Relationship between β-nitropropionic acid content of crownvetch and toxicity in nonruminant animals, *J. Anim. Sci.,* 42, 616, 1976.

83. **Williams, M. C., Yost, G. S., and Stermitz, F. R.,** Miserotoxin, a toxic compound in *Astragalus michauxii, Phytochemistry,* 16, 1438, 1977.

84. **Bell, M. E.,** Toxicology of karaka kernal, karakin and β-nitropropionic acid, *N. Z. J. Sci.,* 17, 327, 1974.

85. **Gould, D. H. and Gustine, D. L.,** Basal ganglia degeneration, myelin alterations, and enzyme inhibition induced in mice by the plant toxin 3-nitropropanoic acid, *Neuropath. Appl. Neurobiol.,* 8, 377, 1982.

86. **Pass, M. A., Majak, W., and Yost, G. S.,** Lack of a protective effect of thiamine on the toxicity of 3-nitropropanol and 3-nitropropionic acid in rats, *Can. J. Anim. Sci.,* 68, 315, 1988.

87. **Majak, W. and McLean, A.,** Nitrotoxin metabolism in livestock. Letter to the Editor, *J. Am. Vet. Med. Assoc.,* 179, 412, 981.

88. **Gilchrist, E. W.,** Letter to the Editors, *Can. J. Anim. Sci.,* 53, 627, 1973.

89. **Hamilton, B. F., Gould, D. H., Wilson, M. P., and Hamar, D. H.,** Enzyme and structural alterations in brains of rats intoxicated with 3-nitropropionic acid, *Fed. Proc.,* 43(Abstr.), 380, 1984.

90. **Gould, D. H., Wilson, M. P., and Hamar, D. W.,** Brain enzyme and clinical alterations induced in rats and mice by nitroaliphatic toxicants, *Toxicol. Lett.,* 27, 83, 1985.

91. **Nicholson, H. H.,** The treatment of timber milk-vetch poisoning among cattle and sheep, *Can. J. Anim. Sci.,* 43, 237, 1963.

92. **Majak, W., Neufeld, R., and Corner, J.,** Toxicity of *Astragalus miser* v. *serotinus* to the honeybee, *J. Apic. Res.,* 19, 196, 1980.

93. **Palmer-Jones, T.,** Nectar from karaka trees poisonous to honeybees, *N. Z. J. Agric.,* 117, 77, 1968.

94. **Byers, R. A., Gustine, D. L., and Moyer, B. G.,** Toxicity of β-nitropropionic acid to *Trichoplusia ni, Environ. Entomol.,* 6, 229, 1977.

95. Janzen, D. H., Juster, H. B., and Bell, E. A., Toxicity of secondary compounds to the seed-eating larvae of the bruchid beetle *Callosobruchus maculatus, Phytochemistry,* 16, 223, 1977.

96. Pasteels, J. M., Braekman, J. C., Daloze, D., and Ottinger, R., Chemical defense in chrysomelid larvae and adults, *Tetrahedron,* 38, 1891, 1982.

96a. Pasteels, J. M., Daloze, D., and Rowell-Rahier, M., Chemical defense in chrysomelid eggs and neonate larvae, *Physiol. Entomol.,* 11, 29, 1986.

97. Hutchins, R. F. N., Sutherland, O. R. W., Gnanasunderam, C., Greenfield, W. J., Williams, E. M., and Wright, H. J., Toxicity of nitro compounds from *Lotus pedunculatus* to grass grub *Costelytra zealandica* (Coleoptera: Scarabaeidae), *J. Chem. Ecol.,* 10, 81, 1984.

98. Greenwood, D. R., Metabolism of karakin, a nitro toxin from *Lotus pedunculatus,* on ingestion by larvae of the grass grub, *Costelytra zealandica, N. Z. J. Zool.,* 11, 451, 1984.

99. Carlson, J. R. and Breeze, R. G., Ruminal metabolism of plant toxins with emphasis on indolic compounds, *J. Anim. Sci.,* 58, 1040, 1984.

100. Carter, J. H., McLafferty, M. A., and Goldman, P., Role of the gastrointestinal microflora in amygdalin (laetrile)-induced cyanide toxicity, *Biochem. Pharmacol.,* 29, 301, 1980.

101. Evans, S. V., Fellows, L. E., and Bell, E. A., Glucosidase and trehalase inhibition by 1,5-dideoxy-1,5-imino-D-mannitol, a cyclic amino alditol from *Lonchocarpus sericeus, Phytochemistry,* 22, 768, 1983.

102. Niwa, T., Inouye, S., Tsuruoka, T., Koaze, Y., and Niida, T., "Nojirimycin" as a potent inhibitor of glucosidase, *Agric. Biol. Chem.,* 34, 966, 1970.

103. Majak, W. and Cheng, K.-J., Identification of rumen bacteria that anaerobically degrade aliphatic nitrotoxins, *Can. J. Microbiol.,* 27, 646, 1981.

104. Kido, T., Yamamoto, T., and Soda, K., Microbial assimilation of alkyl nitro compounds and formation of nitrite, *Arch. Microbiol.,* 106, 165, 1975.

105. Majak, W., Cheng, K.-J., and Hall, J. W., Enhanced degradation of 3-nitropropanol by ruminal microorganisms, *J. Anim. Sci.,* 62, 1072, 1986.

106. Gustine, D. L., Moyer, B. G., Wangsness, P. J., and Shenk, J. S., Ruminal metabolism of 3-nitropropanoyl-D-glucopyranoses from crownvetch, *J. Anim. Sci.,* 44, 1107, 1977.

107. Cheng, K.-J., Phillippe, R. C., Kozub, G. C., Majak, W., and Costerton, J. W., Induction of nitrate and nitrite metabolism in bovine rumen fluid and the transfer of this capacity to untreated animals, *Can. J. Anim. Sci.,* 65, 647, 1985.

108. Allison, M. J., Cook, H. M., and Jones, R. J., Detoxication of 3-hydroxy-4-(1H)-pyridone, the goitrogen of mimosine by rumen bacteria from Hawaiian goats, Abstr., Report on the XVII Conf. Rumen Function, Chicago, 1983, 21.

108a. Hammond, A. C., Allison, M. J., Williams, M. J., Bates, D. B., Prine, G. M., and Adams, E. L., Inoculation of cattle grazing *Leucaena leucocephala* in central Florida with 3, 4-DHP degrading ruminal bacteria, Abstr. Conf. Rumen Function, Chicago, 19, 20, 1987.

109. Scheline, R. R., Drug metabolism by the gastrointestinal microflora, in *Extrahepatic Metabolism of Drugs and Other Foreign Compounds,* Gram, T. E., Ed., SP Medical & Scientific Books, New York, 1980, chap. 17.

110. Angermaier, L. and Simon, H., On the reduction of aliphatic and aromatic nitro compounds by *Clostridia,* the role of ferredoxin and its stabilization, *Hoppe Seylers Z. Physiol. Chem.,* 364, 961, 1983.

111. Murphy, M. J., Ray, A. C., Jones, L. P., and Reagor, J. C., 1,3-Butanediol treatment of ethylene glycol toxicosis in dogs, *Am. J. Vet. Res.,* 45, 2293, 1984.

112. Warmke, H. E., Freyne, R. H., and Morris, M. P., Studies on palatability of some tropical legumes, *Agron. J.,* 44, 517, 1952.

113. Nordfeldt, S., Henke, L. A., Morita, K., Matsumoto, H., Takahashi, M., and Younge, O. R., Feeding tests with *Indigofera endecaphylla* Jacq. (creeping indigo) and some observations on its poisonous effects on domestic animals, *Univ. Hawaii Agric. Exp. Stn. Tech. Bull.,* 15, 1952.

114. Hutton, E. M., Tropical pastures, in *Advances in Agronomy,* Vol. 22, Brady, N. C., Ed., Academic Press, New York, 1970, 1.

114a. Alyward, J. H., Court, R. D., Haydock, K. P., Strickland, R. W., and Hegarty, M. P., *Indigofera* species with agronomic potential in the tropics, *Aust. J. Agric. Res.,* 38, 177, 1987.

115. Bryant, H. T., Hammes, R. C., Blaser, R. E., and Fontenot, J. P., Evaluation of acceptability by beef cattle of crownvetch grazed at several stages of maturity, *J. Anim. Sci.,* 45, 939, 1977.

116. Jung, G. A., Brann, D. E., and Fissel, G. W., Environmental and plant growth stage effects on composition and digestibility of crownvetch stems and leaves, *Agron. J.,* 73, 122, 1981.

117. Phillips, R. L., Cyanogenesis in *Lotus* species, *Crop Sci.,* 8, 123, 1968.

118. Williams, M. C., Purposefully introduced plants that have become noxious or poisonous weeds, *Weed Sci.,* 28, 300, 1980.

119. Parker, R. and Williams, M. C., Factors affecting miserotoxin metabolism in timber milkvetch, *Weed Sci.,* 22, 552, 1974.

120. **Cronin, E. H. and Williams, M. C.,** Chemical control of timber milkvetch and effects on associated vegetation, *Weeds,* 12, 177, 1964.
121. **Cronin, E. H., Williams, M. C., and Olsen, J. D.,** Toxicity and control of kelsey milkvetch, *J. Range Manage.,* 34, 181, 1981.
121a. **Williams, M. C. and Ralphs, M. H.,** Effect of herbicides on miserotoxin concentration in Wasatch milkvetch (*Astragalus miser* var. *oblongifolius*), *Weed Sci.,* 35, 746, 1987.
122. **Parker, R.,** Effect of Various Nitrogen Sources on the Metabolism of Miserotoxin in *Astragalus miser* Dougl. ex Hook, Ph.D. thesis, Utah State University, Logan, 1973.
123. **Majak, W. and Wikeem, B.,** Miserotoxin levels in fertilized *Astragalus miser* var. *serotinus, J. Range Manage.,* 39, 130, 1986.
124. **Majak, W., McLean, A., Pringle, T. P., and van Ryswyk, A. L.,** Fluctuations in miserotoxin concentrations of timber milkvetch on rangelands in British Columbia, *J. Range Manage.,* 27, 363, 1974.
125. **Majak, W., Williams, R. J, van Ryswyk, A. L., and Brooke, B. M.,** The effect of rainfall on Columbia milkvetch toxicity, *J. Range Manage.,* 29, 281, 1976.
126. **Majak, W., Parkinson, P. D., Williams, R. J., Looney, N. E., and van Ryswyk, A. L.,** The effect of light and moisture on Columbia milkvetch toxicity in lodgepole pine forests, *J. Range Manage.,* 30, 423, 1977.
127. **Majak, W.,** Timber milkvetch toxicity, *Can. Agric.,* 24, 24, 1978.
128. **McLean, A. M.,** Poisonous plants, in *Range Management Handbook for British Columbia,* McLean, A., Ed., Wayside Press Ltd., Vernon, B.C., 1979, chap. 3.
128a. **Majak, W., Quinton, D. A., Douwes, H. E., Hall, J. W., and Muir, A. D.,** The effect of clipping on the growth and miserotoxin content of Columbia milkvetch, *J. Range Manage.,* 41, 26, 1988.
129. **Iwasaki, T. and Kosikowski, F. V.,** Production of β-nitropropionic acid in foods, *J. Food Sci.,* 38, 1162, 1973.
130. **Kinosita, R., Ishiko, T., Sugiyama, S., Seto, T., Igarasi, S., and Goetz, I. E.,** Mycotoxins in fermented food, *Cancer Res.,* 28, 2296, 1968.

Chapter 6

VICINE, CONVICINE, AND THEIR AGLYCONES — DIVICINE AND ISOURAMIL

Ronald R. Marquardt

TABLE OF CONTENTS

I. INTRODUCTION*

Vicine and convicine are glycosides that are found primarily in faba beans (*Vicia faba* L) which are one of the most important pulse crops in the world, being consumed in large quantities in the Middle East, Far East, and North Africa, particularly Egypt. The mean annual world faba bean production was 6 million tons in 1977 with China producing about two thirds of the world's faba beans while 10 to 15% are grown in the West Asia and North Africa regions.[1,2] Except for the presence of vicine and convicine and relatively low concentrations of several other antimetabolites,[3] faba beans are an excellent source of protein with their aminoacid balance complimenting that of cereals.[4]

Vicine and convicine are compounds which are hydrolyzed by intestinal microflora[5,6] to highly reactive free-radical generating compounds,[7,8] divicine and isouramil. Divicine and isouramil have been strongly implicated as the causative agents in favism,[9-28] a hemolytic disease in humans, particularly young males, that have a deficiency of erythrocytic glucose-6-phosphate dehydrogenase (G-6PD) activity.[29-31] These free-radical generators may also cause other adverse effects including lipid peroxidation,[32] altered fat[33,34] and mitochondrial metabolism,[36] and possibly diabetes.[35] Some of these adverse effects may be neutralized by increasing the dietary concentration of free-radical scavenging compounds such as vitamins A, C, and E,[34,37-40] and through the use of chelating agents such as EDTA or desferoxamine.[40,41] The adverse effects of these compounds can also be reduced or eliminated by developing glycoside-free cultivars of faba beans or by using processing techniques that selectively extract or hydrolyze these compounds.[42-46]

---

* Abbreviations: G6PD, glucose-6-phosphate dehydrogenase; SOD, superoxide dismutase; GSH, glutathione; RBC, red blood cells.

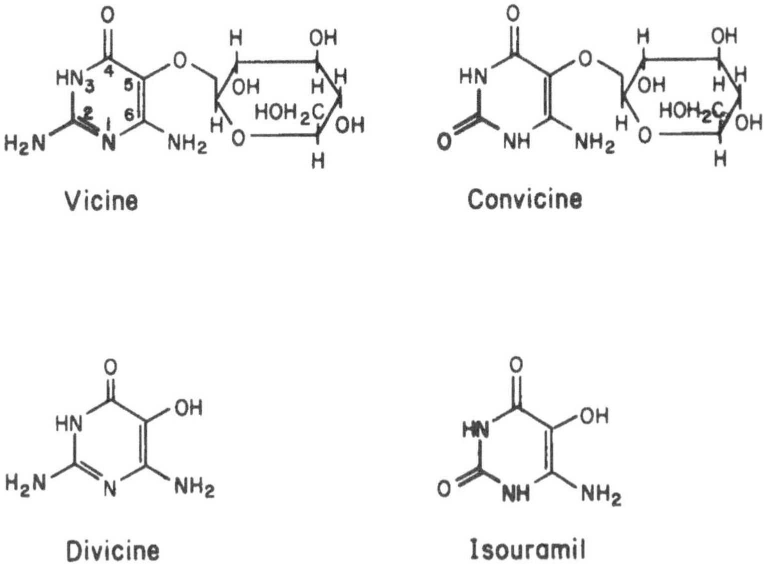

FIGURE 1. Structure of vicine, convicine, and their aglycones.

Vicine and convicine also appear to have certain beneficial properties including the prevention of cardiac arrythmia[47] and under certain conditions are able to inhibit the growth of the malaria parasite, *Plasmodium falciparum.*[41,48-53] The role of vicine and convicine in the seed of *V. faba* has not been established, but it may provide a storage source of nitrogen since it has a high content of nitrogen,[16] or it may protect the seed against certain microorganisms, particularly fungi.[54]

II. PROPERTIES AND CHEMICAL REACTIONS OF VICINE, CONVICINE, DIVICINE, AND ISOURAMIL

A. Structure and Some Properties of Vicine

Vicine [2,6-diamino-5-(β-D-glucopyranosyloxy)-4(1H)-pyrimidinone; 2,6-diamino-4-oxypyrimidine-5-(β-D-glucopyranoside); 2,6-diamino-4,5-dihydroxypyrimidine 5-(β-D-glucopyranoside); divicine 5-glucoside; vicioside; divicine-β-glucoside] has an empirical formula of $C_{10}H_{16}N_4O_7$, a molecular weight of 304.26, contains 39.47% carbon, 5.30% hydrogen, 18.42% nitrogen, and 36.81% oxygen, and has a structural formula as shown in Figure 1.[55] Vicine decomposes at 243 to 244°C and has an optical rotation $[\alpha]_D^{26}$ of $-11.7°$ at a concentration of 3.9 g/100 ml in 0.2 N NaOH,[55,56] is white in color, and has a needle-like structure (Figure 2).[57]

Vicine can be stored in a neutral or basic (1 N NaOH) solution at either 2 or 30°C for up to 7 d without decomposition but is unstable when stored at a low pH for a similar period of time. Vicine, however, does not decompose to a significant degree in acid when kept at a low temperature for relatively short time periods. For example, the decrease in vicine concentration when stored 24 h was less than 1% in 0.1 N HCl at 2 or 30°C and in 1 N HCl at 2°C, but was 16% in 1 N HCl at 30°C.[57] It is rapidly hydrolyzed to its aglycone, divicine, at high temperatures in the presence of concentrated acids.[9,55]

Vicine is minimally soluble in water at pH levels between 4 and 9 (Figure 3) with the values being 3.0 and 3.6 mg/ml, respectively. It is, however, highly soluble at pH values of greater than 10.5 and less than 1.0. The solubility of vicine at pH 7.0 is also influenced by temperature with solubility values being 1.5, 3.3, and 6.5 mg/ml, respectively, at 2, 25,

FIGURE 2. Vicine (upper) and convicine (lower) crystals observed by using a scanning electron microscope. (From Marquardt, R. R., Muduuli, D. S., and Frohlich, A. A., *J. Agric. Food Sci.*, 31, 839, 1983. With permission.)

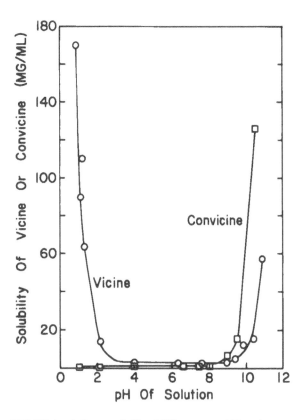

FIGURE 3. Influence of pH at 25°C on the solubility of vicine and convicine. (From Marquardt, R. R., Muduuli, D. S., and Frohlich, A. A., *J. Agric. Food Sci.*, 31, 839, 1983. With permission.)

and 60°C.[57] The solubility of vicine in acetone, absolute ethanol, and methanol at room temperature is approximately 0, 0.4, and 10 mg/ml, respectively. The addition of as little as 1% water to a methanol solution greatly depresses the solubility of vicine in methanol (i.e., from 10 to 1 mg/ml) with minimal solubilities being obtained when the methanol water ratio is 9:1.[58]

The ultraviolet absorption spectrum of vicine is influenced by pH[55,59] (Figure 4) with the maximum molar extinction coefficients (ϵ) at different pH values being shown in Table 1.

B. Structure and Some Properties of Convicine

The structure of convicine [6-amino-5-(β-D glucopyranosyloxy)-2,4(1H,3H)-pyrimidinedione;2,4,5-trihydroxy-6-aminopyrimidine 5-(β-D glucopyranoside)] was established by Bien et al.[60] It has an empirical formula of $C_{10}N_{15}N_3O_8$, a molecular weight of 305.24, and contains 39.35% carbon, 4.95% hydrogen, 13.77% nitrogen, and 41.93% oxygen. It has a structural formula as shown in Figure 1. The plate-like crystals, Figure 2, are yellow in color and decompose without melting at 287°C.[55,57,60]

The stability of convicine in aqueous solutions of different pH is similar but not identical to that of vicine, being relatively stable at a neutral or basic pH and hydrolyzed at low pH values. In general, convicine is more readily hydrolyzed than vicine. The percent decrease in convicine during a 24-h period was 2% in 0.1 *N* HCl at 2°C, 33% in 0.1 *N* HCl at 30°C, 14% at 2°C in 1 *N* HCl, and 93% at 30°C in 1 *N* HCl at 30°C.[57] At high temperatures and high concentrations of acid, it is rapidly hydrolyzed.[60]

The solubility characteristics of convicine in aqueous solutions at different pH levels are

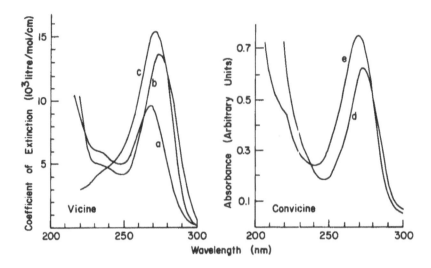

FIGURE 4.   Ultraviolet spectra of vicine in 0.1 *N* NaOH; (b) in 0.1 *M* phosphate buffer, pH 6.8; (c) in 0.1 *N* HCl; and convicine (d) in 0.1 *N* NaOH; and (e) in 0.1 *N* HCl. (From Olsen, H. S. and Anderson, J. H., *J. Sci. Food Agric.*, 29, 323, 1978. With permission.)

Table 1
ULTRAVIOLET ABSORPTION SPECTRA OF VICINE, CONVICINE, AND THEIR AGLYCONES

Compound	Oxidation state	pH	Absorption maxima	Extinction coefficient (Σ) (M^{-1} cm^{-1})	Ref.
Vicine		1	274	16,400	55
		6.8	275;236	13,200;4,400	55
		13	269;235	9,500;5,300	55
Convicine		1	271;220	17,400;6,000	60
		7	271	14,450	60
		13	273	14,450	60
Divicine	Reduced	1	281	12,900	63
	Reduced	1	281	15,850	61
	Reduced	1	281	14,900	64
	Reduced	7	285	9,800	19
	Oxidized		245		19
Isouramil	Reduced	1	280	13,600	63
	Reduced	7	280	14,200	19
	Oxidized	7	255		19

different than that of vicine as convicine has a relatively low solubility at low pH levels whereas vicine is highly soluble at low pH values (Figure 3). Also, in the neutral pH range, convicine is considerably less soluble than vicine. Its solubilities at pH levels between 1.0 and 8.0 are 0.4 and 1.2 mg/ml, respectively. The solubility of convicine is influenced by temperature in a manner similar to vicine with the values being 0.11, 0.26, and 0.72 mg/ml, respectively, at 1, 25, and 60°C in aqueous solutions.[57] The solubility of convicine in chloroform,[56] acetone,[58] absolute ethanol,[58] and methanol[58] at room temperature is approximately 0, 0, 0.03, and 0.08 mg/ml. The ultraviolet absorption spectrum is also affected by pH in a manner similar to that of vicine (Figure 4) with extinction coefficients being similar but not identical to those of vicine (Table 1).[60]

C. Structure and Physical Properties of Divicine and Isouramil

Divicine [2,6-diamino-5-hydroxy-4(3H)-pyrimidineone; 2,6-diamino-4,5-dihydroxypyr-imidine; 2,6-diamino-1,6-dihydro-4,5-pyrimidinedione; 2,6-diamino-4,5-pyrimidinediol; 2,4-diamino-5, 6-dihydroxypyrimidine] is the aglycone of the glycoside vicine and has an empirical formula of $C_4H_6N_4O_2$, a molecular weight of 142.12, and contains 33.0% carbon, 4.26% hydrogen, 39.43% nitrogen, and 22.52% oxygen.[56] The pale yellow[61,62] needles decompose at 300°C,[61] and 1 g dissolves in 100 ml boiling water and in about 350 ml cold water.[56] It is soluble in 10% KOH and has a structural formula as shown in Figure 1.[55,56] The molar extinction coefficient at two pH levels as determined by different researchers[19,61,63,64] is presented in Table 1.

Isouramil [6-amino-5-hydroxy-2,4(1H,3H) pyrimidinedione; 2,4,5-trihydroxy-6-amino-pyrimidine; 6-amino-2,4,5-(1H,3H,6H) pyrimidinetrione] is the aglycone of the glycoside convicine and has an empirical formula of $C_4H_5N_3O_3$, a molecular weight of 143.10, and contains 33.57% carbon, 3.52% hydrogen, 29.36% nitrogen, and 33.54% oxygen.[60,65,66] It has a structure as shown in Figure 1[65,66] and is pale yellow in color.[58,60] Isouramil is insoluble in hot water, alcohol, or acetic acid. It dissolves in sodium hydroxide, ammonia, and dilute hydrochloric acid.[65] The molar extinction coefficient at two pH levels is presented in Table 1.[19,63]

D. Chemical Reactions of Divicine and Isouramil

Divicine and isouramil, as well as the corresponding glycosides, react with Folin-Ciocalteu phenol reagent yielding a blue color similar to that produced by tyrosine and tryptophan. Both divicine and isouramil vigorously reduce alkaline solutions of 2, 6-dichlorophenolin-dophenol, phosphomolybdate, or phosphotungstate. They also produce an intense blue color reaction with an ammoniacal ferric chloride solution which is indicative of the presence of an enolic hydroxyl group.[16,55,56,63,65]

Divicine and isouramil in the absence of oxygen (i.e., under nitrogen)[14,19,64] or in the presence of reducing reagents such as ascorbate,[14] sodium borohydride,[14,19] cysteine,[14,19] glutathione (GSH),[14,19] dithiolthreitol (DDT),[19] mercaptoethanol,[64] sodium dithionite $(Na_2S_2O_4)$,[62] NADH,[67] and NADPH[67] are stable at either a low or a neutral pH and exhibit single absorption peaks with absorbance maxima and extinction coefficients as shown in Table 1.[19,61,63,64] These compounds can be maintained in their fully reduced form at pH 7.4 and 27°C for at least 30 min by 0.1 M mercaptoethanol, even when saturated with 100% oxygen.[64] Exposure of an aqueous solution of the two compounds to oxygen in the absence of a reducing reagent results in a rapid decline in their absorbency peaks with the concomitant appearance of new absorption bands at 255 nm (isouramil)[19] and 245 nm (divicine).[19,55] This hypochromatic effect, which is readily reversed by the addition of reducing reagents,[14,19,62,64,67] can be attributable to formation of the oxidized species of the respective pyrimidine molecules[14,19] (Figure 5).

The half-lives of divicine and isouramil in an oxygen-saturated solution at 26°C and at pH 7.0, but in the absence of reducing reagent, are 0.5 and 3 min, respectively. The compounds are relatively stable at a low pH ($t^{1}/_2 > 20$ h) and reach near maximal decomposition rates at pH 7.3 for divicine and 7.75 for isouramil (Figure 5).[19] Bailey et al.[62] had previously reported that the half-life of divicine at 27°C in an air-saturated buffer (pH 7.4) to be 0.6 min. Both of the above half-lives are similar to the autooxidation rate of divicine at a neutral pH that can be estimated from the decay curve of divicine that was published by Bendich and Clements.[55] The oxidative decomposition of the two aglycones is also markedly enhanced by the presence of traces of transition metals (Cu^{2+} and Fe^{3+})[55,68] and are protected by chelating agents such as EDTA.

Prolonged exposure to oxygen in the absence of a reducing reagent results in total obliteration of the characteristic peaks.[19] This latter change is no longer reversible by reduction

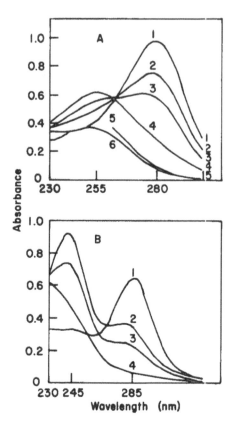

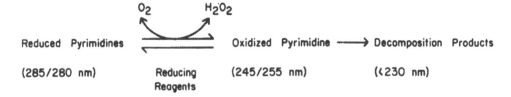

FIGURE 5.   Ultraviolet absorption spectra of a solution of isouramil (A) and divicine (B) incubated at room temperature under nitrogen or in air. (A) A 0.07-mM solution of isouramil in 0.05 M phosphate buffer (pH 7) was used. Curve 1, under nitrogen (no change throughout a 60-min period); curves 2 to 6, in air for 1, 2, 5, 25, and 35 min, respectively. (B) A 0.064-mM solution of divicine in 0.05 M phosphate buffer (pH 6.5) was used. Curve 1, under nitrogen (no change throughout 60-min period); curves 2, 3, and 4 in air for 2, 8.5, and 30 min, respectively. (From Chevion, M., Navok, T., Glaser, G., and Mager, J., *Eur. J. Biochem.*, 127, 405, 1982. With permission.)

FIGURE 6.   Interconversions of divicine and isouramil in presence of oxidizing and reducing reagents. The numbers in parentheses refer to wavelength absorption maxima of divicine/isouramil.[19]

and is suggestive of the rupture of the pyrimidine ring structure.[19] These spectral transitions are closely similar to that obtained when dialuric acid is oxidized to alloxan.[19,55,69] The spectral changes that occur when isouramil or divicine reacts with oxygen can be summarized in Figure 6.[19]

Chevion et al.[19] demonstrated that the oxidation of GSH by an oxygen-saturated solution was very slow at room temperatures and pH 7. However, the addition of catalytic amounts of isouramil or divicine greatly enhanced its rate of oxidation so that within 15 min 80% of the GSH was oxidized to GSSG with the concomitant accumulation of H_2O_2.

The oxidation of 1 mol of pyrimidine is accompanied by a 1:1 stoichiometry with respect to oxygen disappearance, the loss of 2 mol of hydrogen atoms from divicine or isouramil, and the stoichiometric (1:1) formation of 1 mol of hydrogen peroxide. Oxidized divicine and isouramil are converted nonenzymatically to their reduced forms (Figure 6) by 2 mol of GSH, 1 mol of NAD(P)H + H^+, or other reducing compounds with the concomitant formation of GSSG, NAD(P), or oxidized reductant.

Agents other than oxygen are also able to interact with the pyrimidines. Isouramil reduces ferricytochrome C,[19] ferricyanide,[19] o-ferriphenanthroline,[19] and methemoglobin[70] to the corresponding ferrous compounds. The reactions occur within a few seconds, and two ferric ions are reduced by each isouramil molecule. This reaction forms the basis of one of the tests for these compounds, as discussed subsequently.[19]

The oxidized forms of the pyrimidines not only are reduced by different reducing reagents, but also interact directly and nonenzymatically with them. GSH, for example, forms GSH-pyrimidine adducts that have absorption maxima at 305 nm[12,14,19,67] and at 320 nm[67] Benatti et al.[67] isolated two distinct GSH-pyrimidine adducts using reverse-phase high performance liquid chromatography (HPLC) and demonstrated that the GSH-divicine adducts accounted for 2.3% (305-nm peak) and 5.5% (320-nm peak), respectively, of the reacted GSH. Similar adducts are formed between alloxan and GSH but not between its reduced form (dialuric acid) and GSH.[69,71]

Bendich and Clements[55] pointed out in their comprehensive study that some of the distinctive features of divicine and its congeners — that is, their powerful reducing activity, spectral characteristics, and molecular instability — show a striking resemblance to those of ascorbic acid and reductic acid (2, 3 dihydroxy-2-cyclopenten-1-one). These authors concluded that a common structural denomination underlying these properties is a carbonyl-conjugated enediol (A) or aminoenol (B) system. Compounds with these structural arrangements serve to characterize a broad class of compounds designated by the general name reductones.[72] Consequently, all the characteristic properties of the aglycone are abolished by substitution of the hydroxyl group at C-5, such as that represented by the glycosidic linkage present in vicine and convicine. Thus, unlike the free aglycones, the glycosides show no reducing ability, do not undergo oxidation in the presence of oxygen, resist boiling in aqueous solution, and their ultraviolet spectra differ significantly from those of their constituent pyrimidines.[19,55,60]

(A)   (B)

E. Free Radical Intermediates From Divicine and Isouramil

The production of free radicals by divicine was predicted by Flohe et al.[7] and confirmed by Albano et al.[8] The latter researchers[8] detected the free radical of divicine by election spin resonance (ESR) spectroscopy. The ESR has a hyperfine structure consisting of 15 equally spaced lines with field splitting of 0.5 G. The intensity of the ESR signal was

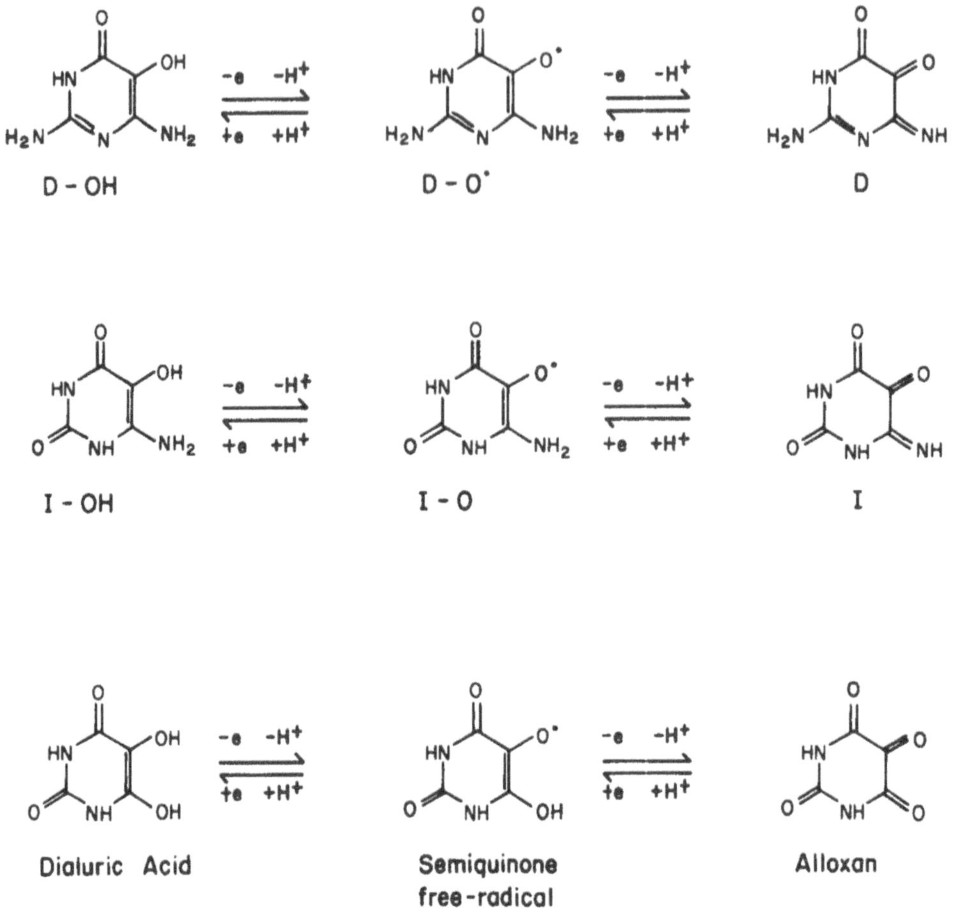

FIGURE 7.   Stepwise one-electron oxidation of divicine, isouramil, and dialuric acid. D-OH (hydroquinone, reduced divicine); D-O (semiquinone free radical); D (quinone, oxidized form); I, isouramil. Dialuric acid and alloxan undergo a similar type oxidation-reduction.

strongly dependent on pH, being maximal at pH 9.0 and low at pH 5.0. The signal was completely suppressed upon the addition of 1 mM GSH. No ESR signal was generated by vicine or by hydrolyzed vicine in the complete absence of oxygen, indicating the release of the aglycone and its reaction with oxygen are essential for the radical formation. A similar pattern would also occur with isouramil. A summary of the one-electron oxidation-reduction of divicine or isouramil[8,27] is presented in Figure 7. The reaction of these compounds with oxygen is similar to the analogous formation of the semiquinone free radical and the quinone, alloxan, from dialuric acid.[69]

III. QUANTITATION AND ISOLATION/SYNTHESIS OF VICINE, CONVICINE, AND THEIR AGLYCONES

A. Quantitation

Several spectrophotometric and chromatographic methods have been utilized for the quantitation of vicine and convicine in plant material and animal tissues. The very first method was that of Higazi and Read[73] which is based on the reaction of vicine and convicine and their corresponding aglycones with Folin-Ciocalteu phenol reagent to yield a blue color similar to that produced by tyrosine or tryptophan. This procedure has been criticized by

Chevion and Navok[74] on the basis that (1) it is nonspecific since aromatic amino acids such as tyrosine, tryptophan, and dihydroxyphenylalanine, present in large quantities in the bean, are not completely removed during extraction and (2) that the recovery of glucosides in the assay solution is low.

A direct ultraviolet spectrophotometric method was developed by Collier[75] and is based on the observation that the molar absorbance coefficient of vicine and convicine in an acid solution are not only similar, but are very high relative to that of other acid-extractable compounds in faba beans. It has been suggested, however, that up to 16% of the absorbance of a protein-free extract of faba beans may be caused by compounds other than vicine and convicine, and as a result the procedure yields high values. Pitz et al.[76] compared the procedure of Collier to the subsequently described gas chromatography (G-C) procedure and concluded that the ultraviolet method of Collier[75] could be used to reliably predict the total glycoside content of faba beans and that this would be particularly beneficial as a screening test since it is rapid and simple.

A third colorimetric procedure was developed by Kim et al.[77] and is based on the reaction of divicine and isouramil following acid hydrolysis of vicine and convicine with a titanium reagent to form a colored complex. Although the authors demonstrated that compounds such as nucleosides and nucleotides do not interfere with the method, they did not provide evidence that an extract of faba beans was free of other compounds that would affect the accuracy of the procedure. Also, the absorbance of the divicine-titanium complex is approximately 10% lower than for the isouramil complex, and as a result it would not be possible to obtain a precise molar absorbency coefficient for samples that contained different proportions of vicine and convicine.

A fourth procedure, developed by Chevion and Navok[74] involved differential extraction of the glucosides vicine and convicine with acetic acid followed by the enzymatic hydrolysis with β-glucosidase under anaerobic conditions. The two aglycones, isouramil and divicine, anaerobically reduce two molecules of O-ferrophenanthroline to O-ferrophenanthroline. The reaction is followed spectrophotometrically at 515 nm. Although the method is designed to quantitate vicine and convicine, they did not present good evidence for the accuracy of the method and the degree of recovery of total glycoside. All of the above procedures, appear to have some limitations particularly with regard to specificity. They, nevertheless, provide a means of estimating total glycosides and may be useful if equipment required for the subsequently described procedures is not available.

Procedures using thin-layer chromatography (TLC), GC, and HPLC have also been developed for the quantitation of both vicine and convicine in plant and in some cases animal tissues. The TLC procedure of Olsen and Andersen[59] involves the extraction by ethanol of vicine and convicine, evaporation of ethanol, separation using TLC, removal of the compound from the plate, and quantitation using ultraviolet spectrophotometry. The procedure is somewhat involved and seems to lack precision but is otherwise reliable.

The GC procedure of Pitz and Sosulski[78] appears to be an accurate and reliable method but is rather tedious. The method requires prior derivatization of the glycosides, which is not only time consuming, but can lead to incomplete derivatization or the formation of optical isomers.

Three HPLC techniques that have been developed for the quantitation of vicine and convicine include two procedures using a reverse-phase column,[79-80] and one using a Li-Chrosorb® NH₂ column.[81] The three procedures are superior to previously published methods, since the methods are highly sensitive, rapid, and accurate, and are able to clearly resolve vicine and convicine. Sample preparation techniques, however, are considerably different for the three HPLC procedures, with the procedure of Lattanzio et al.[80] being the most involved. The extraction procedure of Marquardt and Frohlich[79] appears to be the simplest.

Divicine and isouramil can also be resolved and quantitated using cation exchange[64] or reverse-phase HPLC.[79] Special precautions must be made to exclude traces of oxygen, as these compounds are rapidly degraded in the presence of oxygen, to carry out the analysis at a low pH, or to use antioxidants in the eluting buffer.

B. Isolation of Vicine and Convicine

Vicine was discovered and first isolated by Ritthausen and Kreusler[82] from vetch seeds (*V. sativa*), using an isolation procedure which involved extraction with dilute H_2SO_4 or 80% ethanol and precipitation with $HgSO_4$; followed by removal of Hg^{++} with H_2S and crystallization from dilute ethanol. The yield of pure material was about 0.35%.[83] Lin and Ling[9] modified the above procedure by extracting faba beans under reflux with 65% ethanol, removal of ethanol from the extract by evaporation, extraction of the oily liquid with ether, and precipitation of vicine from the lower layer of liquid by one of two ways. The first procedure was similar to Ritthausen and Kreusler.[82] In the second method, vicine was precipitated with lead acetate in an alkaline solution and then treated as described for the first method. The average yield of white crystals from the second procedure was 0.5%. A somewhat similar procedure was employed by Bendich and Clements.[55] Olaboro et al.[84,85] using faba-bean protein concentrate developed a procedure for the isolation of large quantities of vicine.

Convicine was also initially isolated from vetch seeds by Ritthausen[86] using procedures similar to those for vicine. Bien et al.[60] isolated convicine from fresh beans of *V. faba* after they were soaked in 95% ethanol at room temperature for 3 d. The filtrate was extracted with ether, and the aqueous layer was concentrated under vacuum. Crystallization of convicine started when the solution was kept in the refrigerator. In a subsequent paper Bien et al.[87] developed a method for the synthesis of convicine from 6-acetamido-5-hydroxyuracil with tetra-*O*-acetyl-β-D-glucopyranosyl bromide.

Marquardt et al.[57] have published a procedure for the simultaneous isolation of pure vicine and convicine from an air-classified fraction of faba beans. The procedure involves the extraction of the two glycosides in an aqueous solution from faba-bean protein concentrate, the addition of acetone to the mixture, and the recovery and concentration of the supernatant to a minimal value. Mixed vicine and convicine crystals which form within a period of from 1 to 3 d, are harvested by centrifugation. Vicine is obtained from the harvested mixed crystals by extraction at a low pH, since vicine is highly soluble while convicine is highly insoluble at this pH. Vicine can be recrystallized by adjusting the pH of the solution to neutrality, a pH where its solubility is minimal. The residual and insoluble convicine from the mixed crystals can be readily solubilized in a basic solution and then recrystallized by lowering the pH of the solution to neutrality. The procedure requires that the starting material be faba-bean protein concentrate which can only be prepared by air classification of a pin-milled sample of dehulled faba beans. Air classification not only increases the protein concentration of the sample by more than a factor of two, but also that of vicine and convicine. This method yields relatively high quantities of analytically pure vicine and convicine and is simple to carry out. The method was subsequently modified by Arbid and Marquardt so that vicine and convicine could be isolated from dehulled faba beans rather than faba-bean protein concentrate.[88] The yield of products, however, is considerably reduced. As discussed subsequently, the use of faba beans containing high concentrations of glycosides would also facilitate their isolation.

C. Preparation of Divicine and Isouramil

The glycoside nature of vicine was recognized by Ritthausen[89] who succeeded in isolating the aglycone, divicine, from this compound.[90,91] Divicine can be prepared from vicine by either acid,[9,90-93] enzymatic hydrolysis with β-glucosides,[13,94] or it can be synthesized chemically.[61-63,95,96] An elegant but tedious synthetic approach was utilized by Davoll and Laney.[63]

In this method ethyl tetrahydropyran-2-yloxcyanoacetate was condensed with guanidine in the presence of sodium ethoxide, and the product tetrahydropyranyl ether of divicine was converted to divicine by acid hydrolysis. Divicine has also been chemically synthesized from 2,5,6-triamino-4-pyrimidinone (TP) by oxidation followed by acid-catalyzed hydrolysis of the quinoid TP-5 imine and subsequent reduction with $Na_2S_2O_4$. The yield of the procedure was 74%.[62]

Isouramil was first prepared by Ritthausen[97] by acid hydrolysis of convicine and was first synthesized by Davidson and Bogert.[65] These authors used isobarbituric acid as the starting material which was subjected to nitrosation or coupled with *p*-chlorophenyl diazonium salt, and the reaction products were reduced with $(NH_4)_2S$ to yield isouramil. McOmie and Chesterfield[95] modified the procedure by replacing the reducing reagent, ammonium sulfide, with sodium dithionate and thereby greatly increased the yield, as well as the final purity of the product. This synthetic route was further facilitated by the relatively simple synthesis of isobarbituric acid from 5-bromouracid.[98,99]

The procedure of Davoll and Laney[63] can also be used for the syntheses of isouramil. The method is the same as that for the synthesis of divicine except guanidine was replaced by urea. In addition, isouramil can be prepared from convicine by enzymatic hydrolysis with β-glucosidase.[13,94] Recently methods have been developed by Marquardt and co-workers for the production of stable isolates of both divicine and isouramil. The availability of an off the shelf supply of these compounds should greatly facilitate future studies (unpublished data).

Dehydro-isouramil hydrochloride [4-imino-2,5,6(1H, 3H)-pyrimidinetrione hydrochloride, iso-DHU] has been prepared from isouramil by chlorination of isouramil[35] using the procedure described for dehydrouramil hydrate hydrochloride.[100] A characteristic feature of isoDHU is that it readily forms covalent hydrates, and in aqueous solutions it exists entirely in the hydrated form.

IV. OCCURRENCE OF VICINE AND CONVICINE IN THE SEED OF DIFFERENT PLANTS AND FACTORS AFFECTING THEIR CONCENTRATION IN FABA BEANS

Vicine was first isolated from the seeds of common vetch (*V. sativa* L.),[82,86] and subsequently both vicine and convicine were isolated from *V. faba*.[86,89,97] Vicine was also reported to be present in beet juice and peas[55,101] and in *Lupinus albus*.[102]

More recently, Pitz et al.[103] have reexamined the occurrence of vicine and convicine in the seeds of some *Vicia* species and other pulses using gas-liquid chromatography — mass spectrometry analysis of the glycosides. The results of their thorough analysis suggested that vicine and convicine were only associated with the *Vicia* species and are not present in beet roots or the other common legumes including peas and *L. albus* as had been previously reported.[55,101,102] They reported that the average dry seed concentration of vicine and convicine, respectively, to be 0.72 and 0.27% in *V. faba minor* L., (faba beans), 0.71 and 0.19% in *V. faba major* L. (broad beans), and 0.75 and 0.08% in *V. narbonensis* L. (a possible wild progenitor of faba beans).[103] More recently Dutta et al.[104] reported that alcoholic extracts of seeds of *Momordica charantia* contained vicine.

Both environmental and genetic factors seem to affect the concentration of vicine and convicine in the seed of *V. faba*. Fresh green seeds contain much higher levels of vicine and convicine than mature seeds.[76,105,106] Pitz et al.[76] reported that the mean vicine content of the green pods of four cultivars of faba beans was 1.94% (dry basis) and that of convicine was approximately 0.9%. The pods were harvested 3 to 4 weeks (July 17) after flowering when the moisture content of the seeds averaged 81%. On each successive weekly sampling date, the mean moisture content of the cultivars was progressively less, and there was a corresponding decrease in the mean vicine and convicine content on a dry seed basis. By August 21, when the mean moisture content was 21%, the content of each glucoside leveled off at approximately 0.6% vicine and 0.3% convicine, and the cultivars had the same values as they did on October 4, when they were harvested. These studies suggest that the glycosides were synthesized very early in the development of the seed, that the concentration within the seed decreases with increasing maturity of the seed (presumably due to deposition of protein and starch within the seed), and that the final concentration of glycosides in the mature seeds are not affected by harvesting date provided the seeds have matured.

Pitz et al.[76] also demonstrated that total glycoside content was not closely associated with grain yield or protein content, but showed a negative correlation with seed weight, maturity, and height. The low coefficient of determination (r^2), however, indicates that there is not a strong relationship between these characters and glycoside content. In another, more recent study Bjerg et al.[106] reported that the effects of different environments on glucoside content were small, but that the stage of seed development was important. Unripe seeds with a dry matter percentage below 50% had much higher glycoside concentrations than mature seeds with values reaching 5% of the dry matter in young, newly formed seeds.

The content of vicine in convicine in different cultivars of the mature seed has been reported by a number of researchers. Engel[107] reported that the vicine content of 64 broadbean (*V. faba* L.) samples ranged from 0.07 to 0.68%, Jamalian and Bassiri[105] reported vicine ranged from 0.23 to 0.61% in a study of 58 cultivars, while Gardiner et al.[108] reported that the total glycoside content of 78 widely diverse gentotypes of *V. faba* L. ranged from 0.69 to 1.26% with a mean value of 0.85 ± 0.11%. A more extensive study with 242 diverse strains of faba beans that were grown in three replicated trials, was carried by Pitz et al.[76] They reported that the content of vicine on a dry-matter basis ranged from 0.44 to 0.82%, convicine from 0.13 to 0.64%, and that total glycoside from 0.62 to 1.25%. The mean contents of vicine, convicine, and total glycosides were similar to 36 other less diverse cultivars and averaged 0.66, 0.31 and 0.97, respectively. The correlation (r^2) between vicine and convicine was only +0.22 (p <0.01) which would indicate that the two compounds are under independent genetic control. In a more recent study Bjerg et al.[106] reported that the major portion of the mature seed from 364 faba bean cultivars/lines obtained from different regions in Europe, Africa, and the Middle East contained more than 1% of vicine, convicine, and dihydroxyphenylalanine glycosides. Expressed as a percent of dry matter, the average vicine and convicine contents are 0.5 and 0.3%, respectively, with vicine values ranging from 0.1 to 1.1% and convicine from 0.1 to 0.7%. The estimated heritability for vicine ($h^2 = 0.45$) was somewhat lower than that found for convicine ($h^2 = 0.66$) and for total glycoside content ($h^2 = 0.61$).

Pitz et al.[76] also demonstrated that within a given sample of seed obtained from the same cultivar there was a large variation among individual seeds in the concentration of vicine and convicine. They suggested that if the concentration of the glycosides of individual seeds are under genetic control, then a breeding program could be established in which one half of the seed could be utilized for the assay of vicine and convicine, and the plumule-half of the seed for reproduction. This would greatly assist the plant breeder in the development of low- and high-glycoside-containing cultivars. Recently, Drs. G. Duc and G. Sixdenier at the Station de Genetique et amelioration des Plantes, Dijon, France, have obtained a cultivar of faba beans which does not contain vicine or convicine (personal communication).

In summary, it would appear that there is considerable variability in the concentration of vicine and convicine among and within cultivars of faba beans, with the variability among the mature seeds being principally attributed to genetic factors. There is also considerable differences in the concentration of total glycosides between mature and immature seeds. The development of faba bean cultivars with low concentrations of vicine and convicine should enhance its nutrition value for humans and animals. The development of high-glycoside-containing cultivars would facilitate the isolation of pure compounds. Isolation of the pure compounds would also be facilitated by the use of the immature seeds, since they have a high concentration of both compounds.

V. BIOSYNTHESIS AND POSSIBLE METABOLIC ROLE OF VICINE AND CONVICINE

Brown and Roberts[109] using labeled precursors demonstrated that the synthesis of the glycosides occurs within the pods and is not translocated into them. The results of their study also indicated that the orotic acid pathway was involved in the formation of the pyrimidine ring of both vicine and convicine. Pitz et al.[76] and Gardiner et al.,[108] as indicated previously, reported that the synthesis of vicine and convicine was independently controlled.

The roles that vicine and convicine have in the plant has not been elucidated, but they may serve as a storage source of nitrogen since they contain a very high concentration of nitrogen.[56] Studies by Bjerg et al.[54] have shown that both vicine and convicine and another compound present in faba beans, dihydroxphenylalanine, inhibit the growth *Botrytis cinerea* and *Ascochyta fabae* which are pathogenic to faba beans. An apathogen to *V. faba, Pyrenophora graminea*, was also affected by these compounds. These results suggest that convicine and vicine may be involved in resistance of faba beans to leaf and pod diseases.

VI. CONCENTRATION OF VICINE AND CONVICINE IN DIFFERENT FOOD PRODUCTS AND EFFECT OF TREATMENT WITH β-GLUCOSIDASE

As indicated, the concentration of vicine or convicine in faba bean seeds is influenced by stage of maturity, environmental factors, and genetic variation.[76,103,108] The glycosides appear to occur only in the seed portion of the plant and not in other parts of the plant.

The seed coat or the testa of faba beans contain approximately 13 to 15% of the total dry matter content of the bean,[110] but little or no vicine[111] or convicine.[58] In agreement with these observations, dehulled faba beans or the cotyledons have glycoside contents that are approximately 13 to 16% higher than those of the whole bean.[76,88] Dry milling of dehulled faba beans yields a protein-rich (75.1% protein) and a starch-rich fraction (8.6% protein) with high (1.5%) and low (0.4%) concentrations, respectively, of total glycosides.[42]

Wet processing methods[44] compared to dry milling techniques have been shown to be highly effective at not only producing a protein-rich fraction (94% protein), but at removing more than 97% of vicine and convicine. This product could be utilized in food preparations where low concentrations of glycosides are required.

Cooking and sprouting of faba beans have a negligible effect on their total glycoside content,[43] whereas extracting whole or dehulled beans with 1% acetic acid[43,46] or treatment with β-glucosidase[45] have been shown to be highly effective. The disadvantage of acid extraction is that it undesirably affects the organoleptic properties of the bean.[58] Arbid and Marquardt[45] used almonds, which are a good source of β-glucosidase activity, to hydrolyze vicine and convicine. This treatment requires the addition of finely ground almonds to faba bean paste (cooked faba beans) followed by incubation under appropriate conditions of time and temperature. Under these conditions the glycosides are completely hydrolyzed to divicine and isouramil, which in turn rapidly decompose to a nontoxic compound. This procedure, which can readily be carried out in the home, is an effective and simple means of eliminating the deleterious effects of these compounds. It could be useful in countries such as Egypt where the faba bean consumption is high, particularly for individuals that are predisposed to favism. An advantage of using almonds is that they do not produce the undesirable taste characteristics associated with acid extraction. Other plant products that contain β-glucosidase activity could also be used.

VII. METABOLIC EFFECTS IN ANIMALS

A. Hydrolysis of Vicine and Convicine in the Gastrointestinal Tract of the Chick, Rat, and Human

Frohlich and Marquardt[5] demonstrated that a small percentage of dietary vicine, but not convicine, was absorbed from the gastrointestinal tract of the young chick. Maximum uptake of vicine was within 3 h, the half-life in blood was estimated to be 4.5 h, and all of the vicine was excreted via the bile or kidney. The aglycones were not hydrolyzed *in vitro* by avian tissue homogenates, endogenous enzymes in faba beans, nor at an appreciable rate by 0.1 *N* HCl at 37°C, but were rapidly hydrolyzed by digesta from the ceca of the bird. The total fecal excretion of vicine and convicine was increased when antibiotics were added to the diet, which together with the above data suggested that vicine and convicine are mainly or only hydrolyzed by microorganisms that are present in the gastrointestinal tract. Similar results were obtained with the rat.[6] Dietary vicine was absorbed in relatively small quantities by the rat, was not hydrolyzed by rat tissues, but was very efficiently excreted via the kidney and probably the bile. The glycosides were rapidly hydrolyzed by digesta from the cecum and large intestine, and only slowly by digesta from the small intestine and stomach. When relatively high concentrations of vicine and convicine were fed, only small quantities of these compounds were excreted in the feces and urine, which suggests that they are nearly completely converted into their aglycones in the gastrointestinal tract. The aglycones were detected in the digesta, but not in the blood since under the anaerobic conditions of the digesta, they are relatively stable; whereas in the blood, where there is an abundant supply of oxygen, they are rapidly converted into nonultraviolet-absorbing compounds.[19] The half-life of divicine and isouramil in the blood appears to be less than 1 min, whereas it can persist in cecal digesta for several minutes.[58]

In another study, Arbid and Marquardt[28] demonstrated that when the two glycosides were injected intraperitoneally into the rat, they were rapidly transported from the interabdominal cavity into the intestine where they were hydrolyzed to divicine and isouramil. Death occurs if the dosage of vicine or convicine is high, and this may be prevented by the oral administration of an antibiotic, neomycin, which presumably inhibits growth of intestinal microflora. Also, the amount of vicine and convicine that is excreted is markedly increased in the antibiotic-treated rats.

The results of Bjerg et al.[112] are consistent with the above observations. Rats that were fed faba bean-containing diets had considerable quantities of vicine in the small intestine with only trace amounts of the compound being detectable in the cecum, large intestine, feces, or blood. Human excreta is also highly effective at hydrolyzing these compounds.[6]

The above studies suggest that vicine and convicine are hydrolyzed to their aglycone (divicine and isouramil) by the anaerobic microflora present in the lower portion of the gastrointestinal tract. The hydrolytic products which are stable in the anaerobic atmosphere of the digesta are presumably absorbed into the blood where, in the presence of oxygen and as discussed subsequently, they undergo decomposition to yield activated oxygen species which, if not neutralized by the free-radical scavaging system, would cause cellular damage. The specific microorganism that is responsible for this hydrolysis and the nature of the β-gluconases has not been established.

Luisada[113] reported that the interval between exposure to the faba bean and the onset of favism symptoms ranges from 5 to just under 24 h. Presumably, this reflects the time required for passage of ingested food into that section of the gastrointestinal tract where hydrolysis of the glycosides can occur.

B. Effect of Vicine and Convicine on the Laying Hen

Faba beans contain a factor that depresses egg size in the laying hen.[33,34,84,85,114-116] This factor is present in faba bean protein concentrate in a concentrated form, is thermostable, is soluble in aqueous solutions, can be obtained as a white crystalline material, and has been identified as being vicine.[84,85,116,117] Convicine, when incorporated into the diet, also induces egg size depression (13%) to approximately the same degree as that obtained with 1% vicine.[58] In contrast, dihydroxyphenylalanine, which is also present in faba beans in relatively high concentrations, has no effect when added to the diet at a concentration of 1%.[58] Vicine, when added to the diet of laying hens at a concentration of 1%, in addition to depressing egg size,[33,34] also depressed egg production rate (13%),[33] total egg mass (18%),[33] yolk weight (12%),[33] yolk mass (20%),[33] packed cell volume (29%),[34] the fertility (59%)[34] and hatchability (75%)[34] of eggs, and the ratio of yolk height/diameter (12%).[33] This last parameter is an indirect measure of the fragility of the membrane. In addition, vicine increased the percent of blood spots on the yolk (300%),[33] the plasma lipid concentration (216%),[33] plasma lipid peroxide concentration (370%),[33] degree of spontaneous hemolysis of erythrocytes (56%),[33] and the weight of the liver (9%).[33] The addition of vitamin E to the vicine-containing diet did not have a significant effect on most of the above parameters except that it dramatically improved the fertility (100%)[34] and hatchability (196%) of eggs.[34]

The results with the laying hen demonstrated that vicine and convicine, but not dihydroxyphenylalanine, have a very pronounced effect on its metabolism. The susceptibility of erythrocytes to hemolysis, the increased incidences of blood in the yolk, and decreased hematocrit concentration are similar to the hemolytic effects observed following the ingestion of faba beans by G6PD-deficient humans.[16,31] This data, therefore, supports the proposal that vicine and convicine or their aglycones are the causative agents of favism in humans.

The markedly elevated plasma lipids and lipid peroxides, together with the depressed amount of yolk which is primarily lipid in composition,[118] indicate that these compounds interfere with transport of fat across the yolk membrane. The above observations together with the protective effects of vitamin E with regard to hatchability and fertility of eggs are consistent with the proposal, as discussed subsequently, that the aglycones of vicine and convicine produce free radicals which not only react with erythrocytes, but also other membrane and tissues to cause tissue damage and loss of functional properties.

The relative insensitivity of the young chick to vicine and convicine in contrast to the laying hen[34] may be attributed to its ability to resynthesize reduced glutathione from NADPH in the presence of the aglycones of vicine and convicine at a sufficiently rapid rate so as to neutralize their toxic effects.[19] In the laying hen, there is a high demand for NADPH due to the high rate of fat synthesis,[119] which presumably results in limited availability of NADPH for other biological functions. In the presence of compounds such as vicine and convicine which induce oxidant stress, NADPH becomes limiting, and, as a result, the laying hen is not able to neutralize these effects through the glutathione peroxidase system. A similar, but more specific condition, occurs in human RBCs. The net effect in both species is an increased sensitivity of the erythrocytes to the effects of vicine and convicine. Other tissues and organs, however, are also affected in the laying hen which may be attributed to a general shortage of NADPH under stressful conditions as compared to the specific limitation of NADPH in the erythrocytes of G6PD-deficient humans. The mechanism of these effects are discussed subsequently.

C. Effect of Vicine, Convicine, Divicine, and Isouramil in the Rat

Lin and Ling[10] demonstrated that the only toxic effect in rats fed 0.6 g/kg diet of vicine, apart from a discrete lowering of food intake and rate of growth, was a loss of hair. Collier et al.[120] reported that a diet high in faba beans and containing 1.0% vicine plus convicine did not impair growth, food intake, or liver weight of weanling rats as compared with control animals. Blood glutathione concentration was slightly elevated in the faba bean-fed rats, but blood hemoglobin and liver glutathione were unaffected. These results suggested that rats could tolerate approximately 1% vicine without deleterious effects. In a more recent study, Bjerg et al.[112] reported that convicine, but not vicine, when added to the diet at a concentration of 0.14%, reduced the biological value of dietary protein.

Yannai and Marquardt[26] attempted to sensitize rats to the effects of vicine or the favism-causing factor by prefeeding a diet deficient in riboflavin, tocopherol, and selenium and high in vitamin E-stripped sunflower oil and/or pretreatment with buthionine sulfoximine (BSO). Such deficiencies or treatments interfere with GSH regeneration and its potential as an antioxidant.[32,39,121-126] BSO dramatically reduced blood GSH levels from 49 to 15 mg/100 ml, whereas the deficient diet had no effect on blood GSH concentrations, possibly due to the relatively short feeding period.[26] Vicine, when incorporated into both diets at a concentration exceeding that in faba beans (2%), had very little or no effect on blood GSH concentration, hematocrit level, or percent spontaneous hemolysis of erythrocytes. The failure of oral intakes of vicine and convicine to elicit apparent effects in the rat may be attributed to doses that were not high enough to overwhelm the rats' detoxification system.[19,27,67]

Although rats appear to be insensitive to orally administered glycosides even when "sensitized", rats are sensitive to the glycosides and their aglycones when injected intraperitoneally or intravenously, in the case of divicine.[26,28,40] Rats injected intraperitoneally with vicine and convicine had, after a few hours, an increased respiration rate, abdominal convolutions, generalized cyanosis, and, when the dosage was sufficiently high, death which appeared to be caused by asphyxiation.[28] Tissues of the dead animal were engorged with dark-brown blood, particularly the heart auricles and lungs. The large intestine contained entrapped gases and a very soft- and watery-appearing digesta. The fecal material also was soft and appeared to contain a high content of water, suggesting a movement of body fluids into the gastrointestinal tract. Most of the mortalities following vicine and convicine injections occurred between 8 and 24 h with very few occurring after 24 h.[28] The LD_{50} dosages 24 h after intraperitoneal administration of vicine, convicine, divicine, and isouramil were 4000, 3300, 149, and 114 mg/kg body weight. Corresponding nontoxic doses for which there were no deaths were 3000, 2000, 100, and 80, while all of the rats died when the doses were 5000, 5000, 250, and 250.[28,40] These results demonstrate that the toxicity of the aglycones are similar to each other and approximately 27 to 29 times more toxic on an equal-weight basis than the parent compounds. On the equal-molar basis, the values would be approximately 14. The parent compounds are presumably less toxic as the interconversion to the aglycones may not be quantitative, or the process is sufficiently slow so that the animal is able to neutralize the effects of the products that are formed by the aglycones. Arbid and Marquardt[28] demonstrated that vicine and convicine, when injected intraperitoneally into rats, were rapidly excreted from the abdominal cavity via the gastrointestinal tract and kidney. Vicine and convicine are then hydrolyzed as discussed previously by the microorganism in the digesta to their aglycones, which in turn were absorbed.

Successive injections of subacute amounts of vicine and convicine (1330 mg/kg) over several days caused a decrease in liver size and an increase in spleen size, the latter effect being attributed to an increased phagocytosis of erythrocytes. Blood lymphocytes and possibly eosinophils decreased, while monocytes and neutrophils increased, suggesting that vicine and convicine stimulated adrenal corticoid production.[28,40]

Yannai and Marquardt[26] observed that rats injected intravenously with divicine exhibited a much longer sleeping time after anesthesia with ether (approximately 10 to 20 min) than animals injected with a saline solution. Also, divicine-treated rats had labored breathing (panting) and blue paws and ears (cyanosis), both symptoms indicating impaired oxygen-carrying capacity of the blood. Mild to severe diarrhea was apparent in all animals. These

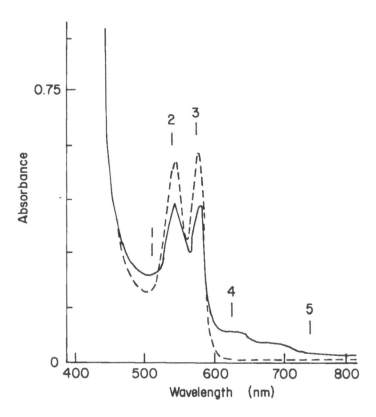

FIGURE 8. Absorption spectra for hemolysates prepared from untreated blood or blood treated with vicine or convicine (---) or divicine (———). Respective absorbance values for the first three treatments which were the same and the divicine treatment were 0.23 and 0.27 at 520 nm (arrow 1), 0.54 and 0.44 at 541.4 nm (arrow 2), 0.58 and 0.43 at 576.4 nm (arrow 3), 0.007 and 0.087 at 629 nm (arrow 4), and 0.004 and 0.025 at 730 mm (arrow 5). From Arbid, M. S. S. and Marquardt, R. R., *J. Sci. Food Agric.*, 37, 539, 1986. With permission.)

symptoms lasted for nearly 1 d and did not occur in rats injected with saline. Blood GSH, hematocrit, and adrenal and spleen weights were also affected in a dose-response manner. Blood GSH appeared to be affected within minutes following divicine injection and then recovered to almost pretreatment values within 24 h. In contrast, hematocrit values were affected only after a longer time period. Hematocrit values relative to the pretreatment period decreased 4% after 1.5 h and 31% after 24 h. Increases in the weights of the adrenal gland and the spleen 24 h following divicine injection, were 144 and 41%, respectively. The decrease in hematocrit was accompanied by little or no change in spontaneous hemolysis of RBCs. In an *in vitro* study[17] in which RBCs were incubated in the presence and absence of divicine, similar results were also obtained in isotonic and hypotonic solution. However, in hypotonic solution (equivalent to 0.45% NaCl), 80% of the normal while only 20% of the divicine-treated RBCs were hemolyzed, which would suggest that divicine under these conditions stabilized or protected the RBCs from hemolysis.

In vitro blood changed in color from bright red to reddish brown within a few minutes of being treated with divicine. A prerequisite of this color change appeared to be the nearly complete depletion of RBCs content of GSH. The ultraviolet absorption spectra of hemolyzed blood following divicine administration *in vivo* or *in vitro* were modified as shown in Figure 8, and these effects were dose related.[28] The effects of vicine, convicine, isouramil, or divicine *in vivo*, or divicine and isouramil, but not vicine or convicine, *in vitro* on the ultraviolet spectrum can be more readily demonstrated by comparing the absorbance ratios at two wavelengths. The ratio of absorbance as established *in vitro* at 576.4/629, for example, was 83 for divicine-treated RBCs and 5 for untreated RBCs. Nearly identical values were obtained *in vivo*[28] Similar spectral changes were observed by Winterbourn et al.[27] which they attributed to the formation of a hemoglobin-peroxide complex and methemoglobin.

Divicine treatment *in vitro* also markedly affects erythrocyte shape (Figure 9).[127] Similarly looking cells were referred to by Mohandas et al.[128] as "echinocytes". These authors explained that the morphological transformations from discocytes to echinocytes occurs due to lipid-soluble anionic agents, which tend to accumulate in the outer half of the membrane bilayer. Such cells were reported by Mezich et al.[129] to form by treatment with menadione within 30 min. After 30 min of incubation the "spiny" cells became "prelytic spheres" and underwent lysis. In contrast, studies by Marquardt and co-workers[58] demonstrated that divicine-treated RBCs retained their shape at both 67 and 100% of physiological concentrations of saline for at least 48 h. The spherical erythrocytes following divicine treatment also had an enhanced sedimentation rate.[58] The more rigid divicine-treated erythrocytes are presumably scavenged by the reticuloendothelial system. As discussed subsequently, the change in red cell membranes in favic individuals appears to be somewhat different.[24]

Erythrocytes from rats that have been treated with divicine also have markedly reduced ability to exchange respiratory gases[58] which may account for the slow recovery from anaesthesia, labored breathing (panting), and blue paws and ears (cyanosis) associated with divicine treatment.[26] Presumably, the animal will die of asphyxiation when a sufficiently high proportion of the erythrocytes are affected.

It may be concluded that divicine and isouramil or their parent compounds, when administered to rats intravenously or intraperitoneally, cause a reduction in the red cell concentration of GSH. GSH in divicine-depleted cells appears to be resynthesized within a relatively short period of time; however, if the concentration of GSH falls below a certain critical level, irreversible cell damage occurs. This is indicated by a change in color of RBCs, altered ultraviolet absorption spectrum of hemolyzed blood [presumably reflecting conversion of hemoglobin into a hemoglobin-peroxide complex (ferryl species) and methemoglobin], increased resistance of RBCs to spontaneous hemolysis (which may suggest cross-linking of protein with its resulting stabilizing effect), altered sedimentation rates of RBCs (which would suggest the shape of the RBC is affected), and finally, changes in shape of erythrocytes from biconcave to a spherical pinocytic form. The more rigid form of erythrocyte is subjected to phagocytosis with splenic sequestration being a major mode of removal of RBCs in the rat.[130] As a result, hematocrit decreases, and the weight of the spleen increases. The functional properties of the erythrocytes are also affected as they have a greatly reduced ability to take up oxygen, as reflected in change in percent oxygen saturation of the RBC[58,131] and the occurrence of cyanosis. These effects occur within a short time of divicine administration, and if the effects are severe, the rat dies of anoxia. Divicine also appears to have other effects as it increases adrenal gland size (indicating the animal is under severe stress), decreases leukocyte counts, and appears to cause an increase in fluidity of the digesta and diarrhea. Studies with the laying hen, as discussed previously, also suggest that these compounds affect other tissues in addition to the RBC. Most of the effects on rat RBCs are similar to those that occur in favic individuals.

Dehydroisouramil (iso-DHU, Figure 7) is an oxidized or quinone form of isouramil. When administered intravenously to rats, it elicited a persistent hyperglycemia as early as 30 min after application, with the morphological changes of the pancreas being considerably more rapid and extensive than those after alloxan injection.[35,100] The more marked potency of iso-DHU *in vivo* may be at least partially attributable to the greater stability of this agent compared to alloxan. Vicine and convicine, in contrast, appear to have no effect or an opposite effect on blood glucose concentration to that of iso-DHU.[28] Additional research, however, must

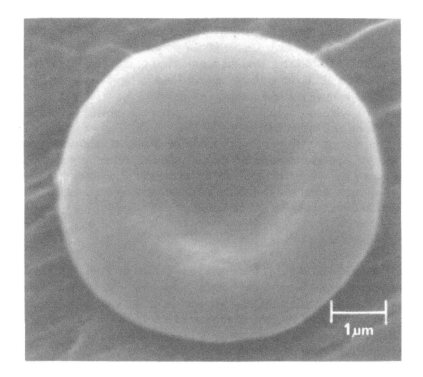

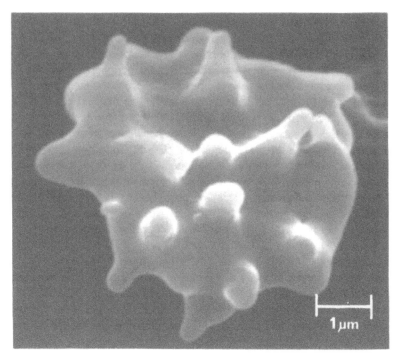

FIGURE 9. Photomicrographs of normal (upper) and divicine-treated erythrocytes obtained from rats. (From Yannai, S. and Marquardt, R. R., *Vet. Human Toxicol.*, 29, 373, 1987. With permission.)

be carried out to clarify the effects of vicine, convicine, and the oxidized and reduced forms of their aglycones on pancreatic function. If it is established that DHU and its divicine analogue are formed in significant quantities in humans that consume faba beans and if one of the target organs is the pancreas, then the consumption of faba beans could have further implications in human health. It has been suggested that in Egypt where the consumption of faba beans is high, there is a high incidence of diabetes.

D. Effect of Vicine on Mice and Dogs

Very little research has been carried out with other species of animals. Lin and Ling[10] reported a transient hemoglobinuria in the urine of puppies 3 h after the oral administration of 0.2 g/kg body weight of vicine. Clark and co-workers[41,51-53] reported that healthy mice were able to withstand injections of divicine at a concentration of 25 mg/kg without any adverse effects in liver, thymus, or splenic tissue.

VIII. EFFECTS ON HUMANS

A. Favism — Description of the Disease and Implications of Vicine and Convicine

The consumption of faba beans (broad bean, *V. faba*) under certain conditions results in the metabolic disease favism, an acute hemolytic anemia.[161] This disease has been recognized since antiquity, but the first authentic description of the disease in the medical literature dates back to the mid 1850s.[132] The disease may be fatal, usually occurs in children (especially males), and is related to a genetic deficiency of glucose-6-phosphate dehydrogenase (E.C.1.1.1.49, G6PD) in erythrocytes.

According to Luisada,[113] the first symptoms of favism occur within 5 to 24 h after ingestion of faba beans. The symptoms consist of malaise, headache, dizziness, nausea, vomiting, chills, pallor, lumbar pain, and fever. Hemoglobinuria appears within 5 to 30 h following exposure to the bean, and jaundice is observed a few hours later. Fever and marked leukocytosis may also occur. Sometimes the red cells in the peripheral blood are distorted in shape, and the hemoglobin may appear to be separated from the membrane.[133] A typical attack lasts from 2 to 6 d. Fatalities usually occur within the first 2 d and rarely during the third day.[113] Although favism may be quite mild, in many cases the anemia is very severe.[134]

The disease has been observed in many different population groups and has a seasonal incidence that coincides with the harvesting of the bean.[135] Contrary to previous reports, it appears to be associated only with the consumption of the green or mature bean and not with the inhalation of pollen.[136] A more detailed review of favism and its relationship to G6PD deficiency have been presented.[16,31] Recent studies, as discussed subsequently, have strongly implicated divicine and isouramil, the aglycones of vicine and convicine, as the causative agents of favism, with a deficiency of G6PD being a predisposing factor.

B. Incidence of Favism

Favism has been observed in many different population groups. It is most prevalent in the islands and coastal regions of the Mediterranean Sea and in the Middle East including Rhodes, Sardinia, Sicily, Cyprus, southern Italy, Greece, Turkey, Lebanon, Israel, Spain, Algeria, Egypt, Sudan, Iraq, and Iran.[113,135-140] The disease is also frequently encountered in China[141] and, as cited in Belsey,[136] in Bulgaria. The incidence of favism in the Middle East varies considerably from one country to another and within each country, and depends on the distribution of the genetic defect, the presence of faba beans in the local diet, and the availability and utilization of medical facilities.[136] An incidence of 5 cases per 1000 population has been estimated in Sardinia.[142] Favism accounts for approximately 1 to 2% of pediatric hospital admissions in Cairo and Alexandria, Egypt.[136]

C. Glucose-6-Phosphate Dehydrogenase Deficiency and Favism

Elucidation of the nature of the inborn error of metabolism underlying the so-called drug sensitivity (that is, the propensity of certain individuals to develop acute hemolysis in response to primaquine and a variety of drugs), provided a major clue for the understanding of the pathogenesis of favism.[142-144] Soon after, Sansone and Segni[145-147] and Szeinberg et al.[148-150] noted that the GSH concentration and stability of red cells of patients with a history of favism were diminished.

Carson et al.[151] revealed that the GSH instability of susceptible erythrocytes was attributable to a deficiency of NADP-linked G6PD. As a result, these cells were unable to maintain an adequate supply of NADPH required for the continuous reduction of GSSG by GSH reductase. It became apparent that all patients with favism were G6PD deficient,[152-155] yet many patients with G6PD deficiency were able to ingest faba beans without experiencing hemolytic episodes. In fact, some individuals with favism had usually ingested faba beans several times without being subjected to the disease.[156] Some factor(s) in addition to G6PD deficiency must, therefore, be required for the development of favism. The trait for thalassemia has been found, for example, to be associated with G6PD deficiency, with a high frequency among certain Caucasian groups.[48] This association has a salutary effect on the enzyme deficiency, for full-blown favism is rare among Caucasions who have thalassemia minor along with G6PD deficiency.[48] The shortened erythrocyte life span in thalassemia minor and the consequently younger red cell population might explain the greater resistance to hemolysis in these particular individuals. This factor would not, however, account for the variable susceptibility of an individual to the disease. Another possibility, as discussed subsequently, is that the nutritional status, particularly that of vitamin E, may also influence the development of the disease.

G6PD deficiency affects about 100 million people of all races throughout the world.[151] The incidence and severity of this genetic defect, however, is widely dissimilar in different ethnic groups.[29,157] Approximately 80 to 100 variants of this deficiency have been identified along with approximately 40 variants with severe G6PD deficiency not associated with chronic anemia, but requiring the additional insult of drugs or other agents for hemolysis to occur.[30,158] The most extensively studied variant of this class is the Mediterranean type of G6PD deficiency with wide distribution all over the Mediterranean, Near East, and even Asia. This variant has been associated with favism.[29,30] In contrast, favism does not occur in North American blacks,[29,31] who have a different variant of G6PD.

G6PD deficiency is an X-linked trait. Males with the mutant gene have only a single red-cell population, and all cells have the mutant type G6PD. Blood destruction will depend on the molecular nature of G6PDH enzyme for each particular variant. Only a proportion of the aging red cell population may be sufficiently deficient in enzyme activity for red cell destruction to occur. Heterozygote females with G6PD deficiency possess two red cell populations: normal and mutant red cells.[31] The ratio of normal to deficient cells most frequently is 1:1, but may range from 1% normal and 99% mutant cells to 1% mutant and 99% normal cells in a few patients. Red cell destruction affects only mutant and not normal cells. As expected, the frequency of females with clinical hemolysis usually is lower than that of affected males, even though the frequency of heterozygotes in the population is almost twice that of affected hemizygotes. Clinically affected females are the few homozygotes and only those heterozygotes who have a preponderance of mutant cells with low enzyme levels.[29,31]

D. Possible Other Effects in Humans

Although the aglycones of vicine and convicine have been most extensively studied relative to their ability to induce favism in susceptible individuals, they also seem to produce other, perhaps subclinical, effects in humans. A hemolytic crisis has in a few cases been associated with cataracts,[159,160] optical atrophy,[161] and a few incidences of mild leukopenia,[162,163] which has also been reported in rats given successive subacute doses of vicine and convicine.[28] The induction of diabetes in rats following administration of the oxidized form of isouramil,[35] which is similar in structure to alloxan (Figure 7), may indicate that the pancreas is a target organ for these compounds. Other studies with rats and the laying hen indicate that liver metabolism[36] and fat metabolism or transport of fats may be affected[33,34] by these compounds. Divicine and isouramil which are free-radical producers[7,8] may be expected to induce metabolic disturbances similar to those produced by other analogous compounds. Additional research, however, must be carried out to establish if these effects occur.

IX. METABOLIC EFFECTS OF DIVICINE AND ISOURAMIL

A. Effect of Divicine and Isouramil on the Concentration of GSH, NADPH, and Ascorbic Acid in Normal and G6PD-Deficient RBCs

Divicine and isouramil affect the metabolism of several compounds, particularly GSH[13,14,16,17,19,28] and NADPH.[17,67] Ascorbic acid also potentiates the effect of divicine or isouramil in that in the oxygenated red cells, an oxygen-mediated recycling of dehydroascorbic acid progressively lowers GSH concentration to near zero, even at a very low concentration of divicine.[13,14,16,17]

Arese et al.[17] demonstrated *in vitro* that there was a parallel and rapid induction of the synthesis of both GSH and NADPH in normal RBCs in the presence of divicine or isouramil. Similar results were obtained with GSH *in vivo* when rats were injected with divicine.[28] This regeneration is accompanied by a short-lived burst of CO_2 production *in vitro*. In the G6PD-deficient cell, however, GSH resynthesis is not stimulated since under *in vivo* conditions only 1/1000 of the potential G6PD activity is expressed.[30,164] As a result enzyme stimulation in the presence of a reduced concentration of NADPH results in a much slower rate of NADPH regeneration in the G6PD-deficient as compared to the normal RBC.

Although NADH is also subjected to oxidation by divicine/isouramil, it does not decrease dramatically in either the normal or the G6PD-deficient RBC, since it is rapidly regenerated by glyceraldehyde 3-phosphate dehydrogenase. Also, there is only a very loose connection between the NADPH/NADP- and the NADH/NAD systems.

B. Nature of the Interaction of Divicine and Isouramil with GSH in the Biological System: Establishment of a Futile Oxidation-Reduction Cycle

As indicated previously the reduced forms of divicine and isouramil are rapidly and nonenzymatically converted in the presence of oxygen to their oxidized form (Figure 6 and 7) with the concomitant formation of H_2O_2. The oxidized form of these compounds, however, is able to react with reducing reagents such as GSH,[14,19,67] NAD(P)H,[67] and ascorbate,[14] which results in the regeneration of the reduce aglycones. This shuttle system results in the net transfer of electrons from various reducing compounds to oxygen to form hydrogen peroxide with divicine and isouramil being the cofactors.[14,19] In a nonbiological system, H_2O_2 would accumulate as it does not react directly with GSH and NADPH,[67] and the shuttle would cease when all reactable reducing compounds were utilized.

In the presence of glutathione peroxidase as would occur *in vivo*, H_2O_2 is converted to H_2O and O_2. The H_2O_2 molecule reacts with two molecules of GSH, while two additional molecules of GSH are dissipated in the reduction of the oxidized form of the pyrimidines. In the overall reaction, therefore, four molecules of GSH are oxidized during a single oxidation-reduction cycle of isouramil or divicine. Glutathione reductase regenerates GSH from GSSG and NADPH + H^+. *In vivo* this futile cycle would continue to operate until the aglycones either spontaneously decomposed or formed irreversible adducts with other components in the cell, or until all of the reducing equivalents were exhausted, after which

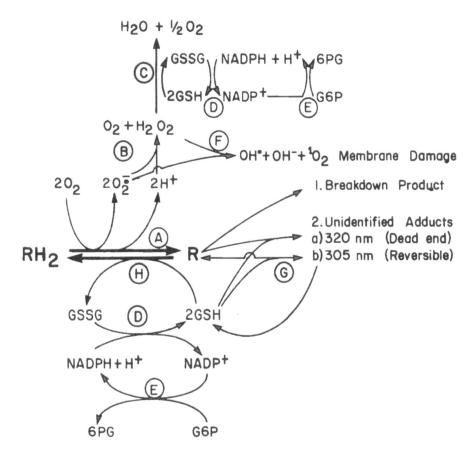

FIGURE 10.   Oxidation-reduction shuttle mechanism for divicine or isouramil (RH_2, reduced pyrimidine; R, oxidized pyrimidine) and other side reactions. A, nonenzymatic autoxidation of RH_2; B, SOD (superoxide dismutase); C, GSH-peroxidase; D, GSH-reductase; E, G6PD (glucose 6-phosphate dehydrogenase); F, nonenzymatic formation of hydroxy and other radicals in presence of metal ions; G, limited spontaneous formation of 320-nm dead-end product and 305-nm reversible product; and H, nonenzymatic reduction of oxidized pyrimidine (R); G6P, glucose-6-phosphate; and 6PG, 6-phosphogluconate; O_2^{-}, superoxide radical; OH·, hydroxyl radical; OH^-, hydroxyl anion; $^1O_2$, singlet oxygen.

irreversible cell damage would occur. The 305-nm divicine (isouramil)-GSH adduct, as discussed previously, is a good substrate for glutathione reductase, whereas the 320-nm adduct does not interact with the enzyme, thereby possibly representing a dead-end species of divicine and isouramil.[67] Based on those observations, the 305-nm adduct may be considered to be still involved in the red-ox cycling of the pyrimidines under *in vivo* conditions, while the 320-nm complex is virtually removed from this cycle and may act as an irreversible trap for divicine and isouramil.

The overall shuttle mechanism as proposed by Chevion et al.[19] and modified to be consistent with the results of Albano et al.,[8] Benatti et al.,[67] and the review by Hebbel[165] is presented in Figure 10. H_2O_2 that is not rapidly removed by the action of GSH reductase is converted in the presence of the superoxide radical (O_2^{-}) and the appropriate cofactors, such as the transition metals, into very reactive species including the hydroxyl radical (OH·) and singlet oxygen ($^1O_2$). These products can cause irreversible cell damage.

C. Role of Superoxide, Hydrogen Peroxide, and Transition Metal Ions and the Corresponding Enzymes on the Autoxidation of Divicine and Isouramil

Winterbourn et al.[27] demonstrated that autoxidation of divicine can occur by at least three different pathways. The normally predominate mechanism is O_2^{-}-dependent and appears to be analogous to the mechanism of adrenaline autoxidation[166] which has the following reaction sequence where RH_2, $RH^{\cdot}$, and R represent the hydroquinone, semiquinone, and quinone forms of divicine, respectively, (Figure 7).

$$RH_2 + O_2 \rightarrow RH^{\cdot} + O_2^{\cdot} + H^+ \quad \text{(slow)} \tag{1}$$

$$RH_2 + O_2^{\cdot} + H^+ \rightarrow RH^{\cdot} + H_2O_2 \quad \text{(fast)} \tag{2}$$

$$RH^{\cdot} + O_2 \rightarrow R + O_2^{\cdot} + H^+ \quad \text{(fast)} \tag{3}$$

After the initial step, Reactions 2 and 3 would constitute a chain and would be responsible for most of the reaction. In the normal biological system SOD would rapidly dismutate all $O_2^{\cdot}$, and the H_2O_2 that is formed would be rapidly converted in the presence of glutathione reductase to H_2O and O_2.

A second pathway, which is prevented by catalase or metal chelators, depends on transition metal ions and H_2O_2. It may be exogenous or produced from the $O_2^{\cdot}$ formed by Reaction 1. Normally this pathway is much slower than Reactions 2 and 3 and is of minor significance. However, it predominates if SOD is present or if the concentrations of transition metal ions, e.g., Cu^{2+} and Fe^{2+}, are micromolar or more. The reaction may involve direct reduction of the metal ions, which readily bind to O-hydroquinone groups of divicine. The rate of autooxidation of divicine or isouramil are accelerated by hemoglobin and H_2O_2 due to the formation of peroxide complexes such as ferrylhemoglobin. Although at an equivalent molar concentration hemoglobin is less effective than Fe^{2+} or Cu^{2+}, in the erythrocyte the high hemoglobin concentration should outweigh this difference.

In both of the above pathways are prevented by SOD, chelating agents, catalase, or glutathione peroxidase, divicine autooxidation would still proceed slowly, presumably via Reaction 1, up to a point where another mechanism takes over, which then results in rapid autooxidation. Details of this third pathway have not been established.

Winterbourn et al.[27] concluded that in the stressed erythrocyte, such as the G6PD-deficient erythrocyte, the impaired function of the GSH-peroxidase pathway due to a shortage of reducing equivalents would result in the accumulation of H_2O_2 due to autoxidation of divicine and isouramil. The H_2O_2 would react with mainly oxyhemoglobin ($HbFe^{2+}O_2$ or $HbFe^{3+}O_2^{\cdot}$), but also methemoglobin ($HbFe^{3+}$) to form the predominantly ferryl species [$(Hb^{IV} OH^-)^{3+}$, a hemoglobin-H_2O_2 complex]. Methemoglobin is also produced by divicine suggesting the following reactions: [$Hb^{IV} OH^-$]$^{3+}$ + RH_2 + $H^+ \rightarrow Hb^{3+}$ (methemoglobin) + $2H_2O$ + $RH^{\cdot}$. Ferrylhemoglobin and the even-more-reactive intermediate formed during the reaction of H_2O_2 with methemoglobin are also capable of oxidizing many electron donors. These oxidized hemoglobin derivatives are, therefore, likely to be major contributors to cell damage in the oxidant-stressed erythrocyte.

The particular oxidation-reduction pathway that occurs partially depends on the relative activities of SOD, GSH peroxidase, and catalase which as discussed above catalyze reactions that are relevant to the steady-state concentration of the potentially toxic oxygen derivatives such as $O_2^-\cdot$ and H_2O_2. These enzymes, in concert, keep at a minimal value the concentration of all active oxygen derivatives, including H_2O_2 and the hydroxy radical ($OH^{\cdot}$) which arises from the secondary interactions of both $O_2^{\cdot}$ and H_2O_2. In favic individuals the activity of G6PD is reduced, and therefore, the ability of GSH reductase and GSH peroxidase to operate at maximal value is greatly impaired,[16,19,27,167] and undesirable secondary reactions occur in the oxidant-stressed cell.

Mavelli et al.[168,169] reported that RBCs of patients with an acute hemolytic anemia, in addition to being deficient in G6PD, had markedly higher SOD activity and less GSH-peroxidase activity than either normal controls or G6PD-deficient patients that were not suffering from a hemolytic crisis. Catalase, which may be involved in a minimal way, was not affected by treatment. The above enzymes were also affected *in vitro* in a similar way when normal RBCs were treated with a high concentration of divicine and ascorbate. The rise of SOD and the drop of GSH peroxidase may assist in perpetuating a consistently increased H_2O_2 flux, thereby making the RBCs of favic subjects more susceptible to per-oxidative injuries associated with the O_2/H_2O_2 sequence of reactions. The effect of divicine on the activities of the two enzymes coupled with the deficiency of GSH in G6PD seem to act synergistically to precipitate a favic crisis and the associate peroxidative reactions.

Divicine and isouramil, instead of causing peroxidative reactions, can directly reduce methemoglobin to ferrous hemoglobin, which is the opposite of the above indicated mechanism. This occurs particularly in cells stabilized against autoxidation, i.e., red cells incubated in a nitrogen atmosphere.[27,70] This reaction may not be metabolically significant because it is largely obscured by the reaction of hemoglobin with H_2O_2.[27]

D. Heinz Body Formation, Protein Aggregation, and Lipid Peroxidation of RBCs

Oxidative stress in the red cell as induced by divicine and isouramil can cause changes in the cell membrane as well as hemoglobin. Methemoglobin which is formed from hemoglobin can be oxidized to reversible and irreversible hemichromes, where hemichromes represent oxidation states of altered iron-porphyrin complexes. Irreversible hemichromes are unstable and rapidly undergo denaturation and precipitation in the membrane as Heinz bodies.[16,22,125,165,170,171]

Protein aggregates have also been shown to form in cells from patients with chronic hemolysis,[172,173] in native thalassemic red cells or thalassemic red cells exposed to oxidative stress,[174] and in cells that have been subjected to oxidant stress[175,176] including divicine.[17,32] Thus, cross-linking of protein is probably due to exposure of sulfhydryls on the membrane proteins which results in the formation of intramolecular disulfide bonds, as the aggregates can be dissociated by mercaptoethanol or dithiolthreitol.[32,173] Cross-linked proteins when subjected to electrophoresis were distributed between band 1 plus 2 (spectrum), 3 (anion channel), and band 4.[177] Cross-linking of protein which occurs in GSH-depleted cells, greatly reduces cell deformity.[176] This may be attributed to an affect on spectrum which is one of the principal determinants of membrane flexibility and is responsible for the shear-elastic properties of the red cell membrane.[177] Palek and Liu[176] demonstrated that reagents able to produce disulfide bonds, such as diamide, produce a 50% decrease in red cell deformability (measured as decreased elongation during a shear stress). Similarly, Arese et al.[17] demonstrated that the polypeptide aggregates that formed 12 to 24 h after G6PD-deficient RBCs were treated with divicine, corresponded with a marked reduction in filterability of the cells.

Lipid peroxidation has been implicated in free radical reactions and membrane alterations associated with aging of cells and tissues.[171] Hochstein and Jain[178] suggested that the increases in intrinsic membrane rigidity are accelerated during aging in RBCs exposed to radical-generating hemolytic agents which may be attributed to the *in vivo* polymerization of membrane proteins consequent to radical-induced peroxidation of membrane lipids. Clear evidence on the role of lipid peroxidation during a favic crisis or as elicited by divicine or isouramil have not, however, been demonstrated.[17,21,23,179] Arese et al.[17] and DeFlora et al.[21] concluded that divicine and isouramil do not exert their hemolytic action by a peroxidating attack on membrane phospholipids since the formation of malondialdehyde was small and much the same in normal and deficient red cells treated with divicine or isouramil. Arese et al.,[17] using four different assays, were not able to repeat the results of Flohe et al.[7] who noticed a short-lived burst of malondialdehyde after isouramil administration. Chevion et al.[23] also were not able to demonstrate the occurrence of lipid peroxidation after isouramil treatment of RBCs utilizing the thiobarbituric acid assay. In addition, neither membrane fluidity nor membrane surface charge as demonstrated by the use of fluorescent probes were affected by isouramil. On the basis of these results they concluded that the membrane and cell effects induced by divicine were not a consequence of lipid peroxidation and alteration in membrane lipid interactions, but that its effects arise from alterations in the interaction and arrangement of membrane proteins. Such an effect might be keeping with the formation of damaging radicals in aqueous compartments in protein domains.

Although it seems unlikely that lipid peroxidation plays a role in the etiology of favism, its effects cannot be completely ruled out, as it has been shown by D'Aquino et al.[32] that divicine in both the vitamin E-riboflavin deficient and normal erythrocytes causes extensive lipid peroxidation with a 30% fall in arachidonic acid concentration. Arese et al.[17] concluded that lipid peroxidation was not associated with hemolytic anemia, but as described subsequently, hemolytic effects may not always be a reliable index of the efficacy of favism-inducing compounds such as divicine or isouramil. Muduuli et al.[34] demonstrated that dietary vicine greatly reduced the fertility of eggs from the laying hen and that the fat-soluble vitamin, vitamin E, provided protection against its effect. This would suggest that lipids may be affected by these compounds. Also, Corash et al.[38] reported that vitamin E supplementation may ameliorate chronic hemolysis in G6PD-deficient individuals. In addition, Marquardt and co-workers[40,58] also demonstrated that vitamin E was highly effective at protecting rats against the toxic effects of divicine. Further studies are required to clarify the role of vitamin E and lipid peroxidation in favism.

E. Rheological Effects and Half-Life of RBCs

Divicine and isouramil, as indicated in the previous section, cause rheological effects.[17,18,23,24,28,127,128,176,177,180,181] Chevion et al.[23] also demonstrated that the *in vitro* incubation of RBCs with isouramil caused a marked decrease in cellular deformability at low shear forces as measured by cone-plate viscometry. At the same time the membrane shear modulus of elasticity, as determined by the dimensions of pressure-induced pseudopods, was increased. These effects were observed in normal RBCs incubated with isouramil in the absence of glucose but not in the presence of glucose. Glucose presumably enabled the cell to regenerate GSH and thereby prevent the adverse effects of isouramil.

Recently it has been shown that RBCs isolated from G6PD-deficient patients suffering from favic crisis had an increased incidence of cross-bonded cells.[24] The degree of cross-bonding in the G6PD-deficient RBCs but not normal cells, also increased when they were treated with divicine, incubated for 7 h, and then subjected to shrinkage in hypertonic plasma.[22,180,181] Cross-bonding of cells, in contrast to effects with certain proteins, as discussed previously, was not reversed by treatment with membrane sulfhydryl-reducing reagents. Heinz bodies were also not necessary for membrane cross-bonding since they were present in only 5% of the RBCs and were evenly distributed. The strength of cross-bonding is such that RBCs swollen in 200-mOsm-buffered saline became almost spherical without pulling apart at the cross-bonded regions. RBCs isolated during the early phases of favic crisis contain as many as 60% cross-bonded RBCs, while these cells are almost absent in the late phases of the crisis, suggesting that they are cleared by sequestration in splenic and other tissues affected by the reticuloendothelial system.

Studies with ^{51}Cr have demonstrated that the half-life of divicine-treated rabbit erythrocytes was markedly reduced.[20] The kinetics of the removal of labeled cells that had been pretreated *in vitro* with isouramil from circulation was biphasic with two first-order values, each with different life spans and relative distribution. Thus, in all treated cells, two populations were identified. The first corresponding to normal undamaged erythrocytes ($t^1/_2$ was approximately 8 to 10 d) and the second corresponding to damaged cells with a half-life of approximately 0.5 d. The relative weight of the fraction of cells characterized by the short life span was dependent on the concentration of isouramil in the incubation system. These effects are in agreement with the change in weight of splenic tissues in rats following injections of divicine.[26]

F. Hemolysis: Paradoxical Effects

The favism-inducing factors, (divicine and isouramil), have been reported to produce contradictory and different effects with regard to erythrocyte stability. These compounds have been reported to have no effect, to increase hemolysis, or, in some cases, to decrease hemolysis of RBCs.

Arese et al.[18] reported a slight increase in mechanical fragility of RBCs when treated with divicine and incubated in their own plasma for 24 h. Chevion et al.[23] also reported that the osmotic fragility of human RBCs that were depleted of glucose was markedly increased when treated with isouramil, but that glucose protected against this effect. Studies with the laying hen demonstrated that the *in vitro* degree of spontaneous hemolysis was variably increased in laying hens that had been fed diets containing vicine and convicine, compared to those fed a normal diet.[32,34] The yolk membrane from the egg was also affected in an analogous manner, suggesting that factors affecting the stability of avian erythrocytes also affect other membranes in a similar manner.

In contrast, several researchers have reported that divicine, isouramil, primaquine, diamine, and other related drugs do not induce overt lysis *in vitro*.[16,18,21,31,32,168] DeFlora et al.[21] demonstrated that nearly complete hemolysis of washed erythrocytes was obtained when incubated in a system containing hydrogen peroxide and sodium azide, but that there were no differences between normal and G6PD-deficient erythrocytes. They concluded that an oxidative mechanism by itself may not be the only determinant of the acute hemolysis that occurs in G6PD-deficient subjects. Mavelli et al.[168] reported that treatment of normal RBCs with high concentrations of divicine and ascorbate produced no hemolysis, but led to hemolysis if the treated red cells were resuspended in homologous plasma. They concluded that a variety of soluble and membrane-bound systems, known to produce active oxygen species, are normally present in plasma and under certain conditions may cause an increase in degree of hemolysis. Factors that are responsible for these effects, however, are not known. Vitamin E status also seems to affect the degree of spontaneous hemolysis of RBCs. D'Aquino et al.[32] demonstrated that isouramil or diamine did not affect the degree of spontaneous hemolysis in the erythrocytes obtained from rats fed a control diet, but that a similar treatment of erythrocytes obtained from riboflavin and vitamin E depleted rats resulted in 100% hemolysis of RBCs. These results indicate that a deficiency of one or both of these vitamins may act synergistically with oxidants such as divicine or isouramil to induce spontaneous hemolysis of the normal erythrocyte, or they may interact synergistically to prevent divicine- or isouramil-induced spontaneous hemolysis of normal erythrocytes. Studies by Marquardt and co-workers,[58,127] as indicated earlier and in contrast to the above reports, have shown that divicine treatment of washed rat erythrocytes not only markedly alters their shape, but also greatly stabilizes them in a hypotonic solution.

G. Effects on Calcium and Potassium Homeostasis

Calcium is an important regulating agent in cell metabolism.[182-184] It is required for the activation of the enzyme protein kinase C which plays a key role in the induction of cellular responses to a variety of ligand-receptor systems and in the regulation of cellular responses to external stimuli. Perturbation of this control system by oxidizing substance presumably could alter the cytoplasmic concentration of calcium and thereby result in an altered type of metabolism.[182-184]

Recently, it has been shown calcium homeostasis in the erythrocyte is severely impaired *in vitro* following divicine treatment[167,185,186] and *in vivo* during a hemolytic crisis.[185] This effect may be attributed in part to the functional failure of the calcium pump (Ca^{2+}-ATPase) and, perhaps to a greater extent, to a large influx of calcium into the erythrocyte through channels that are opened by divicine and isouramil.[167] The increase in calcium content in divicine-treated erythrocytes is associated with a corresponding decrease in the concentration of potassium.[167,185] Shalev et al.[187] have also reported that the exclusion of calcium by the erythrocyte membrane is disrupted by the potent oxidant phenylhydrazine. In addition, Graf et al.[36] have demonstrated that divicine induces calcium release from intact rat-liver mitochondria and proposed that the depleted mitochondrial and probably microsomal Ca^{2+} stores cause an increase in the cytosolic free-Ca^{2+} content. Other compounds including oxaloacetate,[188] hydroperoxides,[189] menadione,[190] and alloxan[191] also induce Ca^{2+} release from rat liver mitochondria. The common denominator for the mode of action of these diverse compounds is the oxidation and hydrolysis of pyrimidine nucleotides.[36] It has been proposed that the mode of action of divicine is threefold:

1.   Divicine leads to the formation of H_2O_2 by autoxidation and redox recycling in the mitochondria, and, as a result, H_2O_2 oxidizes pyrimidine nucleotides.
2.   Divicine can oxidize GSH to GSSG in a cyclic manner resulting in a shift of glutathione redox state.
3.   Divicine may possibly induce intramitochondrial pyrimidine nucleotide oxidation nonenzymatically.[167]

It has been reported that disturbance of cellular Ca^{2+} homeostasis leads to cell damage[192] and eventually to cell death.[193] DeFlora et al.[185] however, concluded that it is unlikely that the remarkable abnormalities of intercellular electrolytes that are observed in favism are the primary cause of hemolysis. They proposed that in spite of the potentially harmful biochemical events these electrolyte imbalances may trigger, additional mechanisms and factors seem to be involved in the process of oxidative hemolysis.

H. Summary

Divicine and isouramil participate in an oxidation-reduction shuttle system in which the net effect is the transfer of electrons from reducing compounds such as GSH, NADP(H), and ascorbic acid through the divicine/isouramil cofactor to O_2 with the resulting formation of H_2O_2. These effects are usually neutralized by the free-radical scavenging system, but can lead to irreversible cell damage if the cell is unable to cope with the rapid depletion of reducing equivalents when challenged with an oxidant. This occurs upon exhaustion of GSH, particularly in G6PH-deficient erythrocytes. Under these conditions there is formation of methemoglobin and, ultimately, Heinz bodies, formation of cross-bonded proteins, and perhaps an increased degree of lipid peroxidation of the erythrocyte membrane. Oxidant stress results in a loss of the structure and function of properties of erythrocytes as evidenced by altered shape, deformability and filterability, susceptibility to spontaneous hemolysis, shortened *in vivo* life span, increased degree of phagocytosis of damaged erythrocytes, transient hemoglobinuria and jaundice, altered flux of calcium and potassium, and finally, loss of the ability of the RBC to exchange respiratory gases. The net effect is that the animal of favic individual will die of asphyxiation if a sufficiently high portion of the RBCs are affected.

X. EFFECT OF ANTIOXIDANTS AND CHELATING AGENTS ON THE TOXICITY OF FAVISM-CAUSING FACTORS

The oxidation-reduction shuttle that is autocatalyzed by divicine and isouramil is counteracted *in vivo* by enzymes and nonenzymatic scavengers and quenchers denoted by the term antioxidants and including compounds such as α-tocopherol (vitamin E), ascorbate (vitamin C), flavonoids, β-carotene, urate, and plasma proteins.[194]

Several experiments have been carried out to determine the influence of nutritional status on the the noxious effects of vicine and convicine. Riboflavin status and, more specifically,

GSH concentrations have been associated with degree of hemolytic response to divicine and isouramil. GSH concentrations depend upon the riboflavin-containing enzyme, GSH reductase, for its regeneration, and therefore the two factors are interrelated. D'Aquino et al.[32] demonstrated that the red cells from riboflavin-deprived rats had a GSH reductase activity which was less than 10% of control rats, but that the concentration of GSH in the red cells was no different than that of the controls. A challenge of the deprived cells with isouramil, however, resulted in a dramatic decrease in the concentration of GSH in RBCs while a similar challenge in control cells resulted in only a moderate decrease in GSH concentrations. Yannai and Marquardt[26] also demonstrated that GSH depletion sensitized rats to divicine. Hematocrit concentration 24 h after divicine administration was depressed by 20% in control rats and 43% in BSO-treated rats. Riboflavin in contrast to GSH deficiency, however, may not be an important secondary factor in the etiology of the diseases. Beutler[158] reported that there was no person among the 350 patients with hemolytic disorders that had a deficiency of GSH reductase. He concluded that riboflavin deficiency which is a common cause of reduced GSH-reductase activity was not usually associated with hemolytic anemia.

Other nutritional factors seem to affect the toxicity of the favism-inducing compounds. Studies with vitamin E supplementation in individuals suffering from a congenital hemolytic anemia, β-thalassemia major, was effective at reducing the abnormally high malondialdehyde levels and increased erythrocyte survival in some of the patients.[37] Corash and co-workers[38,39] reported that vitamin E supplementation in G6PD-deficient individuals decreased chronic hemolysis as evidenced by improved red-cell life span, increased red cell hemoglobin concentration, and decreased reticulocytosis. Other researchers have demonstrated that vitamin E deficiency that occurs in certain pathological conditions results in increased concentrations of malondialdehyde *in vitro*, and increased peroxidative damage *in vivo*, which were responsible for shortened red cell survival times. These hematological abnormalities are alleviated by vitamin E supplementation.[125] Studies with the laying hen[33,34] demonstrated that the vicine-induced depression in fertility and hatchability of eggs was alleviated to a considerable degree with moderate levels of vitamin E supplementation.

The iron chelating agent, desferoxamine, has also been shown to block the antimalarial and the hemolytic activity of divicine.[52] In addition, transition metal chelating agents have been shown to provide complete protection, *in vitro*, against enzyme inactivation caused by favism-inducing agents and the transition metals,[68] presumably by blocking a transition metal ion-driven Fenton-type reaction.[52,125] Meloni et al.,[195] in contrast, reported that desferrioxamine was unsuccessful in the treatment of favism in three humans. They concluded that treatment with desferoxamine should be avoided and that at present blood transfusion must be considered as being the only appropriate treatment. This compound and other free-radical scavenging compounds, however, may only be effective if administered prior to, but not after ingestion of the toxic compounds.

Of considerable interest are the recent studies by Marquardt and Arbid[196] which demonstrated that the toxic effects of the favism-causing factors are greatly modified by certain free-radical scavenging compounds when administered prior to treatment with the favism-causing factors. For example, rats injected intraperitoneally with 25 mg of divicine per 100 g body weight had a GSH concentration of 2 mg per 100 ml blood and a hemoglobin 576 to 628 nm absorbance ration of 6, and all of the rats died within 24 h. Injection of the rats with 100 mg DL-α-tocopherol acetate per 100 g, 1 h prior to divicine injection, result in complete protection against the toxic effects of divicine. The corresponding glutathione concentration, hemoglobin absorbance ratio, and percent mortalities were 23 mg per 100 g, 83, and 0, which were the same as the untreated rats or those injected with only the vitamin. A dose of vitamin E as low as 25 mg provides considerable protection against vitamin E toxicity as only 30% of the rats died, and the GSH concentration and hemoglobin absorbance ratio were only slightly affected. Incorporation of vitamin E into the diet also provides complete protection against the toxic effects of divicine.

Injection of rats with vitamin A (retinol), vitamin C, EDTA, and desferoxamine mesylate (Desferal®) were also protective. These results clearly demonstrate that free-radical scavengers can protect rats against the toxic effects of divicine. These agents appear to prevent divicine-induced depletion of GSH, which seems to be a prerequisite for the formation of an abnormal and nonfunctional RBC.

It may be concluded that certain compounds may be capable of greatly modifying individual susceptibility to favism. These may be one of the important additional factor(s) proposed by Mager et al.,[16] along with G6PD deficiency and exposure to faba beans, that influence the degree and severity of the hemolytic attack. A better understanding of the interactions among these factors may assist in clarifying the bizarre and rather unpredictable mode of occurrence of the disease in susceptible individuals.

XI. MALARIA, FAVISM, AND GLUCOSE-6-PHOSPHATE DEHYDROGENASE ACTIVITY: EFFECT OF DIVICINE AND ISOURAMIL ON MALARIA-INFECTED ANIMALS

A possible link between G6PD deficiency and malaria was first proposed on the basis of epidemiological evidence.[157,197] The universality of this association has been questioned with evidence for[49,198,199] and against[200,201] malaria being sufficient pressure to have selected this otherwise undesirable trait. Although previous workers had assumed that favism would select against G6PD deficiency in malarial areas, Huheey and Martin[202] recalled the observation of Livingstone[203] that the frequency of the genes for G6PD deficiency were highest in the Mediterranean and Middle East, the area where negative selection by favism should be strongest. They hypothesized that faba beans are actively involved in the selection for G6PD deficiency and that rather than being deleterious, "favism" is responsible for the high frequency of G6PD deficiency. They suggested that the stress imposed on the erythrocyte by the favism factors (divicine and isouramil) may act synergistically with an unfavorable G6PD level to eliminate the parasitized erythrocytes. The carriers of the abnormal variant of G6PD would therefore be at a selective advantage with respect to malaria.

Recently, Clark and co-workers[51,52] produced evidence that alloxan was antimalarial and that the mechanism involved metal-independent scission of H_2O_2, an activity expected to be more effective in G6PD-deficient erythrocytes. The recent studies of Golenser et al.[50] and those of Clark et al.[41,53] support the hypotheses of Huheey and Martin.[202] They demonstrated *in vitro* that both G6PD-deficient and normal erythrocytes supported similar growth of the parasite (*Plasmodium falciparum*) that causes malaria in humans. In contrast, G6PD-deficient erythrocytes, after treatment with the favism-inducing agent isouramil, were unable to support the parasites growth, whereas a similar treatment of normal erythrocytes had no effect on the parasites. Clark et al.[41] reported that intravenous injections of divicine into mice infected with *P. vinckei* rapidly killed the parasites and caused hemolysis. Degenerated parasites were frequently observed inside intact, circulating erythrocytes, implying that the parasites were less-well protected from free radical-mediated oxidative damage than were the membranes of their host erythrocytes. Both parasite death and hemolysis following divicine administrations were blocked by the iron chelator, desferrioxamine. This is consistent with effects obtained with alloxan in which desferoxamine prevented the associated transition metal-driven Fenton reactions by removing traces of catalytic nonprotein-bound iron.[204] Clark et al.[41,53] suggested that hydroxy radicals would initiate peroxidative reactions which would not only cause hemolysis but also the killing of the parasite.

Several other activated oxygen generators, such as primaquine,[205] alloxan,[51,206] and *t*-butyl peroxide[53,207] also have antimalarial activity *in vivo*. In addition, vitamin E deficiency in animals reduces the survivability of plasmodia, presumably because more activated oxygen is present in these cells.[208] The concentration of other free radical scavenging compounds would probably have a similar effect to that of vitamin E. These results demonstrate that any factor that will increase the concentration of activated oxygen or free radical production in the host cell will have a negative effect on plasmodial survivability. Malarial parasites may be particularly susceptible to activated oxygen as they are not able to synthesize SOD.[209]

Faba bean consumption under control conditions may therefore provide sufficient protection against malaria for the host to acquire immunity. Selection of faba bean cultivars that are capable of producing high concentrations of vicine or convicine,[76,106,108] harvesting at a very immature state,[76,105,106] mechanical processing,[42] or the consumption of pure vicine or convicine would provide a means of supplying malarial-infected individuals with a relatively inexpensive source of vicine and convicine. Research is required to establish the effectiveness of such a procedure under different nutritional regimens and means to rapidly counteract any possible deleterious side effects.

REFERENCES

1. **Hawtin, G. and Stewart, R.,** The development, production and problems of faba beans (*Vicia faba*) in West Asia and North Africa, *Fabis*, 1, 7, 1979.
2. **Saxena, M. C. and Hawtin, G. C.,** Faba beans in China, *Fabis*, 9, 14, 1984.
3. **Marquardt, R. R.,** Antimetabolites in faba beans: their metabolic significance, *Fabis*, 7, 1, 1983.
4. **Sarwar, G., Sosulski, F. W., and Bell, J. M.,** Nutritive value of field pea and fababean proteins in rat diets, *J. Inst. Can. Sci. Technol.*, 8, 109, 1975.
5. **Frohlich, A. A. and Marquardt, R. R.,** Turnover and hydrolysis of vicine and convicine in avian tissues and digesta, *J. Sci. Food Agric.*, 34, 153, 1983.
6. **Hegazy, M. I. and Marquardt, R. R.,** Metabolism of vicine and convicine in rat tissue: absorption and excretion patterns and sites of hydrolysis, *J. Sci. Food Agric.*, 35, 139, 1984.
7. **Flohe, L., Niebch, G., and Reiber, H.,** Zur Wirkung von Divicin in menschlichen Erytrhocyten, *Z. Klin. Chem. Klin. Biochem.*, 9, 431, 1971.
8. **Albano, E., Tomasi, A., Mannuzzu, L., and Arese, P.,** Detection of a free radical intermediate from divicine of *Vicia faba*, *Biochem. Pharmacol.*, 33, 1701, 1984.
9. **Lin, J.-Y. and Ling, K.-H.,** Studies on favism. I. Isolation of an active principle from faba beans *(Vicia faba)*, *J. Formosan Med. Assoc.*, 61, 484, 1962.
10. **Lin, J.-Y. and Ling, K.-H.,** Studies on favism. II. Studies on the physiological activities of vicine *in vivo*, *J. Formosan Med. Assoc.*, 61, 490, 1962.
11. **Lin, J.-Y. and Ling, K.-H.,** Studies on favism. III. The physiological activities of vicine *in vitro*, *J. Formosan Med. Assoc.*, 61, 579, 1962.
12. **Lin, J.-Y.,** Studies on favism. IV. Reactions of vicine and divicine with sulfhydryl group of glutathione and cysteine, *J. Formosan Med. Assoc.*, 62, 777, 1963.
13. **Mager, J., Glaser, G., Razin, A., Izak, G., Bien, S., and Noam, M.,** Metabolic effects of pyrimidines derived from faba bean glycosides on human erythrocytes deficient in glucose-6-phosphate dehydrogenase, *Biochem. Biophys. Res. Commun.*, 20, 235, 1965.
14. **Razin, A., Hershko, A., Glaser, G., and Mager, J.,** The oxidant effect of isouramil on red cell glutathione and its synergistic enhancement by ascorbic acid a 3,4-dihydroxyphenylalanine. Possible relation to the pathogenesis of favism, *Isr. J. Med. Sci.*, 4, 852, 1968.
15. **Lin, J.-Y. Lee, S.-W., Lin, T.-I., and Ling, K.-H.,** Divicine and favism, *J. Chin. Biochem. Soc.*, 6, 92, 1977.
16. **Mager, J., Chevion, M., and Glaser, G.,** Favism, in *Toxic Constituents of Plant Foodstuffs*, 2nd ed., Liener, I. E., Ed., Academic Press, New York, 1980, 265.
17. **Arese, P., Bosia, A., Naitana, A., Gaetani, S., D'Aquino, M., and Gaetani, G. F.,** Effect of divicine and isouramil on red cell metabolism in normal and G6PD-deficient (Mediterranean variant) subjects. Possible role in the genesis of favism, in *The Red Cell: Fifth Annu. Arbor Conf.*, Brewer, G., Ed., Alan R. Liss, New York, 1981, 725.
18. **Arese, P., Naitana, L., Mannuzzu, L., Turrini, F., Haest, C. W. M., Fisher, T. M., and Deuticke, B.,** Biochemical and micro-rheological modifications in normal and glucose-6-phosphate dehydrogenase-deficient red cells treated with divicine, in *Advances in Red Cell Biology*, Weatherall, D. J., Fiorelli, G., and Gorini, S., Eds., Raven Press, New York, 1982, 375.

19. **Chevion, M., Navok, T., Glaser, G., and Mager, J.,** The chemistry of favism-inducing compounds: the properties of isouramil and divicine and their reaction with glutathione *Eur. J. Biochem.*, 127, 405, 1982.

20. **Chevion, M., Navok, T., and Glaser, G.,** Favism inducing agents: biochemical and mechanistic considerations, in *Advances in Red Cell Biology*, Weatherall, D. J., Fiorelli, G., and Gorini, S., Eds., Raven Press, New York, 1982, 381.

21. **DeFlora, A., Morelli, A., Benatti, U., Lenzerini, L., Meloni, T., Melloni, E., Salamino, F., Sparatore, B., Michetti, M., and Pontremoli, S.,** Metabolic aspects of erythrocyte glucose-6-phosphate dehydrogenase deficiency in, *Advances in Red Cell Biology*, Weatherall, D. J., Fiorelli, G., and Gorini, S., Eds., Raven Press, New York, 1982, 347.

22. **Baker, M. A., Bosia, A., Pescarmona, G., Turrini, F., and Arese, P.,** Mechanism of action of divicine in a cell-free system and in glucose-6-phosphate dehydrogenase-deficient red cells, *Toxicol. Pathol.*, 12, 331, 1984.

23. **Chevion, M., Navok, T., Pfafferott, C., Meiselman, H. J., and Hochstein, P.,** The effects of isouramil on erythrocyte mechanics: implications for favism, *Microcirc. Endothel. Lymphat.*, 1, 295, 1984.

24. **Fischer, T. M., Meloni, T., Pescarmona, G. P., and Arese, P.,** Membrane cross bonding in red cells in favic crises: a missing link in the mechanism of extra vascular haemolysis, *Br. J. Haematol.*, 59, 159, 1985.

25. **Winterbourn, C. C.,** Free radical production and oxidative reactions of hemoglobin, *Environ. Health Perspect.*, 64, 321, 1985.

26. **Yannai, S. and Marquardt, R. R.,** Induction of favism-like symptoms in the rat: effects of vicine and divicine in normal and buthionine sulfoximine-treated rats, *J. Sci. Food Agric.*, 36, 1161, 1985.

27. **Winterbourn, C., Benatti, U., and DeFlora, A.,** Contributions of superoxide, hydrogen peroxide and transition metal ions to auto-oxidation of the favism-inducing pyrimidine aglycone, divicine and its reactions with haemoglobin, *Biochem. Pharmacol.*, 35, 2009, 1986.

28. **Arbid, M. S. S. and Marquardt, R. R.,** Effect of intraperitoneally injected vicine and convicine on the rats: induction of favism-like signs, *J. Sci. Food Agric.*, 37, 539, 1986.

29. **Motulsky, A.,** Hemolysis in glucose-6-phosphate dehydrogenase deficiency, *Fed. Proc. Fed. Am. Soc. Exp. Biol.*, 31, 1286, 1972.

30. **Yoshida, A.,** Hemolytic anemia and G6PD deficiency, *Science*, 179, 532, 1973.

31. **Beutler, E.,** *Hemolytic Anemia in Disorders of Red Cell Metabolism*, Plenum Press, New York, 1978, chap. 2.

32. **D'Aquino, M., Gaetani, S., and Spadoni, M. A.,** Effect of factors of favism on the protein and lipid components of rat erythrocyte membrane, *Biochim. Biophys. Acta*, 731, 161, 1983.

33. **Muduuli, D. S., Marquardt, R. R., and Guenter, W.,** Effect of dietary vicine on the productive performance of laying chickens, *Can. J. Anim. Sci.*, 61, 757, 1981.

34. **Muduuli, D. S., Marquardt, R. R., and Guenter, W.,** Effect of dietary vicine and vitamin E supplementation on productive performance of growing and laying chickens. *Br. J. Nutr.*, 47, 53, 1982.

35. **Rocic, B., Rocic, S., Ashcroft, J. H., Harrison, D. E., and Poje, M.,** Diabetogenic action of alloxan-like compounds: the cytotoxic effect of dehydro-isouramil hydrochloride on the rat pancreatic β-cells, *Diabetes Croat.*, 14, 143, 1985.

36. **Graf, M., Frei, B., Winterhalter, K. H., and Richter, C.,** Divicine induces calcium release from rat liver mitochondria, *Biochem. Biophys. Res. Commun.*, 129, 18, 1985.

37. **Rachmilewitz, E. A., Shifter, A., and Kahane, I.,** Vitamin E deficiency in β-thalassemia major: changes in hematological and biochemical parameters after a therapeutic trial with α-tocopherol, *Am. J. Clin. Nutr.*, 32, 1850, 1979.

38. **Corash, L., Spielberg, S., Bartsocas, C., Boxer, L., Steinherz, R., Sheetz, M., Egan, M., Schlessleman, J., and Schulman, J. D.,** Reduced chronic hemolysis during high-dose vitamin E administration in Mediterranean-type glucose-6-phosphate dehydrogenase deficiency, *N. Engl. J. Med.*, 303, 416, 1980.

39. **Corash, L. M., Sheetz, M., Bieri, J. G., Bartsocas, C., Moses. S., Bashan, N., and Schulman, J. P.,** Chronic hemolytic anemia due to glucose-6-phosphate dehydrogenase deficiency or glutathione synthetase deficiency: the role of vitamin E in its treatment, *Ann. N.Y. Acad. Sci.*, 393, 348, 1982.

40. **Arbid, M. S. and Marquardt, R. R.,** Effect of antimetabolites (divicine) isolated from faba beans *(Vicia faba)* on rat metabolism: protective effects of vitamin E, *Fed. Proc. Fed. Am. Soc. Exp. Biol.*, 44 (Abstr. No. 3079), 937, 1985.

41. **Clark, I. A., Cowden, W. B., Hunt, N. H., Maxwell, L. E., and Mackie, E. J.,** Activity of divicine in *Plasmodium vinckei*-infected mice has implications for treatment of favism and epidemiology of G-6-PD deficiency, *Br. J. Haematol.*, 57, 479, 1984.

42. **Elkowicz, K. and Sosulski, F. W.,** Antinutritive factors in eleven legumes and their air-classified protein and starch fractions, *J. Food Sci.*, 47, 1301, 1982.

43. **Hegazy, M. I. and Marquardt, R. R.,** Development of a simple procedure for complete extraction of vicine and convicine from fababeans (*Vicia faba* L.), *J. Sci. Food Agric.*, 34, 100, 1983.

44. **Arntfield, S. D., Ismond, M. A. H., and Murray, D. E.**, The fate of antinutritional factors during the preparation of a faba bean protein isolate using a micellization technique, *Can. Inst. Food Sci. Technol. J.*, 18, 137, 1985.

45. **Arbid, M. S. S. and Marquardt, R. R.**, Hydrolysis of the toxic constituents (vicine and convicine) in faba bean (*Vicia faba* L.) food preparations following treatment with β-glucosidase, *J. Sci. Food Agric.*, 36, 839, 1985.

46. **Hussein, L., Motawei, H., Nassib, A., Khalil, S., and Marquardt, R. R.**, The complete elimination of vicine and convicine from the faba bean by combinations of genetic selection and processing techniques, *Qual. Plant Plant Foods Hum. Nutr.*, 36, 231, 1986.

47. **Chaumeil, J. C.**, Extraction of vicine from vegetable material and its use in the treatment of cardiac arrythmia, *Chem. Abstr.*, 86, 1111 68f, 1977.

48. **Siniscalo, M., Bernini, L., Latte, B., and Motulsky, A. G.**, Favism and thalassemenia in Sardinia and their relationship to malaria, *Nature (London)*, 190, 1179, 1961.

49. **Roth, E. F., Raventos-Suarez, C., Rinaldi, A., and Nagel, R. L.**, Glucose-6-phosphate dehydrogenase deficiency inhibits *in vitro* growth of *Plasmodium falciparum*, *Proc. Natl. Acad. Sci.*, U.S.A., 80, 298, 1983.

50. **Golenser, J., Miller, J., Spira, D. T., Navok, T., and Chevion, M.**, The effect of a favism inducing agent on the *in vitro* development of *Plasmodium falciparum* in normal and glucose-6-phosphate dehydrogenase deficient erythrocytes, *Blood*, 61, 507, 1983.

51. **Clark, I. A. and Hunt, N. H.**, Evidence for reactive oxygen intermediates causing hemolysis and parasite death in malaria, *Infect. Immun.*, 39, 1, 1983.

52. **Clark, I. A. and Cowden, W. B.**, Antimalarials, in *Oxidative Stress*, Sies, H., Ed., Academic Press, New York, 1985, chap. 7.

53. **Clark, I. A., Mackie, E. J., and Cowden, W. B.**, Injection of free radical generators causes premature onset of tissue damage in malaria-infected mice, *J. Pathol.*, 148, 301, 1986.

54. **Bjerg, B., Heide, M., Knudsen, J. C. N., and Sorensen, H.**, Inhibitory effects of convicine, vicine and dopa from *Vicia faba* on the in vitro growth rates of fungal pathogens, *Z. Pflanzenkr. Pflanzenschutz*, 91, 483, 1984.

55. **Bendich, A. and Clements, G. C.**, A revision of the structural formulation of vicine and its pyrimidine aglycone divicine, *Biochim. Biophys. Acta*, 12, 462, 1953.

56. **Windholz, E. M., Ed.**, *The Merck Index*, 10th ed., Merck and Co., Rahway, NJ, 1983.

57. **Marquardt, R. R., Muduuli, D. S., and Frohlich, A. A.**, Purification and some properties of vicine and convicine isolated from faba beans (*Vicia faba* L.) protein concentrate, *J. Agric. Food Sci.*, 31, 839, 1983.

58. **Marquardt, R. R.**, Unpublished data, 1986.

59. **Olsen, H. S. and Andersen, J. H.**, The estimation of vicine and convicine in fababeans (*Vicia faba* L.) and isolated protein, *J. Sci. Food Agric.*, 29, 323, 1978.

60. **Bien, S., Salemnik, G., Zamir, L., and Rosenblum, M.**, The structure of convicine, *J. Chem. Soc. C*, 496, 1968.

61. **Chesterfield, J. H., Hurst, D. T., McOmie, J. F. W., and Tute, M. S.**, Pyrimidines. XIII. Electrophilic substitution at position 6 and a synthesis of divicine (2,4-diamino-5, 6-dihydroxpyrimidine), *J. Chem. Soc.*, p. 1001, 1964.

62. **Bailey, S. W., Weintraub, S. T., Hamilton, S. M., and Ayling, J. E.**, Incorporation of molecular oxygen into pyrimidine cofactors by phenylalanine hydroxylase, *J. Biol. Chem.*, 257, 8253, 1982.

63. **Davoll, J. and Laney, D. H.**, Synthesis of divicine (2:4 diamino-5:6-dihydroxypyrimidine) and other derivatives of 4:5(5:6)-dihydroxypyrimidine, *J. Chem. Soc.*, p. 2124, 1956.

64. **Bailey, S. W. and Ayling, J. E.**, Cleavage of the 5-amino substituent of pyrimidine cofactors by phenylalanine hydroxylase, *J. Biol. Chem.*, 255, 7774, 1980.

65. **Davidson, D. and Bogert, M. T.**, Isovioluric acid (alloxan-6-oxime), *Proc. Natl. Acad. Sci. U.S.A.*, 18, 490, 1932.

66. **Bien, S., Zamir, L., and Rosenblum, M.**, The structure of convicine, *Isr. J. Chem.*, 4, 30, 1966.

67. **Benatti, U., Guida, L., and DeFlora, A.**, The interaction of divicine with glutathione and pyridine nucleotides, *Biochem. Biophys. Res. Commun.*, 120, 747, 1984.

68. **Navok, T. and Chevion, M.**, Transition metals mediate enzymatic inactivation caused by favism-inducing agents, *Biochem. Biophys. Res. Commun.*, 122, 297, 1984.

69. **Malaisse, W. J.**, Alloxan toxicity to the pancreatic β-cell; a new hypothesis, *Biochem. Pharmacol.*, 31, 3527, 1982.

70. **Benatti, U., Lucrezia, G., Grasso, M., Tonetti, M., DeFlora, A., and Winterbourn, C. C.**, Hexose monophosphate shunt-stimulated reduction of methemoglobin by divicine, *Arch. Biochem. Biophys.*, 242, 549, 1985.

71. **Grankvist, K., Marklund, S., Sehlin, J., and Taljedal, I.**, Superoxide dismutase, catalase and scavengers of hydroxyl radicals protect against the toxic action of alloxan on pancreatic islet cells *in vitro*, *Biochem. J.*, 182, 17, 1979.

72. **Schank, K.,** Reductones, *Synthesis,* p. 176, 1972.
73. **Higazi, M. I. and Read, W. W. C.,** A method for determination of vicine in plant material and in blood, *Agric. Food Chem.,* 22, 570, 1974.
74. **Chevion, M. and Navok, T.,** A novel method for quantitation of favism-inducing agents in legumes, *Anal. Biochem.,* 128, 152, 1983.
75. **Collier, H. B.,** The estimation of vicine in fababeans by an ultraviolet spectrophotometric method, *J. Inst. Can. Sci. Technol.,* 9. 155, 1976.
76. **Pitz, W. J., Sosulski, F. W., and Rowland, G. G.,** Effect of genotype and environment of vicine and convicine levels in fababeans (*Vicia faba* minor), *J. Sci. Food Agric.,* 32, 1, 1981.
77. **Kim, S.-I, Hoehn, E., Eskin, N. H. M., and Ismail, F.,** A simple and rapid colorimetric method for determination of vicine and convicine, *J. Agric. Food Chem.,* 30, 144, 1982.
78. **Pitz, W. J. and Sosulski, F. W.,** Determination of vicine and convicine in fababean cultivars by gas-liquid chromatography, *Can. Inst. Food Sci. Technol. J.,* 12, 93, 1979.
79. **Marquardt, R. R. and Frohlich, A. A.,** Rapid reversed-phase high-performance liquid chromatographic method for the quantification of vicine, convicine and related compounds, *J. Chromatogr.,* 208, 373, 1981.
80. **Lattanzio, V., Bianco, V., and Lafiandra, D.,** High-performance reversed-phase liquid chromatography (HPLC) of favism-inducing factors in *Vicia faba* L., *Experientia,* 38, 789, 1982.
81. **Quemener, B., Gueguen, J., and Mercier, C.,** Determination of vicine and convicine in fababeans by high pressure liquid chromatography, *Can. Inst. Food Sci. Technol. J.,* 15, 109, 1982.
82. **Ritthausen, H. and Kreusler, U.,** Ueber Vorkommen von Amygdalin und eine neue dem Asparagin achnliche Substanz in Wickensamen (Vicia sativa), *J. Prakt. Chem.,* 2, 333, 1870.
83. **Ritthausen, H.,** Ueber Vicin, Bestandteil der Samen von *Vicia sativa,,* *Ber. Dtsch. Chem. Ges.,* 9, 301, 1876.
84. **Olaboro, G., Marquardt, R. R., and Campbell, L. D.,** Isolation of the egg weight depressing factor in faba beans (*Vicia faba* L. var. minor), *J. Sci. Food Agric.,* 32, 1074, 1981.
85. **Olaboro, G., Marquardt, R. R., Campbell, L. D., and Frohlich, A. A.,** Purification, identification and quantification of an egg-weight-depressing factor (vicine) in faba beans (*Vicia faba* L.), *J. Sci. Food Agric.,* 32, 1163, 1981.
86. **Ritthausen, H.,** Ueber vicin und eine zweite stickstoffreiche Substarz der Wickensamen, Convicin, *J. Prakt Chem.,* 24, 202, 1881.
87. **Bien, S., Amith, D., and Ber, M.,** Synthesis of convicine (6 amino-5-β-D glucopyranosyloxyuracil), *J. Chem. Soc. Perkin Trans. 1,* 1089, 1973.
88. **Arbid, M. S. S. and Marquardt, R. R.,** A modified procedure for the purification of vicine and convicine from faba beans (*Vicia faba* L.), *J. Sci. Food Agric.,* 36, 1266, 1985.
89. **Ritthausen, H.,** Vicin ein Glycosid, *Ber. Dtsch. Chem. Ges.,* 29, 2108, 1896.
90. **Ritthausen, H.,** Uber die Zusammensetzung des Vicins, *J. Prakt. Chem.,* 59, 480, 1899.
91. **Ritthausen, H.,** Ueber Divicin, *J. Prakt. Chem.,* 59, 482, 1899.
92. **Levene, P. A.,** On vicine, *J. Biol. Chem.,* 18, 305, 1914.
93. **Levene, P. A. and Senior, J. K.,** Vicine and divicine, *J. Biol. Chem.,* 25, 607, 1916.
94. **Herissey, M. H. and Cheymol, J.,** Sur le Vicioside, *Bull. Soc. Chim. Biol.,* 13, 29, 1931.
95. **McOmie, J. F. W. and Chesterfield, J. H.,** Synthesis of 5-hydroxypyrimidine and a new synthesis of divicine, *Chem. Ind. (London),* 2, 1453, 1956.
96. **Zavyalov, S. I. and Pokhvisneva, G. V.,** Divicine and its 5-0-sulfate in aminolysis reactions, *Izv. Akad. Nauk S.S.S.R. Ser. Khim.,* 10, 2363, 1973.
97. **Ritthausen, H.,** Uber alloxantin als Spaltungsproduct des Convicins aus Saubohnen (*Vicia faba* minor) und Wicken (*Vicia sativa*), *Ber. Dtsch. Chem. Ges.,* 29, 894, 1896.
98. **Wang, S. Y.,** Chemistry of pyrimidines. II. The conversion of 5-bromo to 5-hydroxyuracils, *J. Am. Chem. Soc.,* 81, 3786, 1959.
99. **Wang, S. Y.,** Chemistry of pyrimidines. I. Reaction of bromide with uracil, *J. Org. Chem.,* 24, 11, 1959.
100. **Poje, M., Rocic, B., and Skrabalo, Z.,** β-Cytotoxic action of alloxan and alloxan-like compounds derived from uric acid, *Diabetes Croat.,* 9, 145, 1980.
101. **Jamalian, J., Aylward, F., and Hudson, B. J. F.,** Favism-inducing toxins in broad beans (*Vicia faba*): estimation of the vicine of broad bean and other legume samples, *Qual. Plant. Plant Foods Hum. Nutr.,* 27, 207, 1977.
102. **Pompei, C. and Lucisano, M.,** Le Lupin (*Lupinus albus* L.) comme Source de Proteins pour l'Alimentation Humaine, *Lebensm. Wiss. Technol.,* 9, 289, 1976.
103. **Pitz, W. J., Sosulski, F. W., and Hogge, L. R.,** Occurrence of vicine and convicine in seeds of some *Vicia* species and other pulses, *Can. Inst. Food Sci. Technol. J.,* 13, 35, 1980.
104. **Dutta, P. K., Chakravarty, A. K., Chowdhury, U. S., and Pakrashi, S. C.,** Vicine, a favism-inducing toxin from *Morordica charantia Linn.* seeds, *Indian J. Chem.,* 20B, 669, 1981.
105. **Jamalian, J. and Bassiri, A.,** Variation in vicine concentration during pod development in broad beans (*Vicia faba* L.), *J. Agric. Food Chem.,* 26, 1454, 1978.

106. **Bjerg, B., Knudsen, J. C. N., Olsen, O., Poulsen, M. H., and Sorensen, H.**, Quantitative analysis and inheritance of vicine and convicine content in seeds of *Vicia faba* L., *Z. Pflanzenzuecht.*, 94, 135, 1985.
107. **Engel, A. B.**, Detection of the toxic factors in broad beans *(Vicia faba)* in Progress Report, Rep. No. R3195, World Health Organization Geneva, 1970.
108. **Gardiner, E. E., Marquardt, R. R., and Kemp, G.**, Variation in vicine and convicine concentration of faba bean genotypes, *Can. J. Plant Sci.*, 62, 589, 1982.
109. **Brown, E. G., and Roberts, F. M.**, Formation of vicine and convicine by *Vicia faba*, *Phytochemistry*, 11, 3203, 1972.
110. **Marquardt, R. R., McKirdy, J. A., and Ward, A. T.**, Comparative cell wall constituent levels of tannin-free and tannin-containing cultivars of faba beans *(Vicia faba* L.)., *Can. J. Anim. Sci.*, 58, 775, 1978.
111. **Jamalian, J.**, Favism-inducing toxins in broad beans *(Vicia faba)*. Determination of vicine content and investigation of other non-protein nitrogenous compounds in different broad bean cultivars, *J. Sci. Food Agric.*, 29, 136, 1978.
112. **Bjerg, B., Eggum, B. O., Jacobsen, I., Olsen, O., and Sorensen, H.**, Protein quality in relation to antinutritional constituents in faba beans *(Vicia faba* L.). The effects of vicine, convicine and dopa added to a standard diet and fed to rats, *Z. Tierphysiol. Tierernaehr. Futtermittelkd.*, 51, 275, 1984.
113. **Luisada, L.**, Favism: singular disease affecting chiefly red blood cells, *Medicine (Baltimore)*, 20, 229, 1941.
114. **Davidson, J.**, The nutritive value of field beans *(Vicia faba* L.) for laying hens, *Br. Poult. Sci.*, 14, 557, 1973.
115. **Robblee, A. R., Clandinin, D. R., Hardin, R. T, Milne, G. R., and Darlington, K.**, Studies on the use of fababeans in rations for laying hens, *Can. J. Anim. Sci.*, 57, 421, 1977.
116. **Campbell, L. D., Olaboro, G., Marquardt, R. R., and Waddell, D.**, Use of fababeans in diets for laying hens, *Can. J. Anim. Sci.*, 60, 395, 1980.
117. **Olaboro, G., Campbell, L. D., and Marquardt, R. R.**, Influence of fababean fractions on egg weight among laying hens fed test diets for a short time period, *Can. J. Anim. Sci.*, 61, 751, 1981.
118. **Romanoff, A. L. and Romanoff, A. J.**, *The Avian Eggs*, John Wiley & Sons, New York, 1949.
119. **Griminger, P.**, Lipid metabolism, in *Avian Physiology*, 3rd ed., Sturkie, P. D., Ed., Springer-Verlag, New York, 1976, 253.
120. **Collier, H. B., Aherne, F. X., and Kennelly, J. J.**, Effect of a faba bean diet on growth, liver weight, and a non-protein thiol level of erythrocytes and liver in weanling rats, *Can. J. Anim. Sci.*, 58, 531, 1978.
121. **Tillotson, J. A., and Sauberlich, H. E.**, Effect of riboflavin depletion on the erythrocyte glutathione reductase in the rat, *J. Nutr.*, 101, 1459, 1971.
122. **Hoekstra, W. G.**, Biochemical function of selenium and its relation to vitamin E, *Fed. Proc. Fed. Am. Soc. Exp. Biol.*, 34, 2083, 1975.
123. **Griffith, O. W. and Meister, A.**, Potent and specific inhibition of glutathione synthesis by buthionine sulfoximine (S-n-butyl homocysteine sulfoximine), *J. Biol. Chem.*, 254, 7558, 1979.
124. **Whitting, L. A.**, Vitamin E and lipid antioxidants in free-radical-initiated reactions, in *Free Radicals in Biology*, Vol. 4, Pryor, W. A., Ed., Academic Press, New York, 1980, 295.
125. **Chiu, D., Lubin, B., and Shohet, S. B.**, Peroxidative reactions in red cell biology in *Free Radicals in Biology*, Vol. 5, Pryor, W. A., Ed., Academic Press, Inc., New York, 1982, 115.
126. **Flohe, L.**, Glutathione peroxidase brought into focus, in *Free Radicals in Biology*, Vol. 5, Pryor, W. A., Ed., Acadmic Press, New York, 1982, 223.
127. **Yannai, S. and Marquardt, R. R.**, Effect of divicine, one of its degradation products and hydrogen peroxide on normal and pre-treated erythrocytes, *Vet, Human Toxicol.*, 29, 393, 1987.
128. **Mohandas, N., Greenquist, A. C., and Shohet, S. B.**, Bilayer balance and regulation of red cell shape changes, *J. Supramol. Struct.*, 9, 453, 1978.
129. **Mezick, J. A., Settlemire, C. T., Brierley, G. P., Barefield, K. P., Jensen, W. N., and Cornwell, D. G.**, Erythrocyte membrane interactions with menadione and the mechanism of menadione-induced hemolysis, *Biochim. Biophys. Acta*, 219, 361, 1970.
130. **Schalm, O. W., Jain, N. C., and Carroll, E. J.**, *Veterinary Hematology*, 3rd ed., Lea & Febiger, Philadelphia, 1975, chap. 8.
131. **Arbid, M. S. S. and Marquardt, R. R.**, Favism-like effects of divicine and isouramil in the rat: acute and chronic effects on animal health, mortalities, blood parameters and ability to exchange respiratory gases, *J. Sci. Food Agric.*, 43, 75, 1988.
132. **Aurichio, L.**, Sul favismo, *Rass. Clin. Sci.*, 13, 20, 1935.
133. **Khalil, M., Aziz, Y., Tanious, A., Mahmoud, S., and Gharib, B.**, Study of red cell membrane lipids in glucose-6-phosphate dehydrogenase deficiency anemia, *Gazette Egypt. Paediatr. Assoc.*, 23, 281, 1975.
134. **Sartori, E.**, On the pathogenesis of favism, *J. Med. Genet.*, 8, 462, 1971.
135. **Donoso, G., Hedayat, H., and Khayatian, M.**, Favism, with special reference to Iran, *Bull. W. H. O.*, 40, 513, 1969.

136. **Belsey, M. A.,** The epidemiology of favism, *Bull. W. H. O.,* 48, 1, 1973.
137. **Hassan, M. M.,** Glucose-6-phosphate dehydrogenase deficiency in the Sudan, *J. Trop. Med. Hyg.,* 74, 187, 1971.
138. **Amir-Zaki, L., El-Din, S., and Kubba, K.,** Glucose-6-phosphate dehydrogenase deficiency among ethnic groups in Iraq, *Bull. W. H. O.,* 47, 1, 1972.
139. **Hedayat, Sh., Rahbar, S., Mahbooli, E., Ghaffarpour, M., and Sobhi, N.,** Favism in the Caspian littoral area of Iran, *Trop. Geogr. Med.,* 23, 149, 1971.
140. **Hedayat, Sh., Farhud, D. D., Montazami, K., and Ghadirian, P.,** The pattern of bean consumption, laboratory findings in patients with favism, G-6-P-D deficient, and a control group, *J. Trop. Pediatr.,* 27, 110, 1981.
141. **Du, S.-D.,** Favism in West China, *Chin. Med. J.,* 70, 17, 1952.
142. **Crosby, W. H.,** Favism in Sardinia (Newsletter), *Blood,* 11, 91, 1956.
143. **Beutler, E., Dern, R. J., and Alving, A. S.,** The hemolytic effect of primaquine. VI. An *in vitro* test for sensitivity of erythrocytes to primaquine, *J. Lab. Clin. Med.,* 45, 40, 1955.
144. **Beutler, E., Dern, R. J., Flanagan, C. L., and Alving, A. S.,** The hemolytic effect of primaquine. VII. Biochemical studies of drug-sensitive erythrocytes, *J. Lab. Clin. Med.,* 45, 286, 1955.
145. **Sansone, G. and Segni, G.,** Prime determinazioni del glutatione (GSH) ematico nel favismo, *Boll. Soc. Ital. Biol. Sper.,* 32, 456, 1956.
146. **Sansone, G. and Segni, G.,** Sensitivity to broad beans, *Lancet,* ii, 295, 1957.
147. **Sansone, G. and Segni, G.,** L'instabilita del glutatione ematico (GSH) nel favismo: utilizzazione di um test selettivo. Introduzione al problema genetico, *Boll. Soc. Ital. Biol. Sper.,* 33, 1057, 1957.
148. **Szeinberg, A., Sheba, C., Hirshorn, N., and Bodonyi, E.,** Studies on erythrocytes in cases with past history of favism and drug-induced acute hemolytic anemia, *Blood,* 12, 603, 1957.
149. **Szeinberg, A., Asher, Y., and Sheba, C.,** Studies on glutathione stability in erythrocytes of cases with past history of favism or sulfa drug-induced hemolysis, *Blood,* 13, 348, 1958.
150. **Szeinberg, A., Sheba, C., and Adam, A.,** Selective occurrence of glutathione instability in red blood corpuscles of the various Jewish tribes, *Blood,* 13, 1043, 1958.
151. **Carson, P. E., Flanagan, C. L., Ickes, C. E., and Alving, A. S.,** Enzymatic deficiency in primaquine-sensitive erythrocytes, *Science,* 124, 484, 1956.
152. **Gross, R. T., Hurwitz, R. E., and Marks, P. A.,** An hereditary enzymatic defect in erythrocyte metabolism: glucose-6-phosphate dehydrogenase deficiency, *J. Clin. Invest.,* 37, 1176, 1958.
153. **Vella, F.,** Favism in Asia, *Med. J. Aust.,* 2, 196, 1959.
154. **Larizza, P., Brunetti, P., and Grignani, F.,** Anemie emolitiche enzimopeniche, *Haematologica (Pavia),* 45, 1, 1960.
155. **Harley, J. D.,** Acute haemolytic anaemia in Mediterranean children with glucose-6-phosphate dehydrogenase-deficient erythrocytes, *Aust. Ann. Med.,* 10, 192, 1961.
156. **Kattamis, C. A., Kyriazakou, M., and Chaidas, S.,** Favism. Clinical and biochemical data, *J. Med. Genet.,* 6, 34, 1969.
157. **Motulsky, A. G.,** Metabolic polymorphism and the role of infectious diseases in human evolution, *Human Biol.,* 32, 28, 1960.
158. **Beutler, E.,** Red cell enzyme defects as nondiseases and as diseases, *Blood,* 54, 1, 1979.
159. **Westring, D. W., and Pisciotta, A. V.,** Anemia, cataracts and seizures in patients with glucose-6-phosphate dehydrogenase deficiency, *Arch. Intern. Med.,* 118, 385, 1966.
160. **Harley, J. D., Agar, N. S., and Gruca, M. A.,** Cataracts with a glucose-6-phosphate dehydrogenase variant, *Br. Med. J.,* 2, 86, 1975.
161. **Escobar, M. A., Heller, P., and Trobaugh, F. E., Jr.,** "Complete" erythrocyte glucose-6-phosphate dehydrogenase deficiency, *Arch. Intern. Med.,* 113, 428, 1964.
162. **Beutler, E., Mathai, C. K., and Smith, J. E.,** Biochemical variants of glucose-6-phosphate dehydrogenase giving rise to congenital nonspherocytic hemolytic disease, *Blood,* 31, 131, 1968.
163. **Thigpen, J. T., Steinberg, M. H., Beutler, E., Gillespie, G. T., Jr., Dreiling, B. J., and Morrison, F. S.,** Glucose-6-phosphate dehydrogenase Jackson. A new variant associated with hemolytic anemia, *Acta Haematol.,* 51, 310, 1974.
164. **Brand, K., Arese, P., and Rivera, M.,** Bedeutung und Regulation des Pentosephosphat-Weges in menschlichen Erythrozyten, *Hoppe-Seyler's Z. Physiol. Chem.,* 351, 501, 1970.
165. **Hebbel, R. P.,** Autoxidation and the sickle erythrocyte membrane: a possible model of iron decompartmentalization, in *Free Radicals, Aging and Degenerative Diseases,* Vol. 8, Johnson, J., E., Walford, R., Harmon, D., and Miguel, J., Eds., Alan R. Liss, New York, 1986, 395.
166. **Misra, H. P. and Fridovich, I.,** The role of superoxide anion in the autoxidation of epinephrine and a simple assay for superoxide dismutase, *J. Biol. Chem.,* 217, 3170, 1972.
167. **Benatti, U., Guida, L., Forteleoni, G., Meloni, T., and DeFlora, A.,** Impairment of the calcium pump of human erythrocytes by divicine, *Arch. Biochem. Biophys.,* 239, 334, 1985.

168. **Mavelli, I., Ciriolo, M. R., Rossi, L., Meloni, T., Forteleoni, G., DeFlora, A., Benatti, U., Morelli, A., and Rotilio, G.,** Favism: a hemolytic disease associated with increased superoxide dismutase and decreased glutathione peroxidase activity in red blood cells, *Eur. J. Biochem.,* 139, 13, 1984.

169. **Mavelli, I., Ciriolo, M. R., Rotilio, G., DeSole, P., Castorino, M., and Stabile, A.,** Superoxide dismutase, glutathione peroxidase and catalase in oxidative hemolysis. A study of Fanconi's Anemia Erythrocytes, *Biochem. Biophys. Res. Commun.,* 106, 286, 1982.

170. **Peisach, J., Blumberg, W. E., and Rachmilewitz, E. A.,** The demonstration of ferrihemochrome intermediates in Heinz body formation following the reduction of oxyhemoglobin A by acetylphenylhydrazine, *Biochim. Biophys. Acta,* 393, 404, 1975.

171. **Stern, A.,** Red cell oxidative damage, in *Oxidative Stress,* Sies, H., Ed., Academic Press, New York, 1985, chap. 14.

172. **Allen, D. W., Johnson, G. J., Cadman, S., and Kaplan, M. E.,** Membrane polypeptide aggregates in glucose-6-phosphate dehydrogenase-deficient and *in vitro* aged red blood cells, *J. Lab. Clin. Med.,* 91, 321, 1978.

173. **Johnson, G. J., Allen, D. W., Cadman, S., Fairbank, V. F., White, J. G., Lampkin, B. C., and Kaplan, M. E.,** Red-cell-membrane polypeptide aggregates in glucose-6-phosphate dehydrogenase mutants with chronic hemolytic disease. A clue to the mechanism of hemolysis, *N. Engl. J. Med.,* 301, 522, 1979.

174. **Kahane, I., Shifter, A., and Rachmilewitz, E. A.,** Cross-linking of red blood cell membrane proteins induced by oxidative stress in β-thalassemia, *FEBS Lett.,* 85, 267, 1978.

175. **Haest, C. W. M., Kamp, D., Plasa, G., and Deuticke, B.,** Intra- and intermolecular cross-linking of membrane proteins in intact erythrocytes and ghosts by SH-oxidizing agents, *Biochim. Biophys. Acta,* 469, 226, 1977.

176. **Palek, J. and Liu, S. C.,** Dependence of spectrin organization in red cell metabolism: implications for control of red cell shape, deformability and surface area, *Semin. Hematol.,* 16, 75, 1979.

177. **Fischer, T. M., Haest, C. W. M., Stohr, M., Kamp, D., and Deuticke, B.,** Selective alteration of erythrocyte deformability by SH-reagents. Evidence for involvement of spectrin in membrane shear elasticity, *Biochim. Biophys. Acta,* 510, 270, 1977.

178. **Hochstein, P. and Jain, S. K.,** Association of lipid peroxidation and polymerization of membrane proteins with erythrocyte aging, *Fed. Proc. Fed. Am. Soc. Exp. Biol.,* 40, 183, 1981.

179. **DeFlora, A., Morelli, A., Benatti, U., Pontremoli, S., Melloni, E., Salamino, F., Sparatore, B., Michetti, M., and Meloni, T.,** Membrane lipid components of normal and glucose-6-phosphate dehydrogenase-deficient erythrocytes of asymptomatic and favic subjects, *Acta Biol. Med. Ger.,* 40, 563, 1981.

180. **Fischer, T., Pescarmona, G. P., Bosia, A., Naitana, A., Turrini, F., and Arese, P.,** Mechanism of red cell clearance in favism, *Biomed. Biochem. Acta,* 42, S253, 1983.

181. **Fischer, T. M., Meloni, T., Pescarmona, G. P., and Arese, P.,** Sequestration of red cells in favic crises, *Clin. Hemorheol.,* 3, 228, 1983.

182. **Nishizuka, Y.,** Turnover of inositol phospholipids and signal transduction, *Science,* 225, 1365, 1984.

183. **Nishizuka, Y.,** Studies and perspectives of protein kinase C, *Science,* 233, 305, 1986.

184. **Barnes, D. M.,** How cells respond to signals, *Science,* 234, 286, 1986.

185. **DeFlora, A., Benatti, O., Guida, L., Forteleoni, G., and Meloni, T.,** Favism: disordered erythrocyte calcium homeostasis, *Blood,* 66, 294, 1985.

186. **Bosia, A., Passow, H., Arese, P., Lepke, S., and Mannuzzu, L.,** Effect of divicine on Ca^{++}-stimulated K^+ efflux in normal and G6PD-deficient erythrocytes, *Ital. J. Biochem.,* 29, 393, 1980.

187. **Shalev, O., Leida, M. N., Hebbel, R. P., Jacob, H. S., and Eaton, J. W.,** Abnormal erythrocyte calcium homeostasis in oxidant-induced hemolytic disease, *Blood,* 58, 1232, 1981.

188. **Fishum, G. and Lehninger, A. L.,** Regulated release of Ca^{2+} from respiring mitochondria by $Ca^{2+}/2H^+$ antiport, *J. Biol. Chem.,* 254, 6236, 1979.

189. **Bellomo, G., Matino, A., Richelmi, P., Moore, G. A., Jewell, S. A., and Orrenius, S.,** Pyridine-nucleotide oxidation, Ca^{2+} cycling and membrane damage during tert-butyl hydroperoxide metabolism by rat-liver mitochondria, *Eur. J. Biochem.,* 140, 1, 1984.

190. **Bellomo, G., Jewell, S. A., and Orrenius, S.,** The metabolism of menadione impairs the ability of rat liver mitochondria to take up and retain calcium, *J. Biol. Chem.,* 257, 11558, 1982.

191. **Frei, B., Winterhalter, K. H., and Richter, C.,** Mechanisms of alloxan induced calcium release from rat liver mitochondria, *J. Biol. Chem.,* 260, 7394, 1985.

192. **Bellomo, G., Jewell, S. A., and Orrenius, S.,** Regulation of intracellular calcium compartmentation: studies with isolated hepatocytes and t-butyl hydroperoxide, *Proc. Natl. Acad. Sci. U.S.A.,* 79, 6842, 1982.

193. **Thor, M., Smith, M. T., Hartzell, P., Bellome, G., Jewell, S. A., and Orrenius, S.,** The metabolism of menadione (2-methyl-1, 4-naphthoquinone) by isolated hepatocytes, *J. Biol. Chem.,* 257, 12419, 1982.

194. **Sies, H.,** Oxidative stress: introductory Remarks, in *Oxidative Stress,* Sies, H., Ed., Academic Press, New York, 1985, 1.

195. **Meloni, T., Forteleoni, G., and Gaetani, G. F.,** Desferrioxamine and favism, *Br. J. Haematol.,* 63, 394, 1986.

196. **Marquardt, R. R. and Arbid, M. S. S.,** Protection against the toxic effects of the favism factor (divicine) in rats by vitamin E, A, and C and iron chelating agents, *J. Sci. Food Agric.,* 43, 155, 1988.
197. **Allison, A. C.,** Glucose-6-phosphate dehydrogenase deficiency in red blood cells of East Africans, *Nature (London),* 186, 531, 1960.
198. **Allison, A. C. and Clyde, D. F.,** Malaria in African children with deficient erythrocyte glucose-6-phosphate dehydrogenase, *Br. Med. J.,* i, 1346, 1961.
199. **Luzzatto, L., Usanga, E. A., and Reddy, S.,** Glucose-6-phosphate dehydrogenase deficient red cells: resistance to infection by malaria parasites, *Science,* 164, 839, 1969.
200. **Kidson, C. and Gorman, J. G.,** A challenge to the concept of selection by malaria in glucose-6-phosphate dehydrogenase deficiency, *Nature (London),* 196, 49, 1962.
201. **Martin, D. K., Miller, L. H., Alling, D., Okoye, V. C., Esan, G. J. F., Osunkoya, B. O., and Deane, M.,** Severe malaria and glucose-6-phosphate dehydrogenase deficiency: a reappraisal of the malaria/G-6-PD hypothesis, *Lancet,* i, 524, 1979.
202. **Huheey, J. E. and Martin, D. L.,** Malaria, favism and glucose-6-phosphate dehydrogenase deficiency, *Experientia,* 31, 1145, 1975.
203. **Livingstone, F. B.,** Malaria and human polymorphisms, *Annu. Rev. Genet.,* 5, 33, 1971.
204. **Gutteridge, J. M. C., Richmond, R., and Halliwell, B.,** Inhibition of the iron-catalyzed formation of hydroxyl radicals from superoxide and of lipid peroxidation by desferrioxamine, *Biochem. J.,* 184, 469, 1979.
205. **Cohen, G. and Hochstein, P.,** Generation of hydrogen peroxide in erythrocytes by hemolytic agents, *Biochemistry,* 3, 895, 1964.
206. **Pollack, S., George, J. N., and Crosby, W. H.,** Effects of agents simulating the abnormalities of the glucose-6-phosphate dehydrogenase-deficient red cell on *Plasmodium berghei* malaria, *Nature (London),* 210, 33, 1966.
207. **Allison, A. C. and Eugul, E. M.,** A radical interpretation of immunity to malaria parasites, *Lancet,* ii, 1431, 1982.
208. **Eaton, J. W., Eckman, J. R., Berger, E., and Jacob, H. S.,** Suppression of malaria infection by oxidant-sensitive host erythrocytes, *Nature (London),* 264, 758, 1976.
209. **Fairfield, A. S., Meshnick, S. R., and Eaton, J. W.,** Host superoxide dismutase incorporation by intraerythrocytic plasmodia, in *Malaria and the Red Cell,* Eaton, J. W. and Brewer, G. J., Eds., Alan R. Liss, New York, 1984, 13.

Chapter 7

CALCINOGENIC GLYCOSIDES

Martin Weissenberg

TABLE OF CONTENTS

I. INTRODUCTION

Glycosides having the ability to induce vitamin D intoxication (calcinosis) upon administration into animals are described as calcinogenic principles. They occur in small amounts in the leaves of several plant species, particularly those of the order Solanaceae. Hydrolysis splits off the glycosidic moiety and gives vitamin D_3 sterols as aglycones. Calcinosis was observed in grazing animals as a consequence of the ingestion of calcinogenic plants of wide geographical distribution. The disease is characterized by the deposition of calcium salts in soft tissues, and its symptoms are reminiscent of hypervitaminosis D.

The plants which were found to display calcinogenic activity are listed in Table 1, and it will be noted that calcinogenic glycosides were unequivocally identified only in *Solanum glaucophyllum*[1-12] and in *Cestrum diurnum*.[3,4,7-10,12] No chemical evidence is available for their presence in *Trisetum flavescens*, *S. torvum*, *S. verbascifolium*, and *S. esuriale* although some aqueous vitamin D-like activity could be demonstrated in *T. flavescens* and *S. verbascifolium*.[11] Moreover, the active principles of *S. torvum*, *S. verbascifolium* and *S. esuriale* have not yet been identified, while vitamin D_3 derivatives were found in *T. flavescens*.[11,11a]

Since this chapter is concerned with calcinogenic glycosides, the plants which bear them, *S. glaucophyllum* and *C. diurnum*, will be treated preferentially in the following survey. Nevertheless, a summary of calcinosis incidences and symptoms is also provided in Table 1 for the other species, and the interested reader may usefully consult the included key references to the sources of the data for detailed information. It should be remembered in this connection that similar calcinosis conditions observed in various parts of the world could not always be associated with the grazing of specific plants,[4,8,10,12] whereas other suspected plants which had been grazed proved ineffective in producing calcinosis experimentally in test animals.[10]

S. glaucophyllum is the main plant source of calcinogenic glycosides, and implicitly the factor responsible for a calcinotic disease of livestock prevalent in Argentina and in Brazil, and which causes considerable economic losses to the local cattle industry.[1,8,10,12] During the 1960s, research on the etiology of the disease was undertaken by Carrillo and colleagues at the Balcarce Experimental Station of the Argentinian National Institute of Agricultural Technology (INTA), under sponsorship of the Food and Agriculture Organization (FAO) of the United Nations. The early work reported in local bulletins and journals and summarized by Carrillo[1] led to the identification of *S. glaucophyllum* as the cause of the cattle disease and to a series of subsequent investigations which initiated the modern research in the field of calcinogenic plants. Substantial contributions were concurrently made in Argentina by the groups of R.L. Boland at the Southern National University at Bahia Blanca, R.C. Puche at the National University of Rosario, C.A. Mautalen at the French Hospital in Buenos Aires, and F.M. Rossi at the Institute of Oncology in Buenos Aires. Parallel investigations by R. H. Wasserman and his group at Cornell University in Ithaca, NY, led, inter alia, to the successful identification of $1\alpha,25$-dihydroxy-cholecalciferol ($1\alpha,25$-$(OH)_2D_3$) as an aglycone of the glycosides occurring in *S. glaucophyllum* and *C. diurnum*. Other active research groups in the field of calcinogenic plants were those of H. R. Camberos and G. K. Davis at the University of Florida at Gainesville, K. M. L. Morris at the City of London Polytechnic, D. J. Humphreys at the Royal Veterinary College in London, M. Peterlik at the University of Vienna, and G. Dirksen, H. Zucker, W. A. Rambeck, and their colleagues at the Veterinary Faculty of the Ludwig-Maximilians-University of Munich. Notable contributions were also made by leading groups in vitamin D metabolites research including those of H. F. DeLuca at the University of Wisconsin, Madison (with the cooperation of M. F. Holick, J. L. Napoli, and P. H. Stern), M. R. Haussler at the University of Arizona at Tucson, A. W. Norman at the University of California at Riverside, and D. E. M. Lawson at the University of Cambridge.

Research on calcinogenic plants or plant-induced calcinoses has previously been either briefly summarized[3,4,9,11] or reviewed in considerable detail.[8,9,10,12] The aim of this chapter is to survey the present state of knowledge of calcinogenic glycosides, placing an increased emphasis on their isolation, identification, mechanism of action, and potential pharmacological interest.

II. CHEMISTRY

A. Isolation and Identification

Attempts at the isolation and identification of the pure calcinogenic glycosides from both *S. glaucophyllum* and *C. diurnum* were beset with difficulties. The active principles occur

Table 1
CALCINOGENIC PLANTS

Naturally occurring calcinosis

Genera and species	Family	Grazing animal	Country	Disease	Physiological effects[a]	Active compound	Ref.
Solanum glaucophyllum Desf.	Solanaceae	Cattle, sheep, horses	Argentina	Enteque seco	CST, HCa, HP	1,25-(OH)$_2$D$_3$ glycosides; possibly also: 25-OH D$_3$ glycoside;	1—12
			Brazil	Espichamento	CST	Vitamin D$_3$ glycoside	
Cestrum diurnum L.	Solanaceae	Cattle, horses	Florida (U.S.)	Calcinosis	CST, HCa, HP	1,25-(OH)$_2$D$_3$ glycoside	3,4,7—10,12
Trisetum flavescens (L.) Beauv.	Gramineae	Cattle, sheep	Austria	Pasture disease	CST, HCa, HP	Vitamin D$_3$	4,7,8,10—12
		Cattle	Germany	Enzootic calcinosis	CST, HCa, HP	possibly also: 1,25-(OH)$_2$D$_3$-like water-soluble factor	11a
S. torvum Sw.	Solanaceae	Cattle	Papua, New Guinea	Enzootic calcinosis	CST, HCa, HP	Unknown	10,12
S. verbascifolium L.	Solanaceae	Cattle	Argentina, Africa		CST, HP	Unknown	11,12
S. esuriale Lindl.	Solanaceae	Sheep	Australia	Humpy back	CST	Unknown	10,12

[a] CST, calcification of soft tissues; HCa, hypercalcemia; HP, hyperphosphatemia.

only in minute concentration in the plant leaf, and they are accompanied by chemically similar substances, as well as by pigments and phenolic derivatives. In addition, they are sensitive to acids and to oxidizing agents. Hence, the active glycosides could not so far be isolated in pure form and characterized in spite of considerable efforts. Yet, the structural elucidation of the aglycone and carbohydrate moieties could be achieved. The progress throughout the isolation and purification work was guided and monitored mainly by bioassays.

1. *S. glaucophyllum*
a. *Extraction*

Early investigations on the extraction of the active principle pointed to its solubility in water, methanol,[13-15] ethanol and isopropanol,[15] and aqueous alcohols.[16] The activity is reportedly not extracted with diethyl ether, petroleum ether, benzene,[14] acetone, or methanol-chloroform (2:1),[17] nor is it extractable from aqueous solutions into ether, chloroform, or benzene,[15] despite conflicting results suggesting activity in ether[18] or methylene chloride[20a] extracts. Careful studies of extraction with alcohol-water mixtures indicated good solubility up to a concentration of 80% ethanol[17] or methanol.[19] Plant extraction with water at room temperature was found preferable to reflux with alcohols and neither prolonged periods of reflux, nor reextraction, improved the yield.[20] The apparent insolubility of the active compound in nonpolar solvents was tentatively exploited for defatting and partial removal of pigments either by direct plant treatment or by liquid-liquid extraction of the aqueous solutions of plant extracts with ether, petroleum ether, chloroform, or benzene.[15,21] Small-scale extractions for analytical or biological purposes were usually performed by shaking dried and finely powdered leaves (optionally defatted by prior treatment with nonpolar solvents) in water at a 5:1,[21-24] 10:1,[25-28] or 18:1[29] ratio (v/w) of extractant to plant material. The extraction time was typically 24 h at ambient temperature, although periods of 5 to 60 min at 37 to 40°C[21-24] or overnight at 4°C[27,29] have also been suggested. Reextraction was sometimes proposed.[21,22,28,30] After filtration, the aqueous extract was lyophilized to yield a dark-brown residue (30 to 40% of the original mass).[28,29] In a modified version, treatment of the aqueous filtrate with ethanol was carried out[21,22,30] for partial precipitation of proteins, followed by centrifugation and evaporation to dryness under reduced pressure. Alternatively, ether- or petroleum ether-washed leaf material was extracted with methanol,[31-34] yielding, after filtration and solvent removal, a viscous residue (about 11% extractable material).[34] Finally, the residue either was taken up in water to give solutions containing the equivalent of 0.5 to 4 g of dry leaf per milliliter for biological experiments, or was further subjected to purification.

For larger operations, the actual process of extraction involving water, alcohols, or aqueous alcohols had to be scaled up to an optimal size in order to handle kilogram quantities of plant material and large volumes of solvent. Water was the better solvent; however, filtration and concentration of aqueous extracts were slow and tedious. On the other hand, alcoholic extracts were significantly more contaminated with pigments and resins. The optimized extraction process was carried out with 70% aqueous ethanol under efficient stirring (and facultative reflux) for 24 h. After filtration, ethanol was distilled off under reduced pressure, and the remaining aqueous solution was further treated with nonpolar solvents, thus including this cleanup in the process.[20] The aqueous extract was then subjected to further purification.

b. *Purification*

Preliminary purification of the crude extracts was usually required before chromatography. It involved selective extraction with nonpolar solvents to remove lipids, alkaloids, and pigments,[20,21] dialysis against water to discard proteins, polysaccharides, and pigments,[21] and desalting by gel filtration on Sephadex® G-25[29] or Bio-gel® P-2 polyacrylamide gel columns[35] to eliminate calcium, magnesium, and phosphorus ions. Treatment with ion-exchange resins revealed activity in the neutral fraction only.[15,21]

Attempts were made to remove the accompanying phenolic derivatives which interfere with the isolation process. An early investigation of the phenolic constituents of *S. glaucophyllum* leaves led to the separation and identification of hydroquinone, quercetin, kaempferol, and eight phenolic glycosides.[36] However, treatment of an aqueous leaf extract with lead acetate, which was supposed to precipitate the phenolic compounds, showed an unexpected recovery of the activity in the precipitate.[22,46] A reinvestigation of this reaction under slightly different conditions led to the conclusion that the activity is distributed between both the precipitate and the supernatant, apparently due to partial coprecipitation, and/or interaction, of saponins with lead acetate.[20] At any rate, the surprising result of the lead acetate treatment was attributed at the time to a *phenolic character* of the active principle,[37,170] a view which could not be confirmed subsequently.

In yet another approach,[31,32] partial purification of the leaf extract was attempted by successive fractionation with dilute hydrochloric acid and chloroform and then, following alkalinization of the aqueous phase, with chloroform again, and with chloroform-ethanol(3:2). The remaining active aqueous extract was afterward subjected to paper chromatography which pointed to the presence of two active compounds.[31]

Further purification was done by column chromatography on silicic acid,[26,38] by gel filtration on Sephadex® columns (G-15[15,21,22] followed by G-10[21,22] and LH-20 followed by G-100[34]) or by ion-exchange chromatography of the borate complexes of glycosides on Amberlite® CG-400.[24] Final purification was accomplished by paper chromatography[15,21,22] or by preparative thin-layer chromatography (TLC) on silica gel G plates.[26,38] Both techniques showed ultraviolet (UV) fluorescence to be associated with the purified active material, and the separation was guided by bioassays and UV absorbancy. Homogeneity of the active fraction could not be established unequivocally, and the small quantities obtained and their degree of purity were not sufficient to permit a chemical characterization. Adequate amounts of purified material could, however, be secured in order to allow further investigation of the aglycone and carbohydrate moieties.

c. Aglycone Structure

Partially purified aqueous extracts were submitted to enzymatic hydrolysis with either β-glucosidase[23,24,28,38] or a mixed preparation of glycosidases from the liver of the sea worm, *Charonia lampas*.[39-41] The latter was found to achieve a more efficient glycolytic cleavage due to the presence of a wider range of glycosidases including glucosidases, mannosidases, galactosidases, etc. with both α and β stereospecificities.[40] After 4 to 24 h at 37°C under nitrogen, the released sterols were extracted with chloroform-methanol mixtures and purified by chromatography on Sephadex® LH-20,[23,38-41] followed successively by silicic acid, microCelite®, and Celite® columns.[38-41] The fractionation guided by bioassays revealed that the active principle was removed from the crude aqueous extract into the chloroform phase after hydrolysis and migrated exactly with authentic labeled metabolites. The progress of the purification was also monitored through UV absorption by subtracting background absorbance from columns and solvents.[39,40] A control experiment run in parallel with a blank sample did not display any activity, thus, demonstrating that the lipophilic activity originated in the aqueous leaf extract and not in the crude glycosidase preparation.[40] In an investigation conducted with 3 μg of purified material obtained from 24 g of dried leaf powder, Haussler and colleagues secured unequivocal chromatographical and spectral evidence to identify it as 1α, 25-$(OH)_2D_3$ (Figure 1, *1d*),[40,41] although the small yield, again, did not allow chemical characterization. Accordingly, the purified fraction displayed an UV absorption spectrum identical to that of authentic 1α,25-$(OH)_2D_3$, and its analysis by direct probe mass spectrometry and combined gas chromatography/mass spectrometry (GC/MS) yielded a parent molecular ion of m/e 416, and a fragmentation pattern indistinguishable from that of synthetic 1α,25-$(OH)_2D_3$.[39-41]

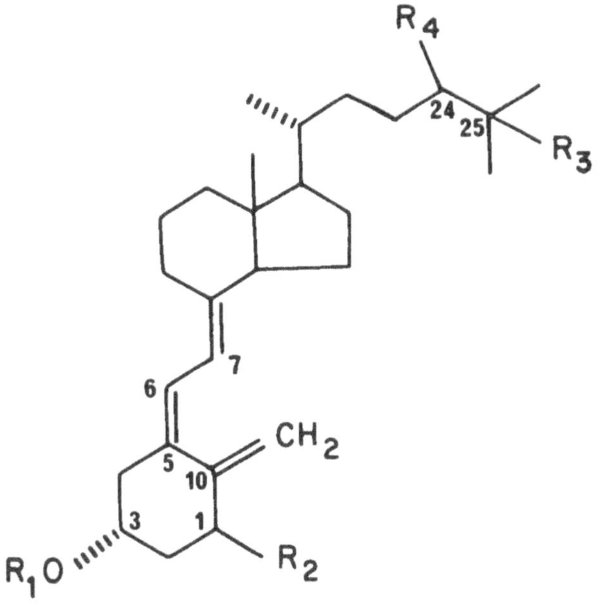

FIGURE 1. Basic structure of vitamin D_3 and related sterols. (a) R_1 = R_2 = R_3 = R_4 = H; (b) R_1 = R_3 = R_4 = H, R_2 = (α)OH; (c) R_1 = R_2 = R_4 = H, R_3 = OH; (d) R_1 = R_4 = H, R_2 = (α)OH, R_3 = OH; (e) R_1 = H, R_2 = (α)OH, R_3 = OH, R_4 = (R)OH.

In another experiment, further purification of the fractions obtained following hydrolysis and subsequent Sephadex® LH-20 chromatography was carried out by analytical and preparative TLC. As a result, vitamin D_3 (Figure 1, *1a*) and 25-hydroxy vitamin D_3 (25-OH D_3, *1c*) were identified, in addition to 1α,25-(OH)$_2$D$_3$ (*1d*), by comigration with authentic samples on TLC and column chromatography, by displaying the same characteristic UV spectra, and by exhibition of vitamin D activity detected by an assay specific for all those metabolites.[23] These findings were suggestive of the presence of all three sterols in the aglycone portion of the naturally occurring glycosides.

Another set of experiments provided further evidence of the presence of vitamin D sterols in the aglycone moiety. Assuming that the intestine might be an important source of glycosidases capable of modifying the active glycosides,[42] Boland and colleagues launched an investigation on the metabolism of the *S. glaucophyllum* glycosides in rumen fluid.[43-48a] Thus, incubation of aqueous leaf extracts with sheep ruminal fluid *in vitro* at 38°C for 72 h followed by preparative TLC, chromatography on Sephadex® LH-20 columns, and finally high performance liquid chromatography (HPLC) led to the isolation of several lipophilic vitamin D_3 metabolites,[43,44,46] including 25-OH D_3, 1α,25-(OH)$_2$D$_3$, and 1α,24,25-(OH)$_3$D$_3$ (Figure 1, *1e*), along with vitamin D_3 itself.[48a] As concurrent incubation of leaf extracts with β-glucosidase gave only small amounts of *1e* in comparison with *1d*, the view was expressed that rumen microbes convert the 1α,25-(OH)$_2$D$_3$ glycoside of *S. glaucophyllum* into free *1d* and a more polar metabolite suggested to be *1e* on the basis of chromatographic and bioassay evidence.[48a] It would appear from these results that lipophilic secosterols are released, and might conceivably be produced, by the action of ruminal fluid on plant extracts. Previous research has pointed to cleavage of the naturally occurring glycosides *in vivo* by the digestive system of animals or by intestinal bacteria, which would liberate 1α,25-(OH)$_2$D$_3$.[28,38,49,50]

d. Carbohydrate Structure

In the only attempt reported to date at the structure elucidation of the carbohydrate moiety

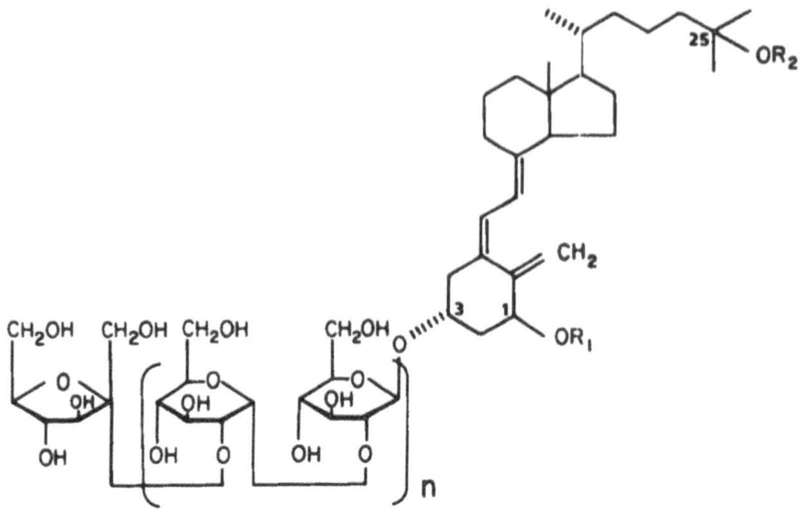

FIGURE 2.   Proposed structure of $1\alpha,25\text{-(OH)}_2\text{D}_3$-glycoside present in *S. glaucophyllum*, with suggested position of a single carbohydrate moiety. $n = 1,2,4$; R_1, R_2 = H or possibly carbohydrate moiety.

of the *S. glaucophyllum* calcinogenic principle, borate complexes of glycosides were prepared from a crude aqueous extract.[24] Subsequent ion exchange chromatography on Amberlite® CG-400 monitored by UV absorption and bioassays revealed the presence of three oligo-saccharides, confirmed also by paper chromatography. Further examination of analytical data established their composition as fructoglucosides with glucose:fructose ratios 8:1, 4:1, and 2:1 (nona-, penta-, and trisaccharide, respectively), the nonasaccharide being the major component. The calculated molecular weights of the fructoglucosides (902, 1126, 1676) were found to agree with previous estimations of 1000 to 2000[16,21,22,29] inferred from gel filtration experiments.[16,21,22] The aglycone $1\alpha,25\text{-(OH)}_2\text{D}_3$ was suggested to be bound at the reducing end of the saccharides.[24]

Parallel incubation of the fructoglucosides with α-amylase and with β-amylase, respectively, indicated that the linkage between glucose is α and that the linkage with the steroid is β. Oxidation with periodic acid of the fructoglucosides showed that fructose is linked to a disaccharide unit (Glc$_p\alpha$1-2Glc) repeating itself one, two, or four times, and known as kojibiose. At this stage no chemical characterization was done, and no spectral evidence was presented to substantiate the assignments. The conclusions arrived at were expressed in the form of a tentative formula (Figure 2), and the evidence thus far cited does not distinguish unambiguously among the possible steroidal positions of conjugation with the carbohydrate unit, or between an α or β linkage of fructose with glucose.[24] These results recall previous reports suggesting that different aglycones[23] might be conjugated to various carbohydrate moieties or sequences,[19,26,31,40,46,51,52] thus affording a series of naturally occurring calcinogenic glycosides.

2. C. diurnum

a. Extraction and Purification

Early work on a phytochemical screening of *Cestrum* species revealed the presence in *C. diurnum* of steroidal saponins, sterols including β-sitosterol, triterpenoids including ursolic acid and β-amyrin, catechol tannins, carbohydrates, and flavonoids. Sucrose, glucose, and fructose were the only sugars detected, and fatty acids were identified as myristic, palmitic, stearic, oleic, and linoleic.[53] The report has also called attention to the cardiotonic properties of the sapogenins from the leaves of *C. diurnum*.[53] Following essentially the same pathway used for *S. glaucophyllum*, a solubility study was conducted with dried leaf powder of *C. diurnum*. Extraction was carried out in parallel with distilled water, absolute ethanol, methanol-chloroform (2:1 v/v), and chloroform, respectively, in a 8:1 v/w ratio of extractant to plant material, and under mechanical stirring. After 2 h at room temperature, the extraction was repeated twice, and the pooled filtrates were either lyophilized or evaporated to dryness under a stream of nitrogen. The dry extracts were redissolved in their respective original solvents, and aliquots were bioassayed for the determination of calcium-binding protein (CaBP). The results showed that the active components were equally soluble in ethanol and in methanol-chloroform (2:1), less soluble in water, and insoluble in chloroform.[54]

For biological experiments, aliquots of extract solutions (lg dry leaf per milliliter) were evaporated to dryness and redissolved in an equivalent volume of propyleneglycol.[54] In a larger scale extraction batch carried out for preparative and identification purposes,[55] the dried leaf powder was first treated with chloroform to remove pigments, then extracted with methanol-chloroform (2:1). The active material was precipitated with acetone, and the resultant pellet was subsequently redissolved in methanol-chloroform (2:1). After preliminary purification on a Sephadex® LH-20 column,[9] the active fraction was subjected to identification.

b. Aglycone Structure

A partially purified extract obtained as described above was incubated with a mixed preparation of glycosidases derived from the liver of the sea worm, *Charonia lampas*, for 16 h at 37°C and pH 5.0, under nitrogen.[55] The reaction mixture was then extracted with chloroform, and the chloroform-soluble material was purified by chromatography on silicic acid followed by Sephadex® LH-20, micro Celite® chromatography, and finally a Celite® liquid-liquid partition column. The pooled peak fractions from the last column exhibited an UV absorption spectrum identical to that of $1\alpha,25$-$(OH)_2D_3$. The purification was monitored by bioassays, and the active compound comigrated with authentic tritiated $1\alpha,25$-$(OH)_2D_3$. Hughes and colleagues managed to secure 1.2 μg of purified material derived from 300 g of dried leaf, and subjected it to GC/MS. The coincident migration of the active lipophilic plant derivative with authentic $1\alpha,25$-$(OH)_2D_3$ on the gas chromatographic column, and the identical fragmentation pattern displayed in their respective mass spectra, indicated the presence of a $1\alpha,25$-$(OH)_2D_3$ glycoside in *C. diurnum*.[55] Further evidence was adduced by analysis of plasma from chicks treated with *C. diurnum*, which revealed the presence of $1\alpha,25(OH)_2D_3$. This observation demonstrated that both *in vitro* and *in vivo* hydrolysis released the same active fragment, indistinguishable from the authentic sample by all chromatographic systems used.[55]

No investigation on the structure of the carbohydrate moiety of *C. diurnum* active factors has been reported to date. However, attention was called at this point to the lesser solubility of *C. diurnum* derivatives in polar solvents as compared with *S. glaucophyllum*, which was accounted for by assuming the presence of fewer carbohydrates units in *C. diurnum* glycosides.[54]

B. Analytical Aspects

Agronomic and genetic work on calcinogenic plants as well as the isolation and purification of their active components should be facilitated by fast assays for vitamin D_3 and its metabolites, even if they would be nonspecific and moderately accurate. A large array of procedures is cited in the literature; however, most of them proved to be time consuming and cumbersome, and thus unsuitable for large screening programs.

The bioassays include curative, prophylactic, and blood calcium absorption or phosphate rise procedures[56] and imply administration of the assayed plant material to test animals and comparison of the biological responses with other test animals treated with known amounts

Table 2
ASSAYS OF CALCINOGENIC ACTIVITY IN PLANTS[a]

	Activity			
Reported	Standard used	$1\alpha,25\text{-}(OH)_2D_3$-equivalent[c] ($\mu g/g$ dry plant)	Assay system[b]	Ref.
0.69 IU/100 mg[d]	D_3		Rachitic line — test in rats	58
200 IU/g	D_3		Rachitic rats bioassay	79
16 IU/200 mg	D_3		Rachitic rats bioassay	78
300 IU/g	D_3		Rachitic chick bioassay	3
10 $\mu g/g$	$1\alpha,25\text{-}(OH)_2D_3$	10	Rachitic chick bioassay	27
24—120 $\mu g/g$	D_3	20—100[e]	Rachitic chick bioassay[f]	57
80 IU/250 mg	D_3		Intestinal Ca absorption in rachitic chicks	33
660 IU/g	D_2		Intestine Ca absorption in rats	59
1 $\mu g/g$	$1\alpha,25\text{-}(OH)_2D_3$	1	Duodenal Ca absorption in rats	60
25 IU/100 mg	D_3		*In vitro* intestinal Ca transport	61
10—60 mg/kg	$1\alpha,25\text{-}(OH)_2D_3$	10—60	*In vitro* intestinal receptor assay	29
30 ng/4.7 mg	$1\alpha,25\text{-}(OH)_2D_3$	0.07	Intestinal chromatin-steroid displacement assay	35
30 pg/26 mg	$1\alpha,25\text{-}(OH)_2D_3$	$1.15 \cdot 10^{-3}$	Rat bone resorption *in vitro*	34
80 ng/60 mg	$1\alpha,25\text{-}(OH)_2D_3$	1.3	Rise of serum phosphate in rat bioassay	11
400 IU/5 g	D_3		Rise of serum phosphate in rat bioassay	62
$1.3 \cdot 10^5$ IU/Kg	D_3		CaBP	17
2.5 $\mu g/g$	D_3		CaBP in chick and rat intestine	63
0.1 $\mu g/g$	$1\alpha,25\text{-}(OH)_2D_3$	0.1	*In vitro* CaBP synthesis	64
$3 \cdot 10^4$ IU/Kg[g]	D_3		*In vivo* CaBP synthesis	54
0.05 $\mu g/100$ mg[g]	$1\alpha,25\text{-}(OH)_2D_3$	0.5	Radioimmunoassay	20b
2—14 $\mu g/100$ mg	$1\alpha,25\text{-}(OH)_2D_3$	20—140	Radioimmunoassay	20b

[a] *S. glaucophyllum*, unless stated otherwise.
[b] Performed with aqueous leaf extracts unless stated otherwise.
[c] The bioactivity of $1\alpha,25\text{-}(OH)_2D_3$ is assumed to be tenfold greater than that of vitamin D_3.[27]
[d] One international unit (IU) is equivalent to 0.025 μg vitamin D_3.
[e] Vitamin D_3 figures (μg) divided by 1.2 gave $1\alpha,25\text{-}(OH)_2D_3$ equivalency.[57]
[f] Performed with leaf powder.
[g] *C. diurnum*.

of standard vitamin D_3 derivatives and with controls. These assays are expensive, and high individual variations in the responses of the test animals are encountered. Competitive protein-binding methods are also frequently used[56] and present advantages like high sensitivity, small sample volumes required, and simultaneous processing of a large number of samples, while the drawbacks are interference from parent derivatives which might compete with the binding proteins, and lesser accuracy. A survey of the literature on the subject is summarized in Table 2. It should be stated that the *in vitro* and *in vivo* assays used by the different groups were later found to give varying results due to differences in either sensitivity and specificity of the methods, or in plant geographical location, climatic conditions, development stage, and sample preparation.[57] All assays were carried out with aqueous leaf extracts with the exception of a rachitic chick bioassay performed with finely powdered leaves.[57] The higher activity found by using the latter assay was admittedly the result of a better sample extraction *in vivo* at the chick intestine, as compared with the conventional extraction of leaves with water. Along with bioassays, a radioimmunoassay using a commercially available antibody was proposed for rapid estimation of vitamin D_3 derivatives in calcinogenic plants, and for monitoring the extraction, isolation, and purification of the active principle therefrom.[20b] The method was found to require small quantities of plant material and crude, dilute aqueous extracts which might be further diluted to the measuring range and directly assayed, without prior purification and hydrolysis steps.[20b]

The relative distribution of individual vitamin D_3 derivatives present in the plant has not yet been quantitatively estimated.

C. Stability

Early work recognized the stability of leaf activity,[6] especially when intoxication of cattle could be reproduced successfully with dry leaves of *S. glaucophyllum* collected 2 years previously in Argentina[65] and Brazil.[66] A reassessment of this inference showed that a leaf sample titrated twice at an interval of 1 year, exhibited no significantly different activity, whereas the higher recorded activity was related to a sample 4 years old.[57]

The thermostability of the calcinogenic principles in *S. glaucophyllum* was tested in feeding experiments with rabbits using the increase of plasma phosphorus levels as an indication of the calcinogenic effect.[67,68] Accordingly, leaves heated for 1 h at 100 to 130°C retained their activity. Further heating at higher temperatures led to a decrease of activity, while deactivation occurred above 180°C.[68,69]

Aqueous extracts of *S. glaucophyllum* (0.5 g leaves per milliliter) were found to be stable up to 12 months when stored in a refrigerator.[21] Reportedly, even evaporation of aqueous extract solutions on a water-bath at 90 to 100°C hardly diminished the activity.[69]

D. Synthetic Analogues

The identification of vitamin D glycosides in calcinogenic plants along with the presence of vitamin D glucuronides in rat bile[70] led to an interest in the synthesis of model vitamin D conjugates. Such synthetic analogues were expected to allow a useful comparison with their naturally occurring counterparts in terms of transport, absorption, and biological activity.

Several groups set up routine procedures for direct glycosilation of vitamin D_3 derivatives, involving mainly reaction with peracetylated α-bromosugars in the presence of silver carbonate. Thus, Kumar and colleagues at the Mayo Clinic in Rochester, obtained vitamin D_3-3β-glucopyranosiduronate (D_3-glucuronide) (Figure 3, *3a*) and described its biological activity.[70] Afterwards, Holick and Holick at the Massachusetts General Hospital in Boston prepared a large series of glycosides biologically active in maintaining calcium and phosphorus metabolism in animals. (Figure 3, R_2 = H, OR_1; R_1 = glucosyl, mannosyl, galactosyl, fructosyl, arabinosyl, xylosyl, sucrosyl, cellobiosyl, etc.)[71] The total number of glycosidic units per compound was not higher than three.[71] Likewise, Fürst et al. of Hoffmann-La Roche & Co. in Basel succeeded in synthesizing several β-D-glucopyranosides (glucosides) and β-D-cellobiosides of vitamin D_3 and its metabolites, including the 1-, 3-, and 25-glucosides of 1α,25-$(OH)_2D_3$ (Figure 3, *3f, 3e,* and *3g,* respectively).[72] The biological activity of these synthetic analogues of the naturally occurring calcinogenic glycosides will be discussed further in Section V.A.2.

III. BIOLOGICAL SIGNIFICANCE IN PLANTS

A. Biosynthesis and Metabolism

Vitamin D_2 has usually been considered the plant vitamin D, as it originated from irradiated yeast. Vitamin D_3 and its metabolites were only recently reported to appear rather widely in the plant world, in contrast to early doubts about their occurrence. The calcinogenic plants mentioned in Table 1 are the main plant source of vitamin D_3 metabolites. Vitamin D_3 itself has also been isolated from noncalcinogenic plants like *Dactylis glomerata* (Orchard grass) and some common meadow grasses,[73,74] as well as from palm[73] and corn[74] extracts, and

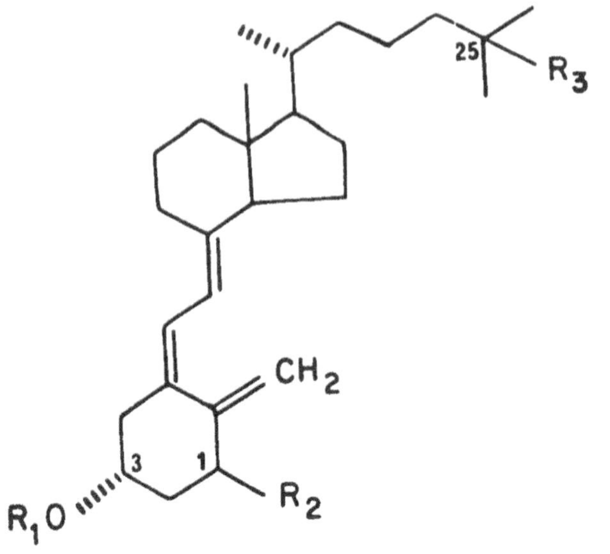

FIGURE 3.   Synthetic vitamin D_3 glycosides and analogues. (a) R_1 = β-D-Glucopyranosiduronyl, R_2 = R_3 = H; (b) R = β-D-Glucopyranosyl, R_2 = R_3 = H; (c) R_1 = β-D-Glucopyranosyl, R_2 = (α)OH, R_3 = H; (d) R_1 = β-D-Glucopyranosyl, R_2 = H, R_1 = OH; (e) R_1 = β-D-Glucopyranosyl, R_2 = (α)OH, R_3 = OH; (f) R_1 = H, R_2 = (α)β-D-Glucopyranosyloxy, R_3 = OH; (g) R_1 = H, R_2 = (α)OH, R_3 = β-D-Glucopyranosyloxy.

other plant species.[75-79] More recent reports on UV light-dependent formation of vitamin D_3 in *T. flavescens* (yellow oat grass),[80] *Medicago sativa* (alfalfa),[81] and phytoplankton[82] had suggested the presence of the natural provitamin 7-dehydrocholesterol. Hence, the inference was drawn[23] that vitamin D_3 is synthesized in plants by photolytic activation of a precursor, in a manner similar to that which occurs in the skin of vertebrates.[82,83] Furthermore, vitamin D_3-25-hydroxylase and 25-OH D_3-1α-hydroxylase activities could be detected in *S. glaucophyllum* leaf homogenates, the former in microsomes and the latter in mitochondria and microsomes.[23] Since the presence of substrates (vitamin D_3 and 25-OH D_3) for both enzymes was demonstrated, a hydroxylation pathway similar to that previously observed in the animal metabolism might presumably be operative in plants.[23] No evidence was yet advanced about the sequential order of hydroxylation that occurs in the plant *in vivo*, while in vertebrates vitamin D_3 is first converted in the liver to 25-OH D_3 which, in turn, is further metabolized by the kidney into 1α,25-$(OH)_2D_3$.[84]

The reactions mentioned heretofore are outlined in Figure 4 and illustrate an admittedly hypothetical pathway for the biosynthesis of vitamin D_3 and its metabolites in plants. The starting point might conceivably be acetyl-CoA, which would undergo further transformations via a mevalonic acid route, through the generally accepted scheme for the biogenesis of Solanaceae steroids[85] including sterols, sapogenins, and steroidal alkaloids. (Incidentally, leaves of several calcinogenic *Solanum* plants listed in Table 1, like *S. glaucophyllum*, *S. torvum*, and *S. verbascifolium*, were found to contain glycosides of the steroidal alkaloid solasodine,[86] whereas *S. glaucophyllum* seeds and tissue cultures obtained therefrom contain solasodine, diosgenin, and sitosterol[87,87a].) No evidence by conversion of labeled precursors to the anticipated final products could be provided so far.

Although the general pathway of aglycone biosynthesis can be envisaged, the route for glycoside conjugation has yet to be investigated, as well as the chronological sequence of hydroxylation steps which occur in the plant *in vivo*.

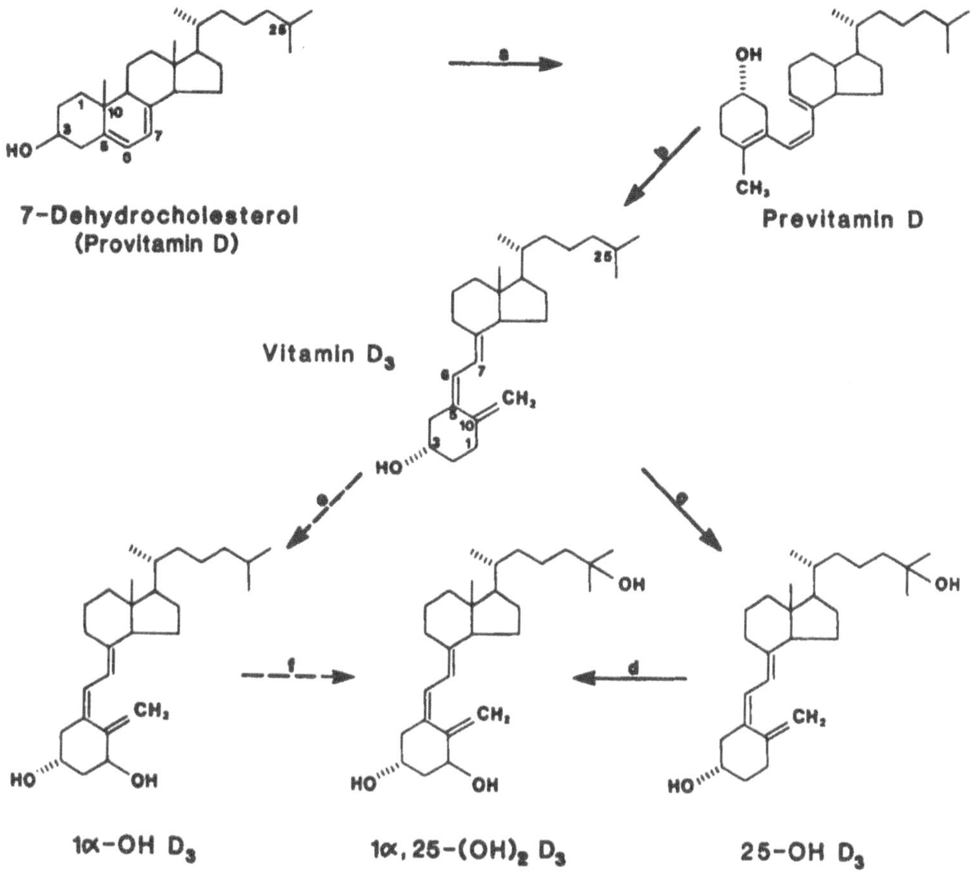

FIGURE 4. Current concept of the pathway of key vitamin D sterols synthesis in *S. glaucophyllum*. Broken line indicates alternative route by which this sterol could be formed. Specific precursor-product relationship has not yet been demonstrated unequivocally. Steps and conditions: (a), UV light irradiation; (b) temperature-dependent step; (c) vitamin D_3-25-hydroxylase; (d) 25-OH D_3-1α-hydroxylase; (e) possible 1α-hydroxylation; (f) possible 25-hydroxylation.

B. Distribution in Plants

The calcinogenic activity of *S. glaucophyllum* and *C. diurnum* was earlier reported to reside chiefly in leaves[1-12,69] and, apparently, none in stems;[13,69] however, a recent distribution study demonstrated substantial occurrence in berries, stems, and roots as well as in cell suspensions.[20a] Most interestingly, notable activity was found also in methylene chloride extracts of berries, stems, and roots (but not of leaves), thus suggesting that free vitamin D sterols might occur in these organs along with their corresponding glycosides.[20a] The importance of geographic location, climatic conditions, and development stage of the plant has been stressed. Thus, higher activity was found in *S. glaucophyllum* plants in the vegetative rather than in reproductive state, and from subtropical areas where calcinosis of cattle is in fact frequently encountered.[57] The influence of various growth and environmental conditions on the relative distribution of the individual vitamin D_3 derivatives present in the plant has not yet been studied.

C. Functions in Plants

The biological significance of vitamin D_3 and its calcinogenic derivatives for plants is not clear. Early speculations were that they may either play a role in the mineral metabolism of the plants or else serve a certain survival function in the plant ecosystem.[40,41] It has also been inferred that they might induce the synthesis of a component of the calcium transport system, manifested as a calcium-dependent ATPase, which could arguably take place in plant cells and result in increased calcium and perhaps also phosphate ion absorption by plants.[88] Incidentally, the *S. glaucophyllum* leaf was reported to contain very high levels of calcium.[89]

Vitamin D_3 derivatives induced physiological responses when supplied to plant systems.[12] Vitamin D_3 itself was found to reduce the calcium-controlled peroxidase secretion in *Beta vulgaris* (sugarbeet) cell culture,[90] and also to inhibit root elongation and stimulate calcium uptake by *Phaseolus vulgaris* roots cultured *in vitro*,[91] presumably owing to changes in calcium ions transport.[91] Further investigation demonstrated that the action of vitamin D_3 on root calcium ion uptake could be mediated by *de novo* protein synthesis, and that the sterol might induce the synthesis in *P. vulgaris* roots[92] of a protein with properties recalling those of plant calmodulin.[12]

Additional plant physiology studies pointed out other potential functions of vitamin D_3 and its derivatives besides the control of cellular calcium level. Thus, it has been assumed that they may be involved in the initiation of sexual reproduction in *Phymatotrichum* fungi species,[93,94] and in culturing plant protoplasts in chemically defined media.[95] Likewise, an endogenous function was proposed for these compounds by considering their activity as growth stimulators either on root formation in *Populus* species[96] or in axenically cultivated marine macroalgae.[97] In the former study, vitamin D_3, $1\alpha,25$-$(OH)_2D_3$ and their corresponding 3-*O*-glucopyranosides markedly promoted adventitious rooting.[96] In the latter case, the most active stimulator among the compounds tested was found to be $1\alpha,25$-$(OH)_2D_3$, thus suggesting that some vitamin D_3 metabolites should be involved in the algal metabolism.[97]

A full appraisal of the functions of vitamin D_3 derivatives in plants has yet to be made; however, it seems likely that they might participate in the plant calcium metabolism, possibly by stimulating the calmodulin synthesis.[12]

IV. TOXICOSES OF ANIMALS

A. Naturally Occurring Calcinosis

Plant poisoning of grazing animals by *S. glaucophyllum* and *C. diurnum* led to the identification of the related calcinogenic glycosides as the causative factor, and prompted the investigation of their biological activity. Consumption of *S. glaucophyllum* leaves has been shown to be the cause of a calcinotic disease of cattle called locally "enteque seco" in Argentina and "espichamento" in Brazil, which feature identical symptoms recalling vitamin D intoxication. A similar form of plant toxicity was later observed in Florida in horses and cattle which grazed leaves of *C. diurnum*. Early descriptions of the symptoms of "enteque seco"[98-100] (also referred to as "enteque ossificans"[100]) and of "espichamento"[101-104] remained seemingly unnoticed until the relationship between the toxicosis and the ingestion of *S. glaucophyllum*[100,105-107] became apparent,[1,13] and a preliminary assessment of the disease etiology could be made.[108-111]

1. Calcinogenic Plants

Calcinogenic glycosides occur in two genera of Solanaceae: *Solanum* and *Cestrum*. The taxonomic keys available to identify the plant species bearing them, *S. glaucophyllum* and *C. diurnum*, are summarized in Table 3. Optimal practices have been developed for their large scale cultivation as potential vitamin D crops, based on studies of germination, vegetative propagation, accessions performance, and environmental factors.[20]

a. S. glaucophyllum

S. glaucophyllum is a deciduous, rhizomatous shrub growing wild typically in areas of low-lying, wet land in the poorly drained eastern and central part of the Buenos Aires

Table 3
IDENTIFICATION OF *S. GLAUCOPHYLLUM* AND *C. DIURNUM*

Character	*S. glaucophyllum* Desfontaines[6,112]	*C. diurnum* Linnaeus[113]
Conspecific name[6,89,114]	*S. malacoxylon* Sendtner *S. glaucum* Dunal *S. glaucum* Bertolini *S. glaucescens* Baile *S. glaucumfrutescens* Larrañaga	
Common name	Duraznillo blanco ("litle peach")[6]	Day blooming jessamine
Height (m)	1—3	Up to 5
Leaf:		
Size (cm)	(9—13) × (1—2)	(6—18) × (2—3.5)
Flower:		
Corolla color	Blue	White
Corolla size (mm)	20—30, across	12—18
Corolla lobes	Short	Blunt; become recurved
Fruit:		
Color	Purplish-black	Dark purple
Size (mm)	7—10, across	6—9
Seed	Many, 5 mm long	2—3, oblong
Chromosome number (2n)	24	16 + 0-6B[247a]

[a] The author is grateful to Prof. Julia F. Morton, University of Miami, Coral Gables, FL, and to Ms. Priscilla Fawcett, Fairchild Tropical Garden, Miami, FL, for their kind help to point out this reference.

province and in the northeastern provinces of Argentina as well as in the Matto Grosso region of Brazil, in Uruguay and Paraguay,[114] and in India.[114a] Accordingly, the incidence of the toxicosis is greatest in flooded areas having pastures of higher feeding value than drier areas.[6,115] The plant has an extensive and deep underground root system which apparently confers considerable resistance to cultivation and weed killers. Dispersal is probably by animals and birds eating the berries or seeds. When a seedling is formed, vegetative propagation develops, and a large array of *S. glaucophyllum* can thus be established. Fallen leaves retaining their toxicity for several months (see Section II. C) mix intimately with neighboring herbage on the underlying pasture (thus improving palatability) and are likely to be eaten accidentally by the grazing animal.[6] Extensive information on the growth habit, distribution, and ecology of *S. glaucophyllum* has been presented in detail[6] (Figure 5).

b. C. diurnum

This plant is native to the West Indies and naturalized in subtropical and tropical regions like Hawaii, California, Texas, and southern Florida, where it is found in neglected fields and pastures[113], and also in India.[114a] The plant flowers and fruits at least twice a year, and it grows rapidly from seeds. Dispersal is also by birds eating fruits and dropping seeds[113] (Figure 5).

2. Calcinosis Symptoms

Calcinosis is regarded as the pathological deposition of calcium salts in soft tissues[7] that

FIGURE 5. Calcinogenic plants cultivated in the experimental fields of the Agricultural Research Organization, The Volcani Center, Bet Dagan, Israel.[20] (a) *S. glaucophyllum;* (b) *C. diurnum.* (Courtesy Dr. Arieh Levy, Agricultural Research Organization, The Volcani Center.)

Table 4
STUDIES OF NATURALLY OCCURRING CALCINOSIS

Grazing animal	S. glaucophyllum	C. diurnum
	Ref.	
Cattle	1,6,13,65,66,69,98, 100,105—107,109, 111,116—124,248	125
Sheep	1,13,69,99,104	
Horses	1,13,69	1 3

occurs in hypervitaminosis D or as a result of the consumption of calcinogenic plants. Table 4 summarizes the literature on the naturally occurring calcinosis induced by the calcinogenic glycoside-bearing plants *S. glaucophyllum* and *C. diurnum* in grazing animals.

Intake by cows of 50 g of air-dried leaves of *S. glaucophyllum* (roughly equivalent to 200 g of fresh leaves, 50 fresh leaves, or 0.1% of the diet) induced the toxicosis in 8 to 10 weeks.[6] Dry plant given by stomach tube to 410-kg heifers and 45-kg Romney ewes in single doses of 5 g and 0.5 g, respectively, produced symptoms for several days.[13] Symptoms were found to be more severe in the bull than in the horse, suggesting prolonged exposure or greater ingestion of *C. diurnum.*[125] The effect of the plant is likely to be cumulative.[6]

The characteristic features of the disease will be described next.

a. General Condition

Early clinical symptoms of the toxicosis include progressive loss of appetite and weight, kyphosis,[121] stiffness of the forelimbs, back arching, tucked-in abdomen, and painful gait.[1,6,13]

With the progress of disease, the animals become emaciated, excitable, and tired, and may show signs of acute cardiac and pulmonary insufficiency.[13] The joints cannot be extended completely, the thoracic limbs bow outward, the shoulders are stiff and deformed.[13] As a result, the animals move with an arched back, carrying their weight on the forepart of the hooves.[1,6,13] Similar clinical symptoms of chronic wasting and lameness were observed in *C. diurnum* intoxication.[113,125]

b. Hypercalcemia and Hyperphosphatemia

Ingestion of calcinogenic plants causes perturbation of calcium and phosphorus metabolism in the grazing animals, resulting in hypercalcemia and hyperphosphatemia[1,6,13,107,125] and, consequently, in an ion product of Ca × P in blood markedly higher than normal. Modification in blood phosphorus concentration has been conveniently used for detection and determination of calcinogenic activity in plants.[11,15,21,24,43,46,62,65,67,68]

c. Calcification of Soft Tissues

The abnormally elevated ion product of Ca × P in plasma leads eventually to the deposition of these ions in soft tissues.[6] Post-mortem examination of affected animals revealed mineralization of soft tissues, especially in the cardiovascular and respiratory systems[1,13,69,100,104,107,113,119,125] and often involving the aorta.[13,119,124] Deposition of calcified materials was also observed in lungs,[1,13,99] kidneys,[1,13] and elastic tissues like arteries, tendons, and ligaments.[1,113,125] The arterial calcification precedes that of lung, the latter being observed only in animals with a calcification of the aorta above 150 μmol Ca per gram of wet tissue.[124]

The mineral deposited in arterial and lung tissues appeared to be a poorly crystallized apatite. Following incineration, the diffraction patterns of carbonate apatite and magnesium whitlockite, respectively, were found.[124]

Relevant data on the histopathology of the naturally occurring calcinosis, particularly with respect to the lesions of the affected organs, have previously been presented in great detail.[8,13]

d. Skeletal Changes

The high levels of calcium and inorganic phosphorus led also to changes of the bone structure characterized by increased bone formation[122] and osteosclerosis.[6] Also recorded were fibrosis of the marrow, endosteal and peristeal hyperostosis, and damaged formation of ground substance of osteoblasts and fibroblasts.[123] Generalized osteopetrosis was related to hypoparathyroidism and hypercalcitoninism,[113] whereas osteonecrosis was believed to result from direct action by the *C. diurnum* active factor.[125]

e. Hematological Changes

Anemia and low erythrocyte counts, hematocrit, or hemoglobin concentration, have been reported following *S. glaucophyllum* grazing by cattle.[98,111] Suspension of plant intake led to the recovery of normal hematological parameters.[111]

f. Miscellaneous

Studies on the concentration of cholesterol (total, free, and esterified)[120] and of sialic acid[118] in serum of both affected and healthy cattle did not show any conclusive discrepancy. The possibility of bacterial[116,117] and parasite[1] infections, or dietary mineral imbalance, was ruled out as well.[1,116,117]

3. Economic Significance

Calcinosis produced by the calcinogenic glycosides-containing plants *S. glaucophyllum* and *C. diurnum* in grazing animals causes serious economic damage in cattle farming and

Table 5
STUDIES OF EXPERIMENTALLY INDUCED
CALCINOSIS AND PHYSIOLOGY

Experimental animal	Ref.	
	S. glaucophyllum	*C. diurnum*
Chick	17,29,33,38,47,50,63,64,126—141	54,55,133,142
Mouse	143,177	
Rat	18,19,28,30,32,34,35,42,44, 45,49,51,52,58—61,63,89,139, 144—178	179
Hamster	149	
Guinea pig	25,151,152,180—185	
Rabbit	14,27,31,42,68,69,149,186—193	
Sheep	31,42,69,110,151, 152,182,194—200	
Pig	201—204	205
Cow	1,31,66,69,107,122, 123,206—218,249	
Horse	69	

losses in meat and milk production. Statistics on the incidence of the disease in Argentina and Brazil have shown that in certain fields as much as 5,[103] 10,[13] and 10 to 30%[6] of the cattle were affected by *S. glaucophyllum* consumption. In Argentina, the toxicosis was reported as the cause of death or slaughter of approximately 300,000 cattle each year, as well as an unknown number of sheep.[25] Intoxicated animals were often detected before death and then sent for slaughter; hence, their value was diminished because of reduced weight and poor meat quality.[6] Another setback was the required retention of a large part of heifer calves as herd replacements.[6] Likewise, as a result of the toxicosis, many pastures remained unused in summer.[6] An estimation of the annual loss in revenue resulting from *S. glaucophyllum*-induced calcinosis in Argentina in 1977 was approximately $20 million (U.S.)[10]

So far no antidote to the intoxication has been available. If the animals are removed in the early stage of disease from the affected areas, they may recover.[13] Transfer to safer pastures at later stages may result in weight improvement, but the tissue calcification persists.[6,13] Leaving the animals in problem areas led to a worsening of their condition, and eventually to death.[1,13]

A joint project of the FAO of the United Nations and the INTA in Buenos Aires studied measures for the field control of the toxicosis.[69] Accordingly, the possibility of removal of calcinogenic plants from pastures, mechanical cultivation, treatment with herbicides, or better drainage of the area, was envisaged and is discussed extensively elsewhere.[6]

B. Experimentally Induced Calcinosis

The clinical symptoms of the disease can be reproduced experimentally in a variety of animals, thus confirming the implication of the calcinogenic plants and their related glycosides in the naturally occurring toxicoses. In Table 5 are listed the references to experimental toxicity and physiology studies carried out with *S. glaucophyllum* and *C. diurnum* in different animals.

The toxic effect has been suggested to depend upon the species of animal treated, the means of administration, the dose of the plant material, and the duration of experiments.[4,8,10] The interplay of these factors might lead to some apparent anomalies commented upon in the pertinent literature.

Differences in animal species have been noted;[126,173] thus, ruminants, rabbits, and guinea pigs were found to be much more sensitive than rats and chickens.[65] In other experiments, the results with rabbits were estimated as remarkable, with hamsters equivocal, and with rats negative.[149] Pigs appeared less susceptible to the action of *S. glaucophyllum* than sheep.[203] Renal calcification was rarely evident in chicks, in contrast to rats and guinea pigs.[10] Young calves proved to be less susceptible than older cattle to the effects of the plant.[210] The sensitivity to *S. glaucophyllum* of several animal species, expressed by plant doses used for a specific strain, has been evaluated.[161] Untreated animals were usually used as the control.[218]

The oral or intraruminal administration of *S. glaucophyllum* in rats, rabbits, sheep, and cattle elicited better responses than intravenous,[4,42,200,211] intramuscular,[42,200] or subcutaneous injection,[42,200] conceivably because the active substance may reach directly its presumed site of action (the cells of the intestinal mucosa), whereas intravenous administration may be followed by rapid excretion.[211] On the other hand, when added to food, the plant dose was found to depend upon intake.[161]

Comprehensive reviews on experimentally produced toxicity and on physiological studies have appeared[8,10,12] and should be referred to for details. The following treatment is intended to summarize briefly the relevant features reported in the extensive literature available on the subject.

1. Toxicity Effects of Calcinogenic Glycosides-Bearing Plants
a. General Condition

Loss of weight in guinea pigs[25] and cattle[208,210] and loss of appetite and stunted growth in calves were usually observed, particularly when high dose levels were employed.[210]

b. Hypercalcemia and Hyperphosphatemia

Chronic administration of calcinogenic plants or extracts thereof was found to disturb the calcium and phosphorus metabolism and, as a result, to produce hypercalcemia and hyper-phosphatemia in various animals[42,193] such as chicks,[33] rats,[151,153] guinea pigs,[25] rabbits,[188-191] sheep,[195,196] and cattle.[1,107,125,206,208,210,211] The increased levels of calcium and phosphorus in blood have been ascribed to increased intestinal absorption of calcium and phosphate[33,60,63,127,150,191,209] from the diet,[149,190] stimulation of bone resorption,[190,191] impaired reabsorption of these ions by the kidney tubules,[161] or a possible interplay of all factors.[8]

c. Mineral Balance

The effect of *S. glaucophyllum* on the mineral balance of calcium, phosphorus, magnesium, and copper was studied in rats,[161] guinea pigs,[180-182] sheep,[182,196,197] and cattle.[211] Calcium, phosphorus, and magnesium absorption increased.[182] Absorption of copper was lower in treated animals, as compared with controls, and its storage in the liver and kidney was less than that of control animals.[182] In other experiments, however, copper levels remained unaffected.[199] Retention of phosphorus was similar to that of the controls, while that of magnesium was lower.[182]

The magnesium absorption was found to decrease in rats fed a normal calcium diet and to increase in those on a low calcium diet.[161] The significance of magnesium in calcinosis has been stressed,[10] and alteration of the mineral balance leading to increased availability of calcium, magnesium, and phosphate may result in deposition of apatite and magnesium whitlockite in soft tissues.[124] These mineral deposits are rich in magnesium, as compared with bone, and their chemical and crystalline composition has already been discussed[124] (see Section IV.A.2.c). Reportedly, inclusion in the diet of phosphate-binding compounds such as aluminum sulfate prevented the absorption of phosphate and reduced the incidence of calcified lesions.[149]

d. Calcification of Soft Tissues

Prolonged administration of plant material to a variety of animals resulted in marked soft tissue calcification, especially of the cardiovascular and respiratory systems[18,25,66,107,110,143,144,148,149,185-187,194,196,205,208,210,215,216,218] as well as musculature, fasciae, tendons, musculature of stomach and part of the large intestine,[185] and kidneys, spleen, and liver.[25] In the lung the characteristic lesions are swelling, fragmentation, and calcification of the elastic fibers of the alveolar wall, accompanied by disappearance of mucopolysaccharide from, and calcification of, the bronchial cartilages and their replacement by fibrous connective tissue.[218] The histopathological features of the experimentally produced calcinosis have been described extensively,[8,10,185,218] and they are similar to those of the naturally occurring toxicosis, discussed previously (Section IV.A.2.c).

Some cardiovascular modifications in calves intoxicated with *S. glaucophyllum* were also reflected in a slightly changed electrocardiogram pattern, which might be indicative of very early lesions.[210]

e. Skeletal Changes and Bone Studies

Toxicity experiments in rats, guinea pigs, and sheep with *S. glaucophyllum* produced changes in the bones defined as osteosclerosis and characterized by an abnormal increase in the density of the bones without macroscopic changes of the anatomic structures.[152] The presence of excess mineral in bone of rabbits as a result of plant intoxication led to osteopetrosis.[27] Further regressive changes in osteocytes resulted in osteonecrosis which, combined with retarded apposition, led subsequently to osteopenia.[27]

During the early stage of *S. glaucophyllum* administration in rats, the rate of bone turnover was depressed.[30] Acute stimulation of cell-level bone formation on trabecular surfaces in *S. glaucophyllum*-treated growing rats may play a role in the hyperostosis seen in naturally occurring calcinosis.[173] At the histological level, long-term administration of *S. glaucophyllum* was reported to induce a higher rate of bone tissue synthesis on trabecular and endosteal surfaces in growing rats on a normal calcium-content diet.[161] Further quantitative histological studies confirmed that *S. glaucophyllum* might be able to mineralize the uncalcified osteoid tissue of vitamin D-deficient rats, inducing no change in the extent of the resorption surfaces.[52] Other features were presented earlier (Section IV.A.2.d).

f. Hematological Changes

Rats fed *S. glaucophyllum* leaves exhibited anemia and a progressive decrease in hematocrit, hemoglobin concentration, and circulating reticulocytes.[163,172,177] Low erythrocyte counts, hematocrit,[207] hemoglobin concentration,[25,207] and packed cell volume[25] have also been reported following experimental administration of *S. glaucophyllum* to cattle and guinea pigs. In sheep, however, packed cell volume and hematocrit remained constant.[199] Upon *S. glaucophyllum* administration in both rats and mice, bone marrow becomes unable to replace red blood cells at an adequate rate. Suspension of plant intake restored the normal hematological parameters.[177]

g. Miscellaneous

The blood glucose and cholesterol concentrations of *S. glaucophyllum*-treated chickens were found to be significantly higher than the corresponding controls, suggesting that the calcinogenic principle may interfere with both protein and carbohydrate metabolism.[140]

Chickens treated with a *S. glaucophyllum* extract had significantly lower serum alkaline phosphatase, alanine aminotransferase, and aspartate aminotransferase activities than the untreated chickens.[140] Decreased serum alkaline phosphatase activity was observed also in experiments with rats[89,160,161] and rabbits,[190] and was attributed to a decrease in the intestinal isoenzyme which was not compensated for by the progressive increase in the bone isoenzyme.[161]

V. PHYSIOLOGICAL AND PHARMACOLOGICAL PROPERTIES

A. Physiological Aspects

1. Structural Characteristics

There is ample evidence on the ability of calcinogenic glycosides to mimic the physiological actions of the main aglycone $1\alpha,25$-$(OH)_2D_3$.[8,10,12] The possible presence of the free aglycone in some particular plant batches has been suspected,[8] because of controversial reports on the activity of ethereal plant extracts[18,148] and on the interaction of unhydrolyzed plant extracts with an intestinal cytosolic receptor[29] held to interact only with free $1\alpha,25$-$(OH)_2D_3$.[38] Recent work has indeed demonstrated activity in methylene chloride extracts of several plant organs[20a] (see Section III.B). The glycoside has been found less active *in vivo* than the free aglycone (arguably due to a possible interference of the carbohydrate moiety with the intestinal absorption) and more active *in vitro*, probably owing to its hydrophilic properties.[174] The suggested presence in plants of a series of glycosides with different aglycones and various carbohydrate sequences could admittedly account for differences in the biological action between plant extracts and $1\alpha,25$-$(OH)_2D_3$.[51,52,61,156] Other views speculated that the *S. glaucophyllum* active principle might act *in vitro* as the unhydrolyzed glycoside conjugate.[34]

The main aglycone, $1\alpha,25$-$(OH)_2D_3$ itself, is considered to be the major active hormonal form of the prohormone, vitamin D_3, from which it originates by an initial 25-hydroxylation in the liver, followed by 1-hydroxylation in the kidney.[84,220] Its main functions are the stimulation of intestinal calcium and phosphate absorption, mediation of bone remodeling, and conservation of minerals in the kidney.[220] A detailed treatment of the vitamin D functions, metabolism, and receptors is beyond the scope of this survey, and series of recent reviews on this subject are available and should be consulted.[9,84,220] Attention will, however, be paid to studies devised to assess the effect of 1-, 3-, and 25-hydroxy substitution pattern, and 5,6- double bond geometry on vitamin D activity. Thus, competitive receptor-binding assays established the importance of the hydroxy groups at C-1 and C-25, and of the natural ring A-configuration (1*S*, 3*R*, 5*Z*) to binding affinity for the intestinal $1\alpha,25$-$(OH)_2D_3$ receptor protein,[221] the most effective being the $1\alpha(S)$-hydroxy group.[221] Likewise, 1α-, 3β-, and 25-hydroxy groups were found to contribute substantially to the *in vitro* bone-resorbing activity of vitamin D_3,[9,222] which reportedly required that at least two of them should be free.[223] These studies led to speculations on whether conjugation to one or more sugar molecules would have an effect equivalent to deletion of a hydroxy group.[34] Some attempts at the elucidation of this point were made with synthetic analogues, and a brief description is given below.

2. Synthetic Analogues

Once the synthetic analogues became available (see Section II.D), their biological activity was investigated. Thus, the D_3-glucuronide (Figure 3, *3a*) was found to be biologically active *in vivo* only after hydrolysis to the free sterol.[70] D_3-glucoside (Figure 3, *3b*) stimulated intestinal calcium absorption and bone calcium mobilization, whereas its 3α-isomer proved somewhat less active, and the corresponding acetate was inactive.[71] In another set of assays, D_3-glucoside was shown to retain virtually all the original activity of the aglycone, while the 1α-OH D_3 3-glucoside (Figure 3, *3c*) retained about 10% of the activity of the free 1α-OH D_3, and 1α-OH D_3 3-cellobioside retained none.[224]

In order to determine whether and how the presence of the β-D-glucopyranoside moiety linked to the 3-, the 1-, or the 25-hydroxy group of $1\alpha,25$-$(OH)_2D_3$ affected the biological activity of the conjugates derived therefrom, a comparison was undertaken among the three corresponding monoglucosides (Figure 3, *3e* to *g*). Preliminary results suggested that the 1- and the 3- glucosides (*3f* and *3e*, respectively) showed no or only little effect on serum calcium, bone weight, CaBP, or calcium deposition in the egg shell, while the 25- glucoside (*3g*) retained more than half the activity of the aglycone, for reasons which are not apparent.[225,226] In prophylactic and therapeutic tests in chickens, only the 25-glucoside had antirachitic activity comparable to that of the free $1\alpha,25\text{-}(OH)_2D_3$, following both intravenous and oral administration.[277] Due to its high potency and solubility in polar solvents, the 25-glucoside was suggested to be of therapeutic interest.[227] However, further investigations[228,228a] showed the three monoglucosides biologically active and equipotent *in vivo*, most likely as a result of hydrolysis to the free aglycone, but less active than the aglycone in this respect.[228a]

3. Physiological Effects of Plant Calcinogenic Glycosides

This subject has been reviewed extensively, and a succint survey will be given below. For more detailed information on this topic, the interested reader is referred to several recent reviews.[8,10,12] The experimental toxicity and physiological studies which describe the symptoms resembling those of the naturally occurring calcinose have been presented in Section IV.B.

a. Stimulation of Intestinal Calcium and Phosphate Absorption

The administration of calcinogenic plants or extracts thereof stimulated the intestinal absorption of calcium in different animal species[17,33,54,60,61,63,64,89,127,129,150,160,189-191,196,209] and increased the intestinal absorption of phosphate.[60,133,139,189,191,196] The stimulatory effect was ascertained following the rise of ^{45}Ca uptake by rat intestinal slices *in vitro*,[150] in chick and rat intestine *in vitro*,[63] and in intact rabbit *in vivo*.[191] Increase in intestinal ^{47}Ca absorption was also observed in intact rats.[155,166] Likewise, *S. glaucophyllum* was found to stimulate ^{32}P absorption in chicks,[133] rats,[153,166] and rabbits,[191,193] notwithstanding a conflicting report.[33] Thus, it has been inferred that the active plant factor and $1\alpha,25\text{-}(OH)_2D_3$ had similar effects on intestinal calcium and phosphate transport.[60,204] The quantitative differences in the effects on calcium and phosphorus metabolism may be dose related.[60]

The plant extracts proved to be a potent stimulator of duodenal calcium absorption in strontium-fed chicks,[127,142] in an embryonic chick organ-cultured intestine *in vitro*,[64] in vitamin D-deficient rats,[60,61,150,179] in nephrectomized rats,[35,59,89,156,159,179] and in thyroparathyroidectomized rats.[147,153,171] Depressed duodenal calcium absorption could be restored by administration of calcinogenic plant extracts in diabetic[158,168] and nephrectomized[89] rats. These studies indicate that the plant active factor could act directly on the intestine, without requiring renal metabolism (i.e., 1α-hydroxylation) or 25-hydroxylation for effects on calcium transport, and actually inhibiting renal 1α-hydroxylation *in vivo*.[29] No changes were observed on the renal handling of calcium and phosphate nor on the intestinal excretion of endogenous calcium in short-term experiments,[59] while a deleterious effect could be seen in long-term experiments.[161]

An inhibitory effect of actinomycin D on intestinal absorption of calcium and phosphate stimulated by *S. glaucophyllum* was found *in vivo* in rats[165,166] and rabbits,[192] and also in organ-cultured chick duodenum.[64] Conversely, prednisolone did not inhibit the stimulatory effect of *S. glaucophyllum* upon intestinal calcium transport, in contrast to the action of the aglycone $1\alpha,25\text{-}(OH)_2D_3$.[51] No relationship between activity of intestinal alkaline phosphatase and calcium or phosphate absorption was observed in experiments done with *S. glaucophyllum* extracts on rachitic chicks treated with ethane-1-hydroxy-1,1-diphosphonate, which was presumed to block the increase of intestinal calcium or phosphate absorption, but to be ineffective against the rise in intestinal alkaline phosphatase activity.[138]

b. Stimulation of CaBP Synthesis

The calcinogenic plant extracts, like their aglycone $1\alpha,25\text{-}(OH)_2D_3$, were found capable of stimulating CaBP synthesis in a number of species: chicks,[127] rachitic chicks,[17] rats,[35] and also in rabbit ileum,[149] in chick mucosa *in vitro,*[63] and in embryonic chick duodenum *in vitro.*[64] They could also restore depressed duodenal CaBP in diabetic rats[168] and overcome the strontium-induced inhibitory effect on CaBP synthesis in chicks.[50,127,142] *S. glaucophyllum*-stimulated CaBP synthesis in the embryonic duodenal organ culture system was completely abolished by actinomycin D.[64]

S. glaucophyllum could apparently compete with $1\alpha,25\text{-(OH)}_2D_3$ for receptor sites on chick intestinal chromatin both *in vivo*[29] and *in vitro.*[29,35] These results suggest that the calcinogenic plant active principle and $1\alpha,25\text{-(OH)}_2D_3$ might act at the same cellular sites and have the same mode of action.

c. *Stimulation of Bone Resorption*

The influence of the calcinogenic plant principle on bone resorption is a controversial issue. Early work reported bone-resorbing activity of the plant extracts *in vivo.*[152,153,161,190,191] These observations were confirmed by *in vitro* studies which demonstrated bone-resorbing effects of *S. glaucophyllum* on mouse calvaria,[35,229,230] on a fetal mouse fibula organ culture system,[231] and on embryonic chick frontal bone.[154] Since conflicting results were concurrently reported for experiments carried out *in vivo*[52,61,173] or *in vitro*[61,160,232] with crude extracts, the inference was drawn that only low doses[34] and/or desalted plant extracts[35,222,231] might have a direct bone-resorbing effect *in vitro*, while high doses inhibited it, arguably due to the presence of an antagonist in some lots of *S. glaucophyllum.*[34] Accordingly, the view has been expressed that the *S. glaucophyllum* factor may act *in vitro* as the unhydrolyzed glycoside conjugate, or else that the enzymes of the bone cells would hydrolyze it to release small amounts of $1\alpha,25\text{-(OH)}_2D_3$, which then should interact with the receptor.[34,231] The effect of calcinogenic plant extracts on bone resorption was therefore suggested to be dose- and time-dependent.[34,52,161]

Bone resorption could also be induced by *S. glaucophyllum* in thyroparathyroidoectomized rats, but not in intact rats, indicating an osteolytic effect of the plant extract in the absence of parathyroid hormone and calcitonin regulation of bone turnover.[171] Depression of bone turnover rate by *S. glaucophyllum* in intact rats was actually demonstrated by calcium kinetics studies.[30]

d. *Hematological Changes*

S. glaucophyllum was found to affect the iron metabolism[172,177] and the phosphorus metabolism in the erythrocyte,[174] the latter process leading to decreased ATP and 2,3-diphosphoglycerate levels, as shown by higher plasma erythropoietin levels and lactate/pyruvate ratio.[174] These results were obtained both *in vivo* and *in vitro* and indicated a direct effect on the red cell.[174]

e. *Effect on Citrate Metabolism*

A time study of the chronic intoxication of growing rats with *S. glaucophyllum* leaves[164,169] showed high plasma citrate levels after 4 d of treatment, even higher levels after 1 week, and normal plasma citrate after 12 d and up to 8 weeks of treatment. The renal ability to oxidize citrate *in vitro* showed a progressive decrease during the three stages. The significant regression between the daily plant intake and urinary citrate was indicative of a direct effect of the plant on citrate metabolism.[169]

Administration of *S. glaucophyllum* extracts was found to increase plasma citrate levels in normal and vitamin D-deficient rats,[193] as well as the citrate content of bone in normal rats.[161] The citrate production was also promoted by *S. glaucophyllum* in embryonic chick frontal bone cultured *in vitro*, indicating a bone resorptive process.[154]

Increased citrate production and mitochondrial ATPase activity and reduction in ATP content of some organs were reported in isolated rat liver mitochondria experiments, suggesting that mitochondria are a major site of action of *S. glaucophyllum.*[154]

f. Effects on Embryos, Fetuses, and Maternal-Fetal Relationship

The $1\alpha,25\text{-}(OH)_2D_3$-like activity of calcinogenic plants was studied also on two very different experimental models: avian embryos and mammalian fetuses. Thus, injection of *S. glaucophyllum* extracts into eggs (at 15[47,129] or 18 d[64] of incubation) induced CaBP synthesis[64] and hypercalcemia[47,129,132] in the chick embryo. Mortality produced by high doses was related to hypercalcemia[129] or, possibly, to plant toxicity.[47]

Oral administration of *S. glaucophyllum* to pregnant cows raised the plasma concentration of calcium and phosphate both in dams and in fetuses.[212-214] When given at the end of gestation, *S. glaucophyllum* increased plasma calcium level in the dam, the fetus, and the neonate,[213] and the calcium content of colostrum.[212,213] The results suggested that the active factor might stimulate the placental transfer of calcium from the dam to the fetus, or else that it may cross the placenta and act directly on the fetus, inducing bone resorption and/or increasing intestinal absorption of calcium from fetal intestine.[212-214] No toxic effects were recorded, presumably because of the low doses used and the short-term treatment.[212-214] The importance of these results for potential use of *S. glaucophyllum* in prophylaxis of parturient hypocalcemia in dairy cows will be discussed later (Section V.B.2.a).

g. Effect on Renal Mineral Control

Experiments done with *S. glaucophyllum* on both normal and nephrectomized rats suggested that the plant biological activity is apparently not dependent on the presence of the kidneys,[89,156,159] notwithstanding a significantly high induced polyuria.[159] Administration of *S. glaucophyllum* extracts to rats did not induce an early effect on the renal handling of calcium and phosphate.[59] Further work, however, substantiated a deleterious effect of the plant extracts on calcium, magnesium, and phosphorus handling by the kidney,[161] arguably related to a reported ATP depletion.[154] Renal clearance values and urinary citrate in rats increased toward the later stage of *S. glaucophyllum* treatment, although the renal ability to oxidize citrate *in vitro* decreased progressively.[169] The potential therapeutic use of calcinogenic plant extracts in patients with chronic renal failure will be discussed further in more detail (Section V.B.1.a).

B. Pharmacological Aspects

The pharmacological properties of the calcinogenic glycosides are presumably related to their main aglycone, $1\alpha,25\text{-}(OH)_2D_3$, while the conjugated hydrophilic carbohydrate groups would confer better solubility features and prevent a too fast liberation of the sterol.[51] Consequently, the intestinal mucosa of the ingesting subjects would be exposed for a relatively longer time to the active metabolite.[51]

Under the trademark Rocaltrol® and in doses of 0.5 to 2.5 μg/d, $1\alpha,25\text{-}(OH)_2D_3$ itself has proved effective in treatment of several disorders of bone and calcium metabolism in humans, such as hypocalcemia of hypoparathyroidism, hypophosphatemic osteomalacia, vitamin D-dependent rickets, and renal osteodystrophy.[233-235] Since the clinical applications of $1\alpha,25\text{-}(OH)_2D_3$ have been the subject of several extensive reviews,[233-237] we shall outline below the potential therapeutic uses of calcinogenic plants in human and veterinary medicine, as well as in animal husbandry. These topics have previously been surveyed briefly.[12,238]

1. Human Medicine
a. Chronic Renal Failure

During a treatment undertaken in 1976, *S. glaucophyllum* plant prepared in a pharmaceutical form by Gador Laboratories, Buenos Aires, was given orally for 6 d in a single daily dose equivalent to 300 mg of dried leaf to four patients with renal failure on prolonged hemodialysis.[239] As a result, calcium absorption was increased in all patients, and serum calcium in three of them. Serum phosphate levels remained unchanged. Similar results were obtained with the equivalent of 1 g aqueous extract of plant leaves during 14 weeks.[240]

An improvement in calcium absorption has also been observed in three uremic patients on hemodialysis, given orally 200 mg *S. glaucophyllum* for 3 d.[156,159] This plant was able to increase the calcium uptake even in patients with terminal renal failure, thus suggesting that its effectiveness is apparently not related to the presence of kidneys.[159]

b. Hypoparathyroidism

In the course of a short-term study reported in 1977, four patients with postsurgical hypoparathyroidism received orally tablets containing the equivalent of 500 mg dried leaves of *S. glaucophyllum*, during 7 to 8 d.[241] The treatment induced a rapid increase in serum calcium levels and in the urinary excretion of calcium and phosphate, whereas serum phosphate tended to decrease as compared with elevated pretreatment levels. No effects were observed on 25-OH D_3 serum levels, urinary excretion of total hydroxyproline, blood hemoglobin, or renal and hepatic functions. Similar results were obtained with plant doses of 1 to 2 g/d.[242]

c. Effect on the Use of Vitamin A

Studies with rachitic rats demonstrated that *S. glaucophyllum* could increase significantly blood calcium levels, hepatic vitamin A content, and tibial ash content,[175] and decrease cytochrome P 450 induction in liver and kidneys,[176] resulting in a more efficient use of retinol for growth. The results suggested a potential therapeutic use of *S. glaucophyllum* in patients with renal lesions that prevent the synthesis of vitamin D metabolites, and consequently of CaBP.[175]

d. Treatment of Bone Disorders

Reports on the ability of *S. glaucophyllum* to cure rachitic lesions in vitamin D- and phosphate-deficient rats suggested its potential therapeutic value in such bone disorders as uremic osteopathy.[58]

2. Veterinary Medicine

a. Prevention of Milk Fever in Dairy Cows

Following positive results in the prophylaxis of parturient paresis (milk fever) in dairy cows by using $1\alpha,25\text{-}(OH)_2D_3$, *S. glaucophyllum* was evaluated as a potentially convenient alternative. Oral administration of plant leaves to cows 1 week before calving prevented the occurrence of hypocalcemia and hypophosphatemia at delivery.[212,215-217] The plant was found far more effective following oral than parenteral administration.[215,216] Single treatments resulted in an initial hypercalcemia and hyperphosphatemia, followed by a marked hypocalcemia and hypophosphatemia. Multiple doses, however, resulted in elevation of these blood parameters with no apparent rebound effect, and increased hypomagnesemia.[217] The treatment did not have an adverse effect on milk production in the first 11 d of lactation.[217] Studies of the systemic tolerance of doses recommended for the control of parturient paresis have drawn attention to eventual calcification risks, especially at high plant doses.[215,216]

b. Effect on Hereditary Pseudo-Vitamin D Deficiency of Pigs

Oral administration of *S. glaucophyllum* powder to rachitic pigs affected with a hereditary form of vitamin D-dependent rickets was effective in curing the disease, as evidenced by increased intestinal calcium absorption, increased phosphate kidney resorption, decreased serum alkaline phosphatase, and improved health of the animals.[202]

c. Effect on Acidotic Chicks

Studies performed on chicks with chronic metabolic acidosis indicated that dietary distribution of *S. glaucophyllum* did not modify the chick response to acidosis. Conversely, a mixture of *S. glaucophyllum* and vitamin D_3 reduced the deleterious effect of acidosis on bone development and enhanced the ash percentage of the tibia to higher values than in nonacidotic chicks. The results suggested a potential therapeutic treatment of some forms of chick osteopathy related to the acid-base equilibrium.[134]

3. Animal Husbandry

The effect of supplementing the diet of laying hens with powdered *S. glaucophyllum* leaves on egg-shell thickness has been studied by several groups.[130,131,136,137,243] A significant increase in shell thickness was apparent for eggs laid on the second and subsequent days of the treatment, but not for those laid during the first 24 h.[130] The percentage of laying decreased from approximately 70 to 60%, arguably due to some toxicity of the plant leaves.[131] No symptoms of hypervitaminosis D were observed.[131] It was suggested[130] that the *S. glaucophyllum* supplement might restore CaBP levels in hens at the end of their laying period,[130,131] and, hence, improve dietary calcium absorption and correct deficiencies in egg-shell calcification.[130,131]

VI. METABOLISM IN HUMANS AND ANIMALS

The calcinogenic activity of the naturally occurring glycosides resides conceivably in the aglycone fragment, with the sugar residues providing favorable solubility and distribution characteristics. The views on the biological action of glycosides as compared with aglycones have been discussed in a preceding section (V.A.1). Further evidence indicated that calcinogenic plant extracts administered *in vivo* gave rise to increased blood levels of $1\alpha,25$-$(OH)_2D_3$[28,38,49] and competed less effectively with radioactive $1\alpha,25$-$(OH)_2D_3$ on interaction with an intestinal receptor than the lipid-soluble moiety released by incubation with β-glucosidase.[38] The observation that the plant extracts are more effective following oral than parenteral administration[4,42,200,211,215,216,246] (see also Section IV.B) suggested the cells of the intestinal mucosa as a source of glycosidase activity.[211] Some additional enzymatic activity might also be provided by bone cells.[34,154,231] Hence, it seems likely that the $1\alpha,25$-$(OH)_2D_3$ released in the animal may result from metabolism of the hydrophilic glycoside at the target tissue (intestine or bone).

Preincubation *in vitro* of calcinogenic plant extracts with ruminal fluid[45,46,65,144,145,200,203] provided an adequate experimental model for simulating the glycosidase activity at the intestinal target tissue. Thus, the use of ruminal fluid from sheep[43,47,48,244] and cows[44] potentiated the effects of the incubated extracts on calcium and phosphorus metabolism in the chick embryo,[47,48] rat,[43-45,144,145,244] guinea pig,[65] rabbit,[43,46] sheep,[200] and pig.[203] The preparative aspects of these studies, outlined chiefly by Boland and colleagues, have already been discussed here (see Section II.A.1.c). The increased biological activity of extracts incubated with ruminal fluid was essentially related to the conversion by rumen microbes of the glycosides mainly into $1\alpha,25$-$(OH)_2D_3$ and $1\alpha,24,25$-$(OH)_3D_3$, the latter metabolite being either released or produced from plant precursors under the action of the complex ruminal microflora.[48a] Finally, the reasonable postulate was advanced that cleavage of the calcinogenic glycosides by intestinal hydrolases,[50] ruminal microbial action,[48a] or bone cells[34] apparently released and/or produced corresponding amounts of aglycones which either might ultimately cause calcinosis in grazing animals[12,50] or else may interact with a receptor *in vitro*.[34] These events are summarized graphically in Figure 6 representing two complementary views[8,10] of an integrated model for the physiological mechanism of calcinogenic glycoside-induced toxicosis. Accordingly, the calcinogenic plant glycosides provide an exogenous supply of $1\alpha,25$-$(OH)_2D_3$, thus by-passing the renal 1α-hydroxylase step. As a result, intestinal CaBP synthesis and plasma calcium and phosphorus levels are increased, and this excessive mineral absorption apparently cannot be physiologically accommodated for except

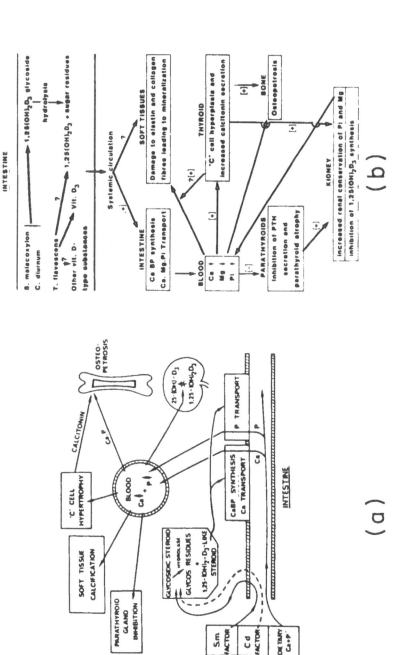

FIGURE 6. Two complementary views of an integrated model for the physiological mechanism of plant-induced calcinosis. (From [a] Wasserman, R. H., in *Dukes' Physiology of Domestic Animals*, 9th ed., Swenson, M. J., Ed., Cornell University Press, Ithaca, NY, 1977, 430; [b] Morris, K. M. L., *Vet. Human Toxicol.*, 24, 34, 1982. With permission.)

by deposition into soft tissues, leading to calcinosis and osteopetrosis. Accompanying features associated with calcinosis and discussed throughout this survey, are depicted in Figure 6 as well.

VII. FUTURE PERSPECTIVES

The economic relevance of the calcinogenic glycosides might be assessed in two ways: as the causal factor of the toxicosis of grazing cattle and, implicitly, as a potentially valuable source of vitamin D_3 sterols.

The considerable research effort aimed at the etiological and toxicological aspects of the calcinosis should be useful for the control of the disease and the reduction of its impact on the economics of cattle farming.

The rapidly increasing use of $1\alpha,25\text{-}(OH)_2D_3$ in clinical and biochemical studies has led to the need for this compound in larger quantities and at reasonable prices. So far, $1\alpha,25\text{-}(OH)_2D_3$ and its derivatives have been prepared by expensive multistep and low-yield syntheses. However, the likelihood of using calcinogenic plants as field crops of vitamin D_3 sterols seems remote at this stage, although at the current price of about \$925 per 25 mg of synthetic $1\alpha,25\text{-}(OH)_2D_3$, they may provide a fair alternative, considering dry leaf yields of approximately 3.3 ton/ha.[20] Furthermore, the reported plant content (20 to 140 mg of $1\alpha,25\text{-}(OH)_2D_3$ per kilogram dry leaves, see Table 2) might conceivably be improved by cultivation studies, tissue culture techniques, or genetic modifications. On the other hand, a concurrent approach was recently devised involving work with synthetic analogues of plant calcinogenic glycosides, which would be of interest both for preparative purposes, and for elucidation of the mechanism of physiological action and of the role played by the carbohydrate moieties and the site of their conjugation to the sterol nucleus.

Considering the laborious and difficult isolation and purification of plant calcinogenic glycosides, an alternative means of therapeutic application mainly in veterinary uses, like prophylaxis of milk fever in dairy cows and improvement of eggshell quality, might arguably be offered by dose-controlled administration of the plant in powdered form or as crude extracts. In that fashion, the calcinogenic plants may represent a fairly inexpensive source of $1\alpha,25\text{-}(OH)_2D_3$.

Most interesting, the metabolism of plant vitamin D_3 sterols seems to parallel that which occurs in animals. However, the taxonomic significance of the calcinogenic glycoside, occurrence, and distribution in plants is still intriguing.

ACKNOWLEDGMENTS

This is contribution No. 1986-E, 1987 series, from The Agricultural Research Organization, The Volcani Center, Bet Dagan, Israel. The work was supported by grant No. I-389-81 from the United States-Israel Binational Agricultural Research and Development Fund (BARD).

REFERENCES

1. **Carrillo, B. J. and Worker, N. A.,** Enteque seco: arteriosclerosis y calcificación metastásica de origin tóxico en animales a pastoreo, *Rev. Invest. Agropecu., Ser. 4,* 4, 9, 1967.
2. Editorial News and Views, Understanding the control of vitamin D synthesis or taking a short cut through South America, *Nature (London),* 253, 87, 1975.
3. **Wasserman, R. H.,** Active vitamin D-like substances in *Solanum malacoxylon* and other calcinogenic plants, *Nutr. Rev.,* 33, 1, 1975.

4. **Humphreys, D. J., Basudde, C. D. K., Varghese, Z., and Moorhead, J. F.,** Plant induced calcinosis in animals, in *Proc. 2nd Workshop Vitamin D,* (Vitamin D and Problems Related to Uremic Bone Disease), Norman, A. W., Schaefer, K., Grigoleit, H. G., von Herrath, D., and Ritz, E., Eds., Walter de Gruyter, Berlin, 1975, 731.
5. **Morris, M.,** A drug from a disease-causing plant?, *New Sci.,* 73, 135, 1977.
6. **Okada, K. A., Carrillo, B. J., and Tilley, M.,** *Solanum malacoxylon* Sendtner: a toxic plant in Argentina, *Econ. Bot.,* 31, 225, 1977.
7. **Wasserman, R. H.,** The nature and mechanism of action of the calcinogenic principle of *Solanum malacoxylon* and *Cestrum diurnum* and a comment on *Trisetum flavescens,* in *Effects of Poisonous Plants on Livestock,* Keeler, R. F., Van Kampen, K. R., and Jones, L. F., Eds., Academic Press, New York, 1978, 545.
8. **Wasserman, R. H. and Nobel, T. A.,** Vitamin D- related compounds as the toxic principle in calcinogenic plants, in *Basic Clinical Nutrition,* Vol. 2, Norman, A. W., Ed., Marcel Dekker, New York, 1980, 455.
9. **Stern, P. H.,** The D vitamins and bone, *Pharmacol. Rev.,* 32, 47, 1980.
10. **Morris, K. M. L.,** Plant induced calcinosis: a review, *Vet. Human Toxicol.,* 24, 34, 1982.
11. **Rambeck, W. A. and Zucker, H.,** Vitamin D-artige Activitäten in calcinogenen Pflanzen, *Zentralbl. Veterinaermed. Reihe A,* 29, 289, 1982.
11a. **Rambeck, W. A., Weiser, H., and Zucker, H.,** A vitamin D_3 steroid hormone in the calcinogenic grass *Trisetum flavescens, Z. Naturforsch. Teil C,* 42, 430, 1987.
12. **Boland, R. L.,** Plants as a source of vitamin D_3 metabolites, *Nutr. Rev.,* 44, 1, 1986.
13. **Worker, N. A. and Carrillo, B. J.,** "Enteque seco", calcification and wasting in grazing animals in the Argentine, *Nature (London),* 215, 72, 1967.
14. **Cassels, B. K., Rossi, F. M., Dallorso, M. E., Daskal, H., and Leiva, A.,** Principios tóxicos del *Solanum malacoxylon* Sendtner. I. Acción de fracciones separadas por solventes selectivos sobre el metabolismo fosfo-cálcico y tejidos blandos del conejo, *Contrib. Cient. Univ. Técn. Estado, Santiago, Chile,* 4, 7, 1971.
15. **Humphreys, D. J.,** Studies on the active principle of *Solanum malacoxylon, Nature (London) New Biol.,* 246, 155, 1973.
16. **Djafar, M. I. and Davis, G. K.,** Some properties of an unknown toxic substance present in the plant *Solanum malacoxylon, Malay. J. Agric. Res.,* 1, 104, 1972.
17. **Wasserman, R. H., Bar, A., Corradino, R. A., Taylor, A. N., and Peterlik, M.,** Calcium absorption and calcium binding protein synthesis in the chick: evidence for a 1,25-dihydroxycholecalciferol-like factor in *Solanum malacoxylon,* in *Calcium-Regulating Hormones,* Talmage, R. V., Owen, M., and Parsons, J. A., Eds., Excerpta Medica, Amsterdam, 1975, 318.
18. **Masselin, J. N., Abadie, G. J., Monesiglio, J. C., and Rossi, F. M.,** Acción del extracto etéreo de Duraznillo blanco *(Solanum malacoxylon), Rev. Invest. Agropecu., Ser. 4,* 6, 1, 1969.
19. **Kraft, D., von Herrath, D., Schaefer, K., Wagner, H., and Ott, E.,** The effect of different *Solanum malacoxylon* extracts on urinary calcium excretion in vitamin D deficient rats, in *Proc. 3rd Workshop Vitamin D* (Vitamin D: Biochemical, Chemical and Clinical Aspects Related to Calcium Metabolism), Norman, A. W., Schaefer, K., Coburn, J. W., DeLuca, H. F., Fraser, D., Grigoleit, H. G., and von Herrath, D., Eds., Walter de Gruyter, Berlin, 1977, 441.
20. **Weissenberg, M., Levy, A., and Wasserman, R. H.,** Studies on calcinogenic Solanaceae plants of potential value as source of active vitamin D-like derivatives, Rep. I-389-81, United States-Israel Binational Agricultural Research and Development Fund (BARD), Bet Dagan, Israel, 1987.
20a. **Weissenberg, M., Levy, A., and Wasserman, R. H.,** Distribution of calcitriol activity in *Solanum glaucophyllum* plants and cell cultures, *Phytochemistry,* 1988, in press.
20b. **Weissenberg, M., Maoz, A., Levy, A., and Wasserman, R. H.,** Radioimmunoassay for rapid estimation of vitamin D derivatives in calcinogenic plants, *Planta Med.,* 54, 63, 1988.
21. **de Boland, A. R., Skliar, M. I., Boland, R. L., Carrillo, B. J., and Ruksan, B.,** A method for the isolation of the active principle of *Solanum malacoxylon, Anal. Biochem.,* 75, 308, 1976.
22. **Skliar, M. I., de Boland, A. R., Boland, R. L., Carrillo, B. J., and Ruksan, B. E.,** Purificacion del principio activo del *Solanum malacoxylon, Medicina (Buenos Aires),* 36, 242, 1976.
23. **Esparza, M. S., Vega, M., and Boland, R. L.,** Synthesis and composition of vitamin D_3 metabolites in *Solanum malacoxylon, Biochim. Biophys. Acta,* 719, 633, 1982.
24. **Vidal, M. C., Lescano, W., Avdolov, R., and Puche, R. C.,** Partial structure elucidation of the carbohydrate moiety of 1,25-dihydroxycholecalciferol glycoside isolated from *Solanum glaucophyllum* leaves, *Turrialba,* 35, 65, 1985.
25. **Camberos, H. R., Davis, G. K., Djafar, M. I., and Simpson, C. F.,** Soft tissue calcification in guinea pigs fed the poisonous plant *Solanum malacoxylon, Am. J. Vet. Res.,* 31, 685, 1970.
26. **Peterlik, M. and Wasserman, R. H.,** 1,25-Dihydroxycholecalciferol-like activity in *Solanum malacoxylon:* purification and partial characterisation, *FEBS Lett.,* 56, 16, 1975.
27. **Dos Santos, M. N., Nunes, V. A., Nunes, I. J., De Barros, S. S., Wasserman, R. H., and Krook, L.,** *Solanum malacoxylon* toxicity: inhibition of bone resorption, *Cornell Vet.,* 66, 565, 1976.

28. Napoli, J. L., Reeve, L. E., Eisman, J. A., Schnoes, H. K., and DeLuca, H. F., *Solanum glaucophyllum* as source of 1α,25-dihydroxy- vitamin D₃, *J. Biol. Chem.*, 252, 2580, 1977.

29. Procsal, D. A., Henry, H. L., Hendrickson, T., and Norman, A. W., 1α,25-Dihydroxyvitamin D₃-like component present in the plant *Solanum glaucophyllum*, *Endocrinology*, 99, 437, 1976.

30. Cabrejas, M., Ladizesky, M., and Mautalen, C., Calcium kinetics in the *Solanum malacoxylon*-treated rat, *J. Nutr.*, 105, 1562, 1975.

31. López, T., Investigaciones sobre la naturaleza química del tóxico del duraznillo blanco *Solanum malacoxylon* Sendt., *Rev. Invest. Agropecu.*, *Ser. 4*, 10, 93, 1973.

32. Rossi, F. M., Cassels, B. K., Daskal, H., Dallorso, M. E., and Leiva, A., Actividad biologica del *Solanum malacoxylon* Sendtner en ratas, *Medicina (Buenos Aires)*, 34, 327, 1974.

33. Cañas, F. M., Ortiz, O. E., Asteggiano, C. A., and Pereira, R. D., Effects of *Solanum malacoxylon* on rachitic chicks. Comparative study with vitamin D₃, *Calcif. Tissue Res.*, 23, 297, 1977.

34. Stern, P. H., Ness, E. M., and DeLuca, H. F., Responses of fetal rat bones to *Solanum malacoxylon* in vitro: a possible explanation of previous paradoxical results, *Mol. Pharmacol.*, 14, 357, 1978.

35. Walling, M. W., Kimberg, D. V., Lloyd, W., Wells, H., Procsal, D. A., and Norman, A. W., *Solanum glaucophyllum (malacoxylon)*: an apparent source of a water-soluble, functional and structural analogue of 1-alpha,25-dihydroxyvitamin D₃, in *Proc. 2nd Workshop Vitamin D* (Vitamin D and Problems Related to Uremic Bone Disease), Norman, A. W., Schaefer, K., Grigoleit, H. G., von Herrath, D., and Ritz, E., Eds., Walter de Gruyter, Berlin, 1975, 717.

36. Rappaportt, I., Giacopello, D., Seldes, A. M., Blanco, M. C., and Deulofeu, V., Phenolic glycosides from *Solanum glaucophyllum*: a new quercetin triglycoside containing D-apiose, *Phytochemistry*, 16, 1115, 1977.

37. Geissman, T. A., as cited in Skliar, M. I., de Boland, A. R., Boland, R. L., Carrillo, B. J., and Ruksan, B., *Medicina (Buenos Aires)*, 36, 242, 1976.

38. Peterlik, M., Bursak, K., Haussler, M. R., Hughes, M. R., and Wasserman, R. H., Further evidence for the 1,25-dihydroxyvitamin D-like activity of *Solanum malacoxylon*, *Biochem. Biophys. Res. Commun.*, 70, 797, 1976.

38a. Haussler, M. R., Wasserman, R. H., Peterlik, M., Bursak, K., McCain, T. A., and Hughes, M. R., Evidence for 1α,25-dihydroxy vitamin D₃-like glycoside as the calcinogenic factor of *Solanum malacoxylon*, *Fed. Proc. Fed. Am. Soc. Exp. Biol.*, 35, 1552, 1976.

39. Wasserman, R. H., Henion, J. D., Haussler, M. R., and McCain, T. A., Calcinogenic factor in *Solanum malacoxylon*: evidence that it is 1,25-dihydroxyvitamin D₃-glycoside, *Science*, 194, 853, 1976.

40. Haussler, M. R., Wasserman, R. H., McCain, T. A., Peterlik, M., Bursak, K. M., and Hughes, M. R., 1,25-Dihydroxyvitamin D₃ glycoside: identification of a calcinogenic principle of *Solanum malacoxylon*, *Life Sci.*, 18, 1049, 1976.

41. Haussler, M. R., Hughes, M. R., McCain, T. A., Zerweck, J. E., Brumbaugh, P. F., Jubiz, W., and Wasserman, R. H., 1,25-Dihydroxy-vitamin D₃: mode of action in intestine and parathyroid glands, assay in humans and isolation of its glycoside from *Solanum malacoxylon*, *Calcif. Tissue Res.*, 22 (Suppl.), 1, 1977.

42. Ladizesky, M., Mautalen, C. A., and Camberos, H., *Solanum malacoxylon*: comparación de su actividad biologica al administrarlo por via oral o parenteral en animales de laboratorio, *Medicina (Buenos Aires)*, 34, 127, 1974.

43. Boland, R. L., Skliar, M. I., de Boland, A. R., Carrillo, B. J., and Ruksan, B. E., Modificacion del principio activo del *Solanum malacoxylon* mediante incubacion con liquido ruminal, *Medicina (Buenos Aires)*, 36, 323, 1976.

44. de Boland, A. R., Skliar, M. I., Gallego, S., Esparza, M., and Boland, R. L., Potentiation of the effects of *Solanum malacoxylon* extracts on rat intestinal phosphate and calcium absorption by incubation with ruminal fluid, *Calcif. Tissue Res.*, 26, 215, 1978.

45. de Boland, A. R., Esparza, M., Gallego, S., Skliar, M. I., and Boland, R. L., Modification by rumen of vitamin D-like activity of *Solanum malacoxylon* in rats, *Acta Physiol. Lat. Am.*, 29, 285, 1979.

46. Skliar, M. I. and Boland, R. L., Formas moleculares multiples del principio activo del *Solanum malacoxylon*, *Medicina (Buenos Aires)*, 39, 38, 1979.

47. Esparza, M. S., Skliar, M. I., Gallego, S. E., and Boland, R. L., Modification by rumen of hypercalcemic activity of *Solanum malacoxylon* on the chick embryo, *Planta Med.*, 47, 63, 1983.

48. Boland, R. L., Skliar, M. I., and Norman, A. W., Detection of 1,24,25-trihydroxyvitamin D₃ in *Solanum malacoxylon* leaf extracts incubated with ruminal fluid, in *Proc. 6th Workshop Vitamin D* (Vitamin D: Chemical, Biochemical and Clinical Update), Norman, A. W., Schaefer, K., Grigoleit, H. G., and von Herrath, D., Eds., Walter de Gruyter, Berlin, 1985, 55.

48a. Boland, R. L., Skliar, M. I., and Norman, A. W., Isolation of vitamin D₃ metabolites from *Solanum malacoxylon* leaf extracts incubated with ruminal fluid, *Planta Med.*, 53, 161, 1987.

49. **Napoli, J. L., Reeve, L. E., Eisman, J. A., Schnoes, H. K., and DeLuca, H. F.**, 1,25-Dihydroxyvitamin D₃ from *Solanum glaucophyllum*, in *Proc. 3rd Workshop Vitamin D* (Vitamin D: Biochemical, Chemical and Clinical Aspects Related to Calcium Metabolism), Norman, A. W., Schaefer, K., Coburn, J. W., DeLuca, H. F., Fraser, D., Grigoleit, H. G., and von Herrath, D., Eds., Walter de Gruyter, Berlin, 1977, 29.

50. **Haussler, M. R., Hughes, M. R., Pike, W. J., and McCain, T. A.**, Radioligand receptor assay for 1,25-dihydroxyvitamin D: biochemical, physiologic and clinical applications, in *Proc. 3rd Workshop Vitamin D* (Vitamin D: Biochemical, Chemical and Clinical Aspects Related to Calcium Metabolism), Norman, A. W., Schaefer, K., Coburn, J. W., DeLuca, H. F., Fraser, D., Grigoleit, H. G., and von Herrath, D., Eds., Walter de Gruyter, Berlin, 1977, 473.

51. **Miravet, L., Carre, M., Fisher-Ferraro, C., and Mautalen, C.**, Action of *Solanum malacoxylon* on intestinal calcium transport in vitamin D-deficient-prednisolone treated rats, *J. Clin. Endocrinol. Metab.*, 45, 1230, 1977.

52. **de Vernejoul, M.-C., Quellie, M. L., Nordin, R. W., Mautalen, C. A., and Miravet, L.**, Action of *Solanum malacoxylon* on bone histology of vitamin D deficient rats, in *Proc. 4th Workshop Vitamin D* (Vitamin D: Basic Research and its Clinical Application), Norman, A. W., Schaefer, K., von Herrath, D., Grigoleit, H. G., Coburn, J. W., DeLuca, H. F., Mawer, E. B., and Suda, T., Eds., Walter de Gruyter, Berlin, 1979, 403.

53. **Karawya, M. S., Rizk, A. M., Hammouda, F. M., Diab, A. M., and Ahmed, Z. F.**, Phytochemical investigation of certain *Cestrum* species. General analysis, lipids and triterpenoids, *Planta Med.*, 20, 363, 1971.

54. **Wasserman, R. H., Corradino, R. A., Krook, L., Hughes, M. R., and Haussler, M. R.**, Studies on the 1α,25-dihydroxycholecalciferol-like activity in a calcinogenic plant, *Cestrum diurnum*, in the chick, *J. Nutr.*, 106, 457, 1976.

55. **Hughes, M. R., McCain, T. A., Chang, S. Y., Haussler, M. R., Villareale, M., and Wasserman, R. H.**, Presence of 1,25-dihydroxy-vitamin D₃ glycoside in the calcinogenic plant *Cestrum diurnum*, *Nature (London)*, 268, 347, 1977.

56. **Rambeck, W. A., Weiser, H., Haselbauer, R., and Zucker, H.**, Vitamin D activity of different vitamin D₃ esters in chicken, Japanese quail and in rats, *Int. J. Vitam. Nutr. Res.*, 51, 353, 1981.

57. **Puche, R. C., Masoni, A. M., Alloatti, D. A., and Roveri, E.**, The antirachitic activity of *Solanum glaucophyllum* leaves, *Planta Med.*, 40, 378, 1980.

58. **Kraft, D., von Herrath, D., Offermann, G., and Schaefer, K.**, The effect of *Solanum malacoxylon* on rachitic bone lesions in the rat, *Naunyn Schmiedebergs Arch. Pharmakol.*, 290, 29, 1975.

59. **Ladizesky, M., Cabrejas, M., Montoreano, R., and Mautalen, C.**, The effect of *Solanum malacoxylon* on the intestinal and renal handling of calcium and phosphate in the rat, in *Proc. 2nd Workshop Vitamin D* (Vitamin D and Problems Related to Uremic Bone Disease), Norman, A. W., Schaefer, K., Grigoleit, H. G., von Herrath, D., and Ritz, E., Eds., Walter de Gruyter, Berlin, 1975, 709.

60. **Walling, M. W. and Kimberg, D. V.**, Effects of 1α,25-dihydroxy-vitamin D₃ and *Solanum glaucophyllum* on intestinal calcium and phosphate transport and on plasma Ca, Mg and P levels in the rat, *Endocrinology*, 97, 1567, 1975.

61. **Uribe, A., Holick, M. F., Jorgensen, N. A., and DeLuca, H. F.**, Action of *Solanum malacoxylon* on calcium metabolism in the rat, *Biochem. Biophys. Res. Commun.*, 58, 257, 1974.

62. **Zucker, H., Kreutzberg, O., and Rambeck, W. A.**, Vitamin D-like action of a steroid from *Trisetum flavescens*, in *Proc. 3rd Workshop Vitamin D* (Vitamin D: Biochemical, Chemical and Clinical Aspects Related to Calcium Metabolism), Norman, A. W., Schaefer, K., Coburn, J. W., DeLuca, H. F., Fraser, D., Grigoleit, H. G., and von Herrath, D., Eds., Walter de Gruyter, Berlin, 1977, 85.

63. **Lawson, D. E. M., Smith, M. W., and Wilson, P. W.**, Relationship of calcium uptake and calcium-binding protein synthesis in chick and rat intestine in response to *Solanum malacoxylon*, *FEBS Lett.*, 45, 122, 1974.

64. **Corradino, R. A. and Wasserman, R. H.**, 1,25- Dihydroxycholecalciferol-like activity of *Solanum malacoxylon* extract on calcium transport, *Nature (London)*, 252, 716, 1974.

65. **Carrillo, B. J.**, The Pathology of Enteque Seco and Experimental *Solanum malacoxylon* Toxicity, Ph.D. thesis, University of California, Davis, 1971.

66. **Tokarnia, C. H. and Döbereiner, J.**, "Espichamento", intoxicacao de bovinos por *Solanum malacoxylon*, no pantanal de mato grosso. II. Estudos complementares, *Pesqui. Agropecu. Bras. Ser. Vet.*, 9, 53, 1974.

67. **Ullrich, W.**, Untersuchungen über die Hitzestabilität des kalzinogenen Wirkstoffes in *Trisetum flavescens* (L.) P.B. und in *Solanum malacoxylon* (Sendtner), Vet. Med. Diss., Munich University, Munich, 1978.

68. **Ullrich, W.**, Untersuchungen über die Hitzestabilität des kalzinogenen Wirkstoffes in *Trisetum flavescens* (L.) P.B. und in *Solanum malacoxylon* (Sendtner), *Berl. Muench. Tieraerztl. Wochenschr.*, 92, 220, 1979.

69. **I.N.T.A. - F.A.O.**, Estudio de las enfermedades y deficiencias nutricionales del ganado vacuno en la Argentina, U.N. Special Fund Project, Final Report, Food and Agriculture Organization of the United Nations, Rome, 1970.

70. **Nagubandi, S., Kumar, R., Londowski, J. M., Corradino, R. A., and Tietz, P. S.,** Role of vitamin D glucosiduronate in calcium homeostasis, *J. Clin. Invest.*, 66, 1274, 1980.

71. **Hollick, S. and Hollick, M. F.,** Vitamin D glycosides and their use, U.S. Patent 4,410,515, 1983.

72. **Fürst, A., Labler, L., and Meier, W.,** Synthese von β-D-glucopyranosiden einiger hydroxylierter Vitamin D- Verbindungen, *Helv. Chim. Acta*, 66, 2093, 1983.

73. **Raoul, Y., Le Boulch, N., Gounelle, J. C., Marnay-Gulat, C., and Ourisson, G.,** Isolement et caractérisation du cholécalciferol des végétaux superieurs, *FEBS Lett.*, 1, 59, 1968.

74. **Raoul, Y., Le Boulch, N., Gounelle, J. C., Marnay-Gulat, C., and Ourisson, G.,** Substances antirachitiques des végétaux. Presence du cholécalciferol, *Bull. Soc. Chim. Biol.*, 52, 641, 1970.

75. **Blazheevich, N. V.,** Vitamin D-like activity in plants of the Solanaceae family (literature review), *Vopr. Pitan.*, 4, 51, 1977; *Chem. Abstr.*, 87, 132590, 1977.

76. **Matsumoto, I.,** Glycosides of active vitamin D in vegetables, *GC-MS News*, 5, 4, 1977; *Chem. Abstr.*, 88, 186047, 1978.

77. **Zucker, H. and Rambeck, W. A.,** Vitamin D in Tieren und Pflanzen, *Prakt. Tieraerzt*, 61, 220, 1980.

78. **Varga, L., Morava, E., and Gergely, A.,** Vitamin D activity of certain plants and vegetable food in rats, Taplalkozastud. Helyzete Feladatai Magyarorszagon, (Magy. Taplalkozastud. Tarsasag Vandorgyulesenek Tud. Anyaga), 8th, Mozsik, G., Javor, T., and Szakaly, S., Eds., Akademiai Kiado, Budapest, 1981, 705; *Chem. Abstr.*, 100, 155602, 1984.

79. **Varga, L., Gergely, A., and Morava, E.,** Vitamin D content of some vegetables, *Elelmez. Ipar*, 36, 142, 1982; *Chem. Abstr.*, 98, 142118, 1983.

80. **Zucker, H., Stark, H., and Rambeck, W. A.,** Light-dependent synthesis of cholecalciferol in a green plant, *Nature (London)*, 283, 68, 1980.

81. **Horst, R. L., Reinhardt, T. A., Russell, J. R., and Napoli, J. L.,** The isolation and identification of vitamin D_2 and vitamin D_3 from *Medicago sativa* (alfalfa plant), *Arch. Biochem. Biophys.*, 231, 67, 1984.

82. **Holick, M. F., Adams, J. S., Clemens, T. L., MacLaughlin, J., Horiuchi, N., Smith, E., Holick, S. A., Nolan, J., and Hannifan, N.,** Photoendocrinology of vitamin D: the past, present and future, in *Proc. 5th Workshop Vitamin D* (Vitamin D, Chemical, Biochemical and Clinical Endocrinology of Calcium Metabolism), Norman, A. W., Schaefer, K., von Herrath, D., and Grigoleit, H. G., Eds., Walter de Gruyter, Berlin, 1982, 1151.

83. **Okano, T., Yasumura, M., Mizuno, K., and Kobayashi, T.,** In vivo and in vitro conversion of 7-dehydrocholesterol into vitamin D_3 in rat skin by ultraviolet rays irradiation, *J. Nutr. Sci. Vitaminol.*, 24, 47, 1978.

84. **Henry, H. L. and Norman, A. W.,** Vitamin D: metabolism and biological actions, *Annu. Rev. Nutr.*, 4, 493, 1984.

85. **Heftmann, E.,** Biogenesis of steroids in Solanaceae, *Phytochemistry*, 22, 1843, 1983.

86. **Weiler, E. W., Kruger, H., and Zenk, M. H.,** Radioimmunoassay for the determination of the steroidal alkaloid solasodine and related compounds in living plants and herbarium specimens, *Planta Med.*, 39, 112, 1980.

87. **Jain, S. C. and Sahoo, S.,** Isolation of steroids and glycoalkaloids from *Solanum glaucophyllum* Desf., *Pharmazie*, 41, 820, 1986.

87a. **Sahoo, S. L. and Jain, S. C.,** Isolation of steroids and glycoalkaloid from seeds and seedlings callus of *Solanum glaucophyllum* Desf. tissue culture, *Cell Chromosome Res.*, 10, 104, 1987.

88. **Heftmann, E.,** Functions of steroids in plants, *Phytochemistry*, 14, 891, 1975.

89. **Walling, M. W. and Kimberg, D. V.,** Calcium absorption by intestine. Stimulation in vitamin D-deficient nephrectomized rats by *Solanum glaucophyllum*, *Gastroenterology*, 69, 200, 1975.

90. **Kevers, C., Sticher, L., Penel, C., Greppin, H., and Gaspar, T.,** The effect of ergosterol, ergocalciferol, and cholecalciferol on calcium-controlled peroxidase secretion by sugarbeet cells, *Physiol. Plant.*, 57, 17, 1983.

91. **Vega, M. A., Santamaria, E. C., Morales, A., and Boland, R. L.,** Vitamin D_3 affects growth and Ca^{2+} uptake by *Phaseolus vulgaris* roots cultured in vitro, *Physiol. Plant.*, 65, 423, 1985.

92. **Vega, M. A. and Boland, R. L.,** Vitamin D_3 activity in in vitro calcium uptake and growth of *Phaseolus vulgaris* roots, in *Proc. 6th Workshop Vitamin D* (Vitamin D: Chemical, Biochemical and Clinical Update), Norman, A. W., Schaefer, K., Grigoleit, H. G., and von Herrath, D., Eds., Walter de Gruyter, Berlin, 1985, 723.

93. **Banieckl, J. F. and Bloss, H. E.,** The basidial stage of *Phymatotrichum omnivorum*, *Mycologia*, 61, 1054, 1969.

94. **Bloss, H. E.,** Observations on species of *Phymatotrichum*, *J. Ariz. Acad. Sci.*, 6, 102, 1970.

95. **Brar, D. S., Rambold, S., Constabel, F., and Gamborg, O. L.,** Isolation, fusion and culture of sorghum and corn protoplasts, *Z. Pflanzenphysiol.*, 96, 269, 1980.

96. **Pythoud, F., Buchala, A. J., and Schmid, A.,** Adventitious root formation in green cuttings of *Populus tremula*: characterization of the effect of vitamin D_3 and indolylbutyric acid, *Physiol. Plant.*, 68, 93, 1986.

97. **Fries, L.**, D-vitamins and their precursors as growth regulators in axenically cultivated marine macroalgae, *J. Phycol.*, 20, 62, 1984.

98. **Lignières, J.**, Contribution à l'étude de la pasteurellose bovine connue en Argentine sous le nom de "Diarrhée" et l'"Enteque", *Bull. Soc. Cent. Med. Vet. Nouv. Ser.*, 16, 761, 1898.

99. **Lignières, J.**, L'arteriosclérose épidemique du mouton, *Rev. Gen. Med. Vet.*, 20, 1, 1911.

100. **Collier, W. A.**, Zur Kenntnis einer als "Enteque" bezeichneten Krankheit der Rinder in der Provinz Buenos Aires, *Zeitschrift für Infektionskrankheiten, parasitare Krankheiten und Hygien der Haustiere*, 31, 81, 1927.

101. **Vasconcelos, A.**, Ossificaçao do pulmao de un bovino, *Rev. Vet. Zool.*, 4, 390, 1916.

102. **Tibirica, P. de Q. T.**, Arteriosclerose bovina, *Ann. Fac. Med. Univ. Sao Paulo*, 2, 311, 1927.

103. **Pardi, M. C. and Dos Santos, J. A.**, Ossificaçao pulmonar e calcificaçao vascular em bovinos do pantanal Mato Grossense, *Veterinaria (Brasil)*, 1, 3, 1947.

104. **de Barros, S., Pohlenz, J., and Santiago, C.**, Zur Kalzinose beim Schaf, *Dtsch. Tieraerztl. Wochenschr.*, 77, 346, 1970.

105. **Eckell, O. A.**, Accion toxica del *"Solanum glaucum"* Dun. (Duraznillo blanco), *Ann. Fac. Med. Vet. (La Plata)*, 5, 9, 1943.

106. **Eckell, O. A.**, Accion toxica del *"Solanum glaucum"* Dun. (Duraznillo blanco), *Rev. Med. Vet. (Buenos Aires)*, 25, 453, 1943.

107. **Döbereiner, J., Tokarnia, C. H., Da Costa, J. B. D., Campos, J. L. E., and Dayrell, M. De S.**, *"Espichamento"*, intoxicaçao de bovinos por *Solanum malacoxylon* no pantanal de mato grosso, *Pesqui. Agropecu. Bras. Ser. Vet.*, 6, 91, 1971.

108. **Molfino, R., Alonso, R., and Riccitelli, J.**, Posibles causas edéficas del "Enteque seco", *Rev. Fac. Agron. Univ. Nac. La Plata*, 31, 53, 1955.

109. **Eckell, O. A., Gallo, G. G., Martin, A. A., and Portella, R. A.**, Observaciones sobre el "Enteque seco" de los bovinos, *Rev. Fac. Cienc. Vet. La Plata Univ. Nac. La Plata*, 2, 193, 1960.

110. **Machado, T. L., Jr. and Salle, C. T. P.**, A *Solanum malacoxylon* na etiologia da calcificaçao dos tecidos moles em ovinos no Rio Grande do Sul, *Bol. Inst. Pesqui. Vet. Desiderio Finamor*, 1, 81, 1972; *Biol. Abstr.*, 56, 29125, 1973.

111. **Bingley, J. B.**, "Enteque seco". Perspectivas bioquímicas, *Bol. Tec. INTA, EEA Balcarce*, No. 29, 1964.

112. **D'Arcy, W. G.**, *Solanum* and its close relatives in Florida, *Ann. Mo. Bot. Gard.*, 61, 819, 1974.

113. **Krook, L., Wasserman, R. H., Shively, J. N., Tashjian, A. H., Jr., Brokken, T. D., and Morton, J. F.**, Hypercalcemia and calcinosis in Florida horses: implication of the shrub *Cestrum diurnum* as the causative agent, *Cornell Vet.*, 65, 26, 1975.

114. **Cabrera, A. L.**, Solanaceae in Flora de la Provincia de Buenos Aires, in Colección Cientifica del INTA, Vol. 4, Part 5a, Instituto Nacional de Tecnologia Agropecuaria, Buenos Aires, Argentina, 1965, 190.

114a. **Deb, D. B.**, Solanaceae in India, in *The Biology and Taxonomy of the Solanaceae*, Linnean Society Symp. Ser. No. 7, Hawkes J. G. and Lester, R. N., Eds., Academic Press, London, 1979, 87.

115. **Tilley, J. M. A.**, Enteque seco in Argentina, Annual Report, Grassland Research Inst., Hurley, U.K., 73, 1967.

116. **Gaggino, O. P. and Carrillo, B. J.**, "Enteque seco". I. Observación de las lesiones de tipo paratuberculosos en bovinos con "Enteque seco", *Rev. Invest. Ganad. INTA*, No. 16, 39, 1963.

117. **Gaggino, O. P. and Carrillo, B. J.**, "Enteque seco". II. Resultados de pruebas intradérmicas dobles con tuberculinas bovina y aviar en bovinos de zonas de "Enteque seco" en el sudeste de la Provincia de Buenos Aires, *Rev. Invest. Agropecu. Ser. 4*, 1, 67, 1964.

118. **Cuerpo, L. and Casal, J. J.**, Valores en suero bovino de ácido siálico. Relación entre bovinos aparentemente normales y con "Enteque seco", *Rev. Invest. Agropecu. Ser. 4*, 2, 85, 1965.

119. **Masselin, J. N. and Chiaravalle, A. M.**, Frecuencia de la esclerosis aórtica en el bovino, *Rev. Invest. Agropecu. Ser. 4*, 2, 117, 1965.

120. **Müler, S. M. and Casal, J. J.**, Valores en suero bovino de colesterol total, libre, esterificado y asociado a alfa y beta lipoproteínas. Relación entre bovinos aparentemente normales y con "Enteque seco", *Rev. Invest. Agropecu. Ser. 4*, 2, 95, 1965.

121. **Bingley, J. B., Landó, E. R., and Carrillo, B. J.**, Enteque seco. Aspectos bioquímicos II. Niveles de fósforo en sangre de bovinos en áreas afectadas, *Rev. Invest. Agropecu. Ser. 4*, 2, 207, 1965.

122. **Döbereiner, J. and Dämmrich, K.**, Skelettveränderungen bei Rindern nach Vergiftungen mit *Solanum malacoxylon* Sendtner, *Verh. Dtsch. Ges. Pathol.*, 58, 323, 1974.

123. **Dämmrich, K., Döbereiner, J., Done, S. H., and Tokarnia, C. H.**, Skelettveränderungen nach Vergiftungen mit *Solanum malacoxylon* bei Rindern, *Zentralbl. Veterinaermed. Reihe A*, 22, 313, 1975.

124. **Puche, R. C., Faienza, H., Valenti, J. L., Zuster, G., Osmetti, G., Hayase, K., and Dristas, J. A.**, On the nature of arterial and lung calcifications induced in cattle by *Solanum glaucophyllum*, *Calcif. Tissue Res.*, 26, 61, 1978.

125. **Krook, L., Wasserman, R. H., McEntee, K., Brokken, T. D., and Teigland, M. B.**, *Cestrum diurnum* poisoning in Florida cattle, *Cornell Vet.*, 65, 557, 1975.

126. **Ross, E., Simpson, C. F., Rowland, L. O., and Harms, R. H.**, Toxicity of *Solanum sodomaeum* and *Solanum malacoxylon* to chicks, *Poult. Sci.*, 50, 870, 1971.

127. **Wasserman, R. H.**, Calcium absorption and calcium-binding protein synthesis: *Solanum malacoxylon* reverses strontium inhibition, *Science*, 183, 1092, 1974.

128. **Basudde, C. D. K. and Humphreys, D. J.**, The effect of the administration of *Solanum malacoxylon* on the chick, *Res. Vet. Sci.*, 18, 330, 1975.

129. **Narbaitz, R. and Carrillo, B. J.**, Production of hypercalcemia in the chick embryo by an extract of *Solanum malacoxylon*, *Rev. Can. Biol.*, 35, 181, 1976.

130. **Morris, K. M. L., Jenkins, S. A., and Simonite, J. P.**, The effect on egg-shell thickness of the inclusion of the calcinogenic plant *Solanum malacoxylon* in the diet of laying hens, *Vet. Rec.*, 101, 502, 1977.

131. **Gallego, S. E., Boland, R. L., Bonino, M., Azcona, J. O., and Villar, J.**, Efecto de la administración de *Solanum malacoxylon* en la calcificación de huevos de gallina, *Rev. Invest. Agropecu. Ser. 1*, 14, 67, 1978.

132. **Narbaitz, R.**, Production of hypercalcemia in the chick embryo by exogenous parathyroid hormone, *Solanum malacoxylon* and 1,25-dihydroxycholecalciferol, in *Endocrinology of Calcium Metabolism*, Copp, D. H. and Talmage, R. V., Eds., Excerpta Medica, Amsterdam, 1978, 371.

133. **Peterlik, M. and Wasserman, R. H.**, Stimulatory effect on 1,25-dihydroxycholecalciferol-like substances from *Solanum malacoxylon* and *Cestrum diurnum* on phosphate transport in chick jejunum, *J. Nutr.*, 108, 1673, 1978.

134. **Sauveur, B.**, Effets du *Solanum malacoxylon* et du phosphore alimentaire chez le poulet en acidose métabolique chronique, *Ann. Biol. Anim. Biochim. Biophys.*, 19, 1789, 1979.

135. **Masoni, A. M., Alloatti, D. A., Roveri, E., and Puche, R.**, Efecto comparativo del 1.25 dihidrocolecalciferol (DHCC) y 1.25 dihidrocolecalciferol-glucósido (DHCCG) sobre el metabolismo mineral en el pollo raquítico, *Medicina (Buenos Aires)*, 39, 796, 1979.

136. **Azcona, J. O., Bonino, M. F., Chang, M. J., Gallego, S. E., and Boland, R. L.**, Effect of calcinogenic substances from *Solanum malacoxylon* on egg shell thickness and performance of laying hens, in *Proc. 6th European Poultry Conf.*, Vol. III, 1980, 197; Ibidem, in *Zootec. Int.*, (2), 12, 1982.

137. **Sauveur, B.**, Effect of dietary 25-(OH)D₃ and *Solanum malacoxylon* on egg shell quality of brown eggs, in *Quality of Eggs, Proc. 1st European Symp.*, Beuving, G., Scheele, C. W., and Simons, P. C. M., Eds., Spelderholt Institute of Poultry Research, Beekbergen, Netherlands, 1981, 194.

138. **Asteggiano, C., Tolosa, N., Pereira, R., Moreno, J., and Canas, F.**, Lack of relationship between activity of intestinal alkaline phosphatase and calcium or phosphate absorption, *Acta Physiol. Lat. Am.*, 31, 77, 1981.

139. **Zucker, H., Geissdoerfer, K., Pettrich, M., and Rambeck, W. A.**, Influence of the calcinogenic plants *Trisetum flavescens* and *Solanum malacoxylon* on intestinal transport of calcium and phosphate, in *Current Advances in Skeletogenesis*, Int. Congr. Ser. 589, Excerpta Medica, Amsterdam, 1982, 201; Ibidem, *Abstr. 5th Int. Workshop Calcif. Tiss.*, Israel, 1982, 98.

140. **Basudde, C. D. K.**, The effect of *Solanum malacoxylon* on serum enzyme activities, blood glucose, and cholesterol levels in chicks, *Poult. Sci.*, 61, 1001, 1982.

141. **Tröger, C.**, Wirkungsvergleich von *Solanum malacoxylon*, *Trisetum flavescens*, 1,25-Dihydroxycholecalciferol, 1,25-Dihydroxyergocalciferol und 24,25-Dihydroxycholecalciferol am rachitischen Hühnerküken, Inaugural dissertation, Ludwig-Maximilians-University, Munich, 1984.

142. **Wasserman, R. H., Corradino, R. A., and Krook, L. P.**, *Cestrum diurnum*: a domestic plant with 1,25-dihydroxycholecalciferol-like activity, *Biochem. Biophys. Res. Commun.*, 62, 85, 1975.

143. **Rossi, F. M., Rojas, A., Glait, H., and de Lustig, E. S.**, Efectos del "*Solanum malacoxylon*" Sendtner sobre el Ca tisular y esquelético en ratón, *Medicina (Buenos Aires)*, 33, 666, 1973.

144. **Gaggino, O. P., Deshpande, P. D., and Tilley, J. M.**, Reproducción experimental de lesiones arterioscleróticas caracteristicas del enteque seco en ratas, *Rev. Invest. Agropecu. Ser. 4*, 4, 123, 1967.

145. **Carrillo, B. J.**, Intoxicacion experimental en ratas con *Solanum malacoxylon*, *Medicina (Buenos Aires)*, 31, 604, 1971.

146. **Rossi, F. M., Cassels, E. K., Daskal, H., Dallorso, M. E., and Leiva, A.**, Actividad biologica del "*Solanum malacoxylon* Sendtner" (SM) en ratas, *Medicina (Buenos Aires)*, 31, 604, 1971.

147. **Campos, C. F. and Mautalen, C. A.**, Efecto del "*Solanum malacoxylon*" sobre la calcemia y la fosfatemia en ratas paratiroidectomizadas, *Medicina (Buenos Aires)*, 31, 605, 1971.

148. **Masselin, J. N., Gaggino, O. P., Monesiglio, J. C., and Parodi, J. J.**, Efectos de la administracion a ratas de extractos etereos de algunos vegetales, *Rev. Invest. Agropecu. Ser. 4*, 8, 117, 1971.

149. **Wase, A. W.**, Effect of *Solanum malacoxylon* on serum calcium and phosphate in laboratory animals, *Fed. Proc.*, 31, 708, 1972.

150. **O'Donnell, J. M. and Smith, M. W.**, Vitamin D-like action of *Solanum malacoxylon* on calcium transport by rat intestine, *Nature (London)*, 244, 357, 1973.

151. **Carrillo, B. J.**, Efecto de la intoxicación de *Solanum malacoxylon* en la morfologia de las células parafoliculares de la tiroides, *Rev. Invest. Agropecu. Ser. 4*, 10, 41, 1973.

152. **Carrillo, B. J.**, Efecto de la intoxicación de *Solanum malacoxylon* en el sistema óseo, *Rev. Invest. Agropecu. Ser. 4*, 10, 65, 1973.

153. **Campos, C., Ladizesky, M., and Mautalen, C.**, Effect of *Solanum malacoxylon* on the serum levels of calcium and phosphate in thyroparathyroidectomized and rachitic rats, *Calcif. Tissue Res.*, 13, 245, 1973.

154. **Puche, R. C. and Locatto, M. E.**, Effects of *Solanum malacoxylon* on embryonic bone in vitro and on isolated mitochondria, *Calcif. Tissue Res.*, 16, 219, 1974.

155. **Ladizesky, M., Cabrejas, M., Labarrere, C., and Mautalen, C.**, *Solanum malacoxylon* and vitamin D: comparison of its antirachitic activity and its effect on the intestinal absorption of calcium in the rat, *Acta Endocrinol. Panam.*, 5, 71, 1974.

156. **von Herrath, D., Kraft, D., Offermann, G., and Schaefer, K.**, *Solanum malacoxylon:* eine therapeutische Alternative für 1,25-Dihydroxycholecalciferol bei urämischen Calciumstoffwechselstörungen, *Dtsch. Med. Wochenschr.*, 99, 2407, 1974.

157. **Puche, R. C., Fernandez, M. C., Locatto, M. E., Ferretti, J. L., and Valenti, J. L.**, Hypostenuria and nephrocalcinosis induced in rats chronically fed leaves of *Solanum glaucophyllum*, *IRCS Med. Sci.*, 3, 125, 1975.

158. **Schneider, L. E., Wasserman, R. H., and Schedl, H. P.**, Depressed duodenal calcium absorption in the diabetic rat: restoration by *Solanum malacoxylon*, *Endocrinology*, 97, 649, 1975.

159. **von Herrath, D., Schaefer, K., Kraft, D., and Offermann, G.**, Effect of *Solanum malacoxylon* on calcium metabolism in experimental uremia and in uremic patients, in *Proc. 2nd Workshop Vitamin D* (Vitamin D and Problems Related to Uremic Bone Disease), Norman, A. W., Schaefer, K., Grigoleit, H. G., von Herrath, D., and Ritz, E., Eds., Walter de Gruyter, Berlin, 1975, 703.

160. **Moorhead, J. F., Humphreys, D. J., Varghese, Z., Basudde, C. D. K., Jenkins, M. V., and Wills, M. R.**, Studies on "vitamin D-like action" of *Solanum malacoxylon*, in *Proc. 2nd Workshop Vitamin D* (Vitamin D and Problems Related to Uremic Bone Disease), Norman, A. W., Schaefer, K., Grigoleit, H. G., von Herrath, D., and Ritz, E., Eds., Walter de Gruyter, Berlin, 1975, 725.

161. **Puche, R. C., Locatto, M. E., Ferretti, J. L., Fernandez, M. C., Orsatti, M. B., and Valenti, J. L.**, The effects of long term feeding of *Solanum glaucophyllum* to growing rats on calcium, magnesium, phosphorus and bone metabolism, *Calcif. Tissue Res.*, 20, 105, 1976.

162. **Ferretti, J. L., Locatto, M. E., Fernandez, M. C., Orsatti, M., and Puche, R. C.**, Estudio etiopatogénico integrado del enteque seco, *Medicina (Buenos Aires)*, 36, 549, 1976.

163. **Orsatti, M. B., Fucci, L. L., Juster, G., Osmetti, G., and Puche, R.**, Estudio sobre la anemia de los animales intoxicados con «*Solanum glaucophyllum*» (SG), *Medicina (Buenos Aires)*, 36, 573, 1976.

164. **Locatto, M. E., Fernandez, M. del C., and Puche, R. C.**, Efectos del *Solanum glaucophyllum* (SG) sobre el metabolismo del citrato en la rata, *Medicina (Buenos Aires)*, 36, 597, 1976.

165. **Ferraro, C., Ladizesky, M., Cabrejas, M., Montoreano, R., and Mautalen, C.**, The effect of actinomycin D on the intestinal absorption of phosphate and calcium in *Solanum malacoxylon* treated rats, in *Phosphate Metabolism, Kidney and Bone*, Avioli, L., Bordier, Ph., Fleish, H., Massry, S., and Slatopolsky, E., Eds., Armour, Paris, 1976, 365.

166. **Ferraro, C., Ladizesky, M., Cabrejas, M., and Mautalen, C. A.**, Intestinal absorption of calcium and phosphate. Effect of actinomycin D upon the stimulation induced by *Solanum malacoxylon*, *Medicina (Buenos Aires)*, 37, 385, 1977.

167. **Borrozzino, G., Facchini, G., Maranesi, M., and Pignatti, C.**, Sulla probabile attivita 1α,25-diidrossicolecalciferolo del *Solanum malacoxylon* essicato, assunto con l'alimentazione, *Boll. Soc. Ital. Biol. Sper.*, 53, 2129, 1977.

168. **Schneider, L. E. and Schedl, H. P.**, Effects of *Solanum malacoxylon* on duodenal calcium binding protein in the diabetic rat, *Endocrinology*, 100, 928, 1977.

169. **Locatto, M. E., Fernandez, M. C., and Puche, R. C.**, Metabolismo del citrato en ratas tratadas con *Solanum glaucophyllum*, *Medicina (Buenos Aires)*, 38, 250, 1978.

170. **Puche, R. C. and Locatto, M. E.**, Sobre el mecanismo de acción del principio activo del *Solanum malacoxylon*, *Medicina (Buenos Aires)*, 33, 665, 1973.

171. **Ladizesky, M., Fainstein Day, P., and Mautalen, C. A.**, Efectos metabolicos del *Solanum malacoxylon* en ratas tiroparatiroidectomizadas, *Medicina (Buenos Aires)*, 38, 341, 1978.

172. **Dusso, A., Alloatti, D., Orsatti, M., and Puche, R.**, Efecto del 1-25 dihidroxicolecalciferol glucosido (DHCCG) sobre el metabolismo del hierro en la rata, *Medicina (Buenos Aires)*, 39, 711, 1979.

173. **Norrdin, R. W., De Barros, C. S. L., Queille, M. L., Carré, M., and Miravet, L.**, Acute effects of *Solanum malacoxylon* on bone formation rates in growing rats, *Calcif. Tissue Int.*, 28, 239, 1979.

174. **Locatto, M. E., Fernandez, M. C., Faienza, H., Orsatti, M. B., Puche, R. C., Boland, R. L., and Skliar, M. I.**, Effect of 1,25-dihydroxycholecalciferol and 1,25-dihydroxycholecalciferol glycoside on 2,3-diphosphoglycerate levels of the rat erythrocyte, *Pfluegers Arch.*, 389, 81, 1980.

175. **Ferrando, R., Henry, N., and Fourlon, C.**, Activité antirachitique directe de *Solanum malacoxylon* desséché. Influence sur l'utilisation de la vitamine A, *Recl. Med. Vet.*, 156, 739, 1980.

176. **Ferrando, R., Truhaut, R., Henry, N., and Fourlon, C.**, Influences comparées de la vitamine D₃ et de *Solanum malacoxylon* sur la calcémie, l'induction du cytochrome P 450 et l'utilisation du rétinol chez le Rat recevant un régime rachitigène, *C.R. Acad. Sci., Ser. 3*, 292, 713, 1981.

177. **Orsatti, M. B., Dusso, A., and Puche, R. C.,** Effect of *Solanum glaucophyllum* feeding on erythropoiesis and iron metabolism in the rat, *Medicina (Buenos Aires)*, 42, 43, 1982.

178. **Pettrich, M.,** Einfluss calcinogener Pflanzen auf den Calcium-und Phosphat-transport im isolierten Rattendarm, Thesis, Ludwig-Maximilians-University, Munich, 1982.

179. **Walling, M. W., Kimberg, D. V., Wasserman, R. H., and Feinberg, R. R.,** Duodenal active transport of calcium and phosphate in vitamin D-deficient rats: effects of nephrectomy, *Cestrum diurnum*, and 1α,25-dihydroxyvitamin D_3, *Endocrinology*, 98, 1130, 1976.

180. **Davis, G. K., Camberos, H. R., and Djafar, M. I.,** Mineral balance and cardiovascular calcification in guinea pigs fed *Solanum malacoxylon*, *Fed. Proc.*, 28, 561, 1969.

181. **Camberos, H. R., Davis, G. K., and Djafar, M. I.,** Balance mineral en cobayos tratados con *Solanum malacoxylon*, *Medicina (Buenos Aires)*, 29, 441, 1969.

182. **Camberos, H. R., Davis, G. K., and Djafar, M. I.,** Action of *Solanum malacoxylon* on the mineral balance of calcium, phosphorus, magnesium, and copper in sheep and guinea pigs, in *Trace Element Metabolism in Animals, Proc. World Assoc. Anim. Prod./Int. Biol. Progr., Int. Symp. 1969*, Mills, C. F., Ed., Churchill Livingstone, London, 1970, 369.

183. **Camberos, H. R. and Davis, G. K.,** Retención de calcio y calcio endógeno en cobayos tratados con "*Solanum malacoxylon*" (SM), *Medicina (Buenos Aires)*, 31, 606, 1971.

184. **Ousavaplangchai, L.,** Vergleichende Untersuchungen zur Histopathologie der Intoxication mit Vitamin D_3 und *Solanum malacoxylon* bei Meerschweinchen, Inaugural-Dissertation, Vienna University, Vienna, 1972.

185. **Köhler, H., Leibetseder, J., Skalicky, M., and Swoboda, R.,** Zur Kalzinose der Rinder in Österreich. IV. Vergleichende experimentelle Untersuchungen über den Ca- und P-Stoffwechsel des Meerschweinchens nach Hypervitaminose D und Intoxication mit *Solanum malacoxylon*, *Zentralbl. Veterinaermed. Reihe A*, 24, 441, 1977.

186. **Rossi, F. M., Dallorso, M. E., Daskal, H., Gaggino, O. P., and Leiva, A.,** Reproducción experimental de enteque seco en conejo. I. Lesiones cardiovasculares, *Gac. Vet.*, 31, 230 and 415, 1969.

187. **Rossi, F. M., Dallorso, M. E., Daskal, H., Leiva, A., and Gaggino, O.,** Enteque seco experimental en conejos. Calcificación de tejidos blandos producida por *Solanum malacoxylon*, *Medicina (Buenos Aires)*, 29, 442, 1969.

188. **Mautalen, C. A., Rossi, F. M., Dallorso, M. E., Daskal, H., and Leiva, A.,** Efecto del *Solanum malacoxylon* sobre el metabolismo fosfo-cálcico en el conejo, *Medicina (Buenos Aires)*, 29, 441, 1969.

189. **Mautalen, C. A. and Rossi, F. M.,** Efecto del "*Solanum malacoxylon*" sobre el metabolismo fosfocalcico en el conejo, *Medicina (Buenos Aires)*, 30, 443, 1970.

190. **Mautalen, C. A.,** Mecanismo de acción del *Solanum malacoxylon* sobre el metabolismo mineral en el conejo, *Rev. Argent. Endocrinol. Metab.*, 17, 1, 1971.

191. **Mautalen, C. A.,** Mechanism of action of *Solanum malacoxylon* upon calcium and phosphate metabolism in the rabbit, *Endocrinology*, 90, 563, 1972.

192. **Basudde, C. D. K. and Humphreys, D. J.,** The effect of the active principle of *Solanum malacoxylon* on rabbits and the inhibition of its action by actinomycin D, *Calcif. Tissue Res.*, 18, 133, 1975.

193. **Basudde, C. D. K. and Humphreys, D. J.,** The vitamin D_3 metabolite-type activity of *Solanum malacoxylon*, *Clin. Endocrinol.*, 5 (Suppl.), 109 S, 1976.

194. **Gaggino, O. P.,** Desarrollo de la lesion arteriosclerotica incipiente en el enteque seco reproducido experimentalmente en ovejas, *Rev. Invest. Agropecu. Ser. 4*, 6, 31, 1969.

195. **Camberos, H. R., Djafar, M. I., and Davis, G. K.,** Experimental cardiovascular calcification in sheep produced by "*Solanum malacoxylon*", Florida Academy of Science, Gainesville, 1969.

196. **Camberos, H. R. and Davis, G. K.,** Acción del *Solanum malacoxylon* sobre el balance mineral en ovinos, *Gac. Vet. Asoc. Lat. Am. Prod. Anim. Mem.*, 3, 31, 1969.

197. **Camberos, H. R. and Davis, G. K.,** Acción del *Solanum malacoxylon* sobre el balance mineral en ovinos, *Gac. Vet. (Buenos Aires)*, 32, 466, 1970.

198. **Camberos, H. R.,** Arteriosclerosis calcificante en ruminantes: enteque seco, *Gac. Vet. (Buenos Aires)*, 33, 120, 1971.

199. **Bingley, J. B., Ruksan, B. E., and Carrillo, B. J.,** Serum calcium fractions in sheep treated with *Solanum malacoxylon*, *Res. Vet. Sci.*, 21, 121, 1976.

200. **Ruksan, B. and Carrillo, B. J.,** Efecto en ovinos del *Solanum malacoxylon* incubado con liquido de rumen, *Medicina (Buenos Aires)*, 36, 318, 1976.

201. **Fox, J. and Care, A. D.,** The effects of hydroxylated derivatives of vitamin D_3 and of extracts of *Solanum malacoxylon* on the absorption of calcium, phosphate and water from the jejunum of pigs, *Calcif. Tissue Res.*, 21 (Suppl.), 147, 1976.

202. **Harmeyer, J., von Grabe, C., and Martens, H.,** Effects of metabolites and analogs of vitamin D_3 in hereditary pseudo-vitamin D deficiency of pigs, in *Proc. 3rd Workshop Vitamin D* (Vitamin D: Biochemical, Chemical and Clinical Aspects Related to Calcium Metabolism), Norman, A. W., Schaefer, K., Coburn, J. W., DeLuca, H. F., Fraser, D., Grigoleit, H. G., and von Herrath, D., Eds., Walter de Gruyter, Berlin, 1977, 785.

203. **Ruksan, B. E., Wells, G. A. H., and Lewis, G.,** *Solanum malacoxylon* toxicity to pigs, *Vet. Rec.*, 103, 153, 1978.

204. **Fox, J. and Care, A. D.,** The effects of hydroxylated derivatives of vitamin D₃ and of aqueous extracts of *Solanum malacoxylon* on the absorption of calcium, phosphate, sodium, potassium and water from the jejunum of pigs, *J. Endocrinol.*, 82, 417, 1979.

205. **Kasali, O. B., Krook, L., Pond, W. G., and Wasserman, R. H.,** *Cestrum diurnum* intoxication in normal and hyperparathyroid pigs, *Cornell Vet.*, 67, 190, 1977.

206. **Garces, N. E., Okada, A. K., Lando, E., Rucksan, B. E., Culot, J. P., Tilley, J. M. A., Deshpande, P. D., and Roberts, R. M.,** Efecto del *Solanum malacoxylon* sobre el calcio y el fósforo inorgánico en sangre de bovinos, *Bol. Tec. INTA*, 54, 20, 1967.

207. **Reinoso Castro, H.,** Enteque seco: reproducción experimental en bovinos, *Rev. Med. Vet. (Buenos Aires)*, 51, 131, 1970.

208. **Carrillo, B. J., Tilley, J. M., Garces, N. E., Gaggino, O. P., Ruksan, B., and Worker, N. A.,** Intoxicación experimental de bovinos con "*Solanum malacoxylon*", *Gac. Vet. (Buenos Aires)*, 33, 468, 1971.

209. **Sansom, B. F., Vagg, M. J., and Döbereiner, J.,** The effects of *Solanum malacoxylon* on calcium metabolism in cattle, *Res. Vet. Sci.*, 12, 604, 1971.

210. **Döbereiner, J., Done, S. H., and Beltran, L. E.,** Experimental *Solanum malacoxylon* poisoning in calves, *Br. Vet. J.*, 131, 175, 1975; and **Zaccardi, E. M.,** *Rev. Fac. Cienc. Vet. La Plata Univ. Nac. La Plata*, 10, 375, 1968.

211. **Kunz, W.,** Über den Einfluss von *Solanum malacoxylon* (Sendtner) auf den Kalzium-, Phosphor-, und Magnesium-Gehalt im Blutserum beim Rind nach parenteraler und oraler Applikation, *Berl. Muench. Tieraerztl. Wochenschr.*, 90, 69, 1977.

212. **Roux, R., Davicco, M.-J., Carrillo, B. J., and Barlet, J.-P.,** *Solanum glaucophyllum* in pregnant cows. Effect on colostrum mineral composition and plasma calcium and phosphorus levels in dams and newborn calves, *Ann. Biol. Anim. Biochim. Biophys.*, 19 (1A), 91, 1979.

213. **Barlet, J.-P., Roux, R., and Davicco, M.-J.,** The effects of *Solanum glaucophyllum* ingestion by pregnant cows on plasma calcium levels in dams, fetuses and neonates and on the mineral composition of colostrum, *Proc. 4th Workshop Vitamin D* (Vitamin D: Basic Research and its Clinical Application), Norman, A. W., Schaefer, K., von Herrath, D., Grigoleit, H. G., Coburn, J. W., DeLuca, H. F., Mawer, E. B., and Suda, T., Eds., Walter de Gruyter, Berlin, 1979, 345.

214. **Barlet, J.-P., Davicco, M.-J., Lefaivre, J., and Carrillo, B. J.,** Fetal blood calcium response to maternal hypercalcemia induced in the cow by calcium infusion or by *Solanum glaucophyllum* ingestion, *Horm. Metab. Res.*, 11, 57, 1979.

215. **Kunz, W. and Hänichen, T.** *Solanum malacoxylon:* Untersuchungen über die Verträglichkeit der zur Prophylaxe der Hypokalzämischen Gebärlähmung empfohlenen Mengen, *Berl. Muench. Tieraerztl. Wochenschr.*, 94, 421, 1981.

216. **Kunz, W. and Hänichen, T.,** *Solanum malacoxylon:* investigations on the systemic tolerance of doses recommended for prophylaxis of parturient paresis, in Metabolic Disorders in Farm Animals, Proc. 4th Int. Conf. Prod. Disease Farm Animals, Munich, 1980, Giesecke, D., Dirksen, G., and Stangassinger, M., Eds., 1981, 211.

217. **McMurray, C. H., Rice, D. A., Gordon, F., and Humphreys, D.,** The potential use of *Solanum malacoxylon* in the prevention of parturient paresis in dairy cows, in Proc. 5th Int. Conf. Prod. Disease Farm Animals, Uppsala, 1983, Swedish University of Agricultural Sciences, Uppsala, 1983, 22.

218. **Done, S. H., Döbereiner, J., and Tokarnia, C. H.,** Systemic connective tissue calcification in cattle poisoned by *Solanum malacoxylon:* a histological study, *Br. Vet. J.*, 132, 28, 1976.

219. **Noseda, R. P., Cumba, S. A., Gimeno, E. J., Bardon, J. C., Borrajo, H., Maida, J. C., and Forastieri, H.,** Reproducción experimental de ⟨⟨Enteque seco⟩⟩ por Duraznillo blanco, su tratamiento y posterior estudio comparativo de animales tratados y no tratados, *Gac. Vet. (Buenos Aires)*, 38, 105, 1976.

220. **Haussler, M. R.,** Vitamin D receptors: nature and function, *Annu. Rev. Nutr.*, 6, 527, 1986.

221. **Paaren, H. E., Mellon, W. S., Schnoes, H. K., and DeLuca, H. F.,** Ring A-stereoisomers of 1-hydroxyvitamin D₃ and their relative binding affinities for the intestinal 1α,25-dihydroxyvitamin D₃ receptor protein, *Bioorg. Chem.*, 13, 62, 1985.

222. **Stern, P. H., Horst, R. L., Gardner, R., and Napoli, J. L.,** 10-Keto or 25-hydroxy substitution confer equivalent in vitro bone-resorbing activity to vitamin D₃, *Arch. Biochem. Biophys.*, 236, 555, 1985.

223. **Stern, P. H., Trummel, C. L., Schnoes, H. K., and DeLuca, H. F.,** Bone resorbing activity of vitamin D metabolites and congeners in vitro: influence of hydroxyl substituents in the A ring, *Endocrinology*, 97, 1552, 1975.

224. **Rambeck, W. A., Weiser, H., and Zucker, H.,** Biological activity of glycosides of vitamin D₃ and 1α-hydroxyvitamin D₃, *Int. J. Vitam. Nutr. Res.*, 54, 25, 1984; **Lo, E. W., Holick, S. A., Ray, R., Zanis, P., McLaughlin, J., and Holick, M. F.,** The design, synthesis and biological evaluation of water soluble derivatives of vitamin D₃, *Clin. Res.*, 32, 521A, 1984.

225. **Rambeck, W. A., Weiser, H., Hennes, U., Meier, W., and Zucker, H.**, Bioactivity of the C(1)-, C(3)- and C(25)-β-D-glucopyranoside of 1α, 25-(OH)₂D₃, *Calcif. Tissue Int.*, 36 (Suppl. 2), S43, 151, 1984.

226. **Rambeck, W. A., Weiser, H., Meier, W., Labler, L., and Zucker, H.**, Biological activity of the three mono-β-D-glucopyranosides of 1,25-dihydroxycholecalciferol, *Int. J. Vitam. Nutr. Res.*, 55, 263, 1985.

227. **Rambeck, W. A., Weiser, H., and Zucker, H.**, Antirachitische Aktivität von Glukosiden des 1,25-Dihydroxyvitamin D₃ und des 1α-Hydroxyvitamin D₃ beim Hühnerküken, *Wien. Tieraerztl. Monatsschr.*, 73, 169, 1986.

228. **Londowski, J. M., Kost, S. B., Gross, M., Labler, L., Meier, W., and Kumar, R.**, Biologic activity of 3β-D-glucopyranosides of vitamin D compounds, *J. Pharmacol. Exp. Ther.*, 234, 25, 1985.

228a. **Londowski, J. M., Kost, S. B., Meier, W., Labler, L., and Kumar, R.**, Biological activity of the C-1, C-3, C-25,β-D-glucopyranosides of 1,25-dihydroxyvitamin D₃, *J. Pharmacol. Exp. Ther.*, 237, 837, 1986.

229. **Lloyd, W., Wells, H., Walling, M. W., and Kimberg, D. V.**, Stimulation of bone resorption in organ culture by salt-free extracts of *Solanum glaucophyllum*, *Endocr. Res. Commun.*, 2, 159, 1975.

230. **Simonite, J. P., Morris, K. L. M., and Collins, J. C.**, Induction of bone resorption in vitro by an extract of *Solanum malacoxylon*, *J. Endocrinol.*, 68, 18P, 1976.

231. **Liskova-Kiar, M. and Proschek, L.**, Influence of partially purified extracts of *Solanum malacoxylon* on bone resorption in organ culture, *Calcif. Tissue Res.*, 26, 39, 1978.

232. **Rossi, F. M.**, Liberación de Ca 40 y Ca 45 en cultivos de hueso tratados con *Solanum malacoxylon* Sendtner, *Medicina (Buenos Aires)*, 32, 735, 1972.

233. **Riggs, B. L.**, Calcitriol in disorders of bone and calcium metabolism, *Clin. Ther.*, 3, 33, 1980.

234. **Haussler, M. R. and Cordy, P. E.**, Metabolites and analogues of vitamin D, *J. Am. Med. Assoc.*, 247, 841, 1982.

235. **Meier, W.**, Die chemische Entwicklung von Rocaltrol, *Nieren. Hochdruckkr.*, 10, 175, 1981.

236. **Chesney, R. W.**, Current clinical applications of vitamin D metabolite research, *Clin. Orthop. Relat. Res.*, 161, 285, 1981.

237. **Simpson, R. U. and Wishaar, R.**, Newly discovered activities for calcitriol (1,25-dihydroxyvitamin D₃): implications for future pharmacological use, *BioEssays*, 4, 65, 1986.

238. **de Vernejoul, M.-C., Mautalen, C. A., and Miravet, L.**, Le *Solanum malacoxylon*: de la plante toxique à l'agent thérapeutique, *Nouv. Presse Med.*, 7, 1941, 1978.

239. **Mautalen, C. A., Ferraro, C., Cabrejas, M., Landi, E., and Gotlieb, D.**, Effects of *Solanum malacoxylon* on calcium metabolism in patients with chronic renal failure, *Calcif. Tissue Res.*, 22 (Suppl.), 534, 1977.

240. **Moos, M., Fischer-Ferraro, C., Cabrejas, M., Mautalen, C. A., Julianelli, V., Locatelli, A., and Cantarovich, F.**, Comparación del efecto del *Solanum malacoxylon* (SM) y del 1α-hidroxicolecalciferol (1αHO-D₃), en pacientes urémicos en hemodialisis, *Medicina (Buenos Aires)*, 36, 549, 1976.

241. **Casco, C., Ferraro, C., Ladizesky, M., Man, Z., Ghiringhelli, W., Cabrejas, M., and Mautalen, C. A.**, Effects of *Solanum malacoxylon* in hypoparathyroidism, in *Proc. 3rd Workshop Vitamin D* (Vitamin D: Biochemical, Chemical and Clinical Aspects Related to Calcium Metabolism), Norman, A. W., Schaefer, K., Coburn, J. W., DeLuca, H. F., Fraser, D., Grigoleit, H. G., and von Herrath, D., Eds., Walter de Gruyter, Berlin, 1977, 759.

242. **Casco, C., Ladizesky, M., Ferraro, C., Man, Z., Ghiringhelli, W., Cabrejas, M., and Mautalen, C.**, Efecto del *Solanum malacoxylon* sobre el metabolismo fosfocálcico y los niveles séricos de 25-hidroxicolecalciferol en el hipoparatiroidismo, *Medicina (Buenos Aires)*, 36, 594, 1976.

243. **Reichenbächer, H.**, Der Einfluss von Goldhafer *(Trisetum flavescens)* und *Solanum malacoxylon* auf die Eischalenqualität, Inaugural Dissertation, Fachbereich Tiermedizin, Munich, 1979.

244. **Rambeck, W. A., Kreutzberg, O., and Zucker, H.**, Ruminal fluid increases the vitamin D activity of *Trisetum flavescens*, in *Proc. 5th Workshop Vitam D*. Vitamin D, Chemical, Biochemical and Clinical Endocrinology of Calcium Metabolism, Norman, A. W., Schaefer, K., von Herrath, D., and Grigoleit, H. G., Eds., W. de Gruyter, Berlin, 1982, 329.

245. **Wasserman, R. H.**, Bones, in *Dukes' Physiology of Domestic Animals*, 9th ed., Swenson, M. J., Ed., Cornell University Press, Ithaca, NY, 1977, 430; and **Wasserman, R. H.**, Physiological regulation of calcium metabolism: the consequences of excess intake of 1,25-dihydroxycholecalciferol from natural sources, *Ann. N. Y. Acad. Sci.*, 307, 442, 1978.

246. **Wasserman, R. H., Peterlik, M., Haussler, M., Hughes, M., Bursac, K. M., and Krook, L.**, The calcinogenic plants: their nature and mechanism of action, *Isr. J. Med. Sci.*, 12, 1492, 1976.

247. **Sobti, S. N., Verma, V., Rao, B. L., and Pushpangadan, P.**, In IOPB chromosome number reports. LXV, *Taxon*, 28, 627, 1979.

248. **Wasserman, R. H.**, Phosphate absorption, vitamin D, and a botanical source of active vitamin D activity, *Proc. Cornell Nutr. Conf. Feed Manuf.*, 1974, 100.

249. **Sansom, B. F.**, Estimation of absorption coefficients for manganese-54 and calcium-47 in dairy cows by whole-body counting, *Assessment Radioactive Contam. Man, Proc. Symp. 1971*, IAEA, Vienna, 1972, 563.

Chapter 8

CARCINOGENIC BRACKEN GLYCOSIDES

Iwao Hirono

TABLE OF CONTENTS

I. INTRODUCTION

Bracken fern, *Pteridium aquilinum,* is widely distributed in many parts of the world and used as a human food in Japan and some other countries. The toxic effect of bracken fern on livestock has attracted the attention of veterinary scientists since the end of the last century.[1,2,3] The predominant feature of cattle bracken poisoning is depressed bone marrow activity, which gives rise to severe leukopenia, thrombocytopenia, and hemorrhagic syndrome. The carcinogenicity of bracken fern was demonstrated most clearly by the experiment of Evans and Mason[3] showing that rats fed a diet containing bracken fern developed multiple intestinal adenocarcinomas. Subsequently, the simultaneous induction of urinary bladder tumors and the occurrence of mammary carcinoma in Sprague-Dawley rats fed a bracken-containing diet has been reported. However, neither the causative principle of cattle bracken poisoning nor bracken carcinogen had been isolated until very recently when our group succeeded in isolating a carcinogenic principle, ptaquiloside, a novel norsesquiterpene glucoside of the illudane type. It was also demonstrated that ptaquiloside is a causative principle of cattle bracken poisoning.

II. TOXICITY OF BRACKEN FERN

The toxic syndromes induced by ingestion of bracken fern are different in horses, cattle, and sheep.[1] Ingestion of bracken fern by horses leads to a thiamine deficiency because the fern contains thiaminase, which can cleave the thiamine molecule. The predominant feature of bracken fern poisoning in cattle is depressed bone marrow activity, which gives rise to severe leukopenia, especially of granulocytes, thrombocytopenia, the hemorrhagic syndrome, and hematuria. Cattle "bracken poisoning" is quite dramatic and nearly always fatal.[2]

III. CARCINOGENIC PROPERTIES OF BRACKEN FERN

Evans and Mason[3] reported in 1965 that rats fed a diet containing bracken fern developed multiple ileal adenocarcinomas. Simultaneous intestinal and urinary bladder tumors were induced in rats fed a diet containing bracken fern until the animals died or were killed.[4] Adenomatous polyps and adenocarcinomas developed predominantly in the ileum. Urinary bladder tumors occurred in 81% of the autopsied rats and were either papillomas or sessile or papillary carcinomas. Hirono et al.[5] suggested that the highest incidence of urinary bladder tumors is obtained if relatively small amounts of the carcinogen are given over long periods. In Japan, young bracken fern fronds in the fiddlehead or crosier stage of growth are used as human food (Figure 1).[5,6] However, when rats were given a diet containing powdered young bracken fern in a proportion of 1 part by weight of bracken fern to 2 parts of basal diet for 4 months, all the rats that survived for more than 7 months after the start of the experiment had ileal tumors. The most common site for intestinal tumors induced by bracken fern was the terminal 20 cm of the ileum. Histologically, the intestinal tumors were not only epithelial tumors, such as adenomas and adenocarcinomas, but also sarcomas.[5,7]

In Japan, young bracken fern is usually used as a human food after its astringent taste has been removed by one of the following treatments.

1.   Fresh bracken fern is immersed in boiling water containing wood ash or sodium bicarbonate, then seasoned; sometimes, it is merely boiled before being eaten.
2.   Fresh bracken is pickled in salt and immersed in boiling water before use (Figure 1).

The carcinogenic activity of processed bracken used as a human food was studied in ACI

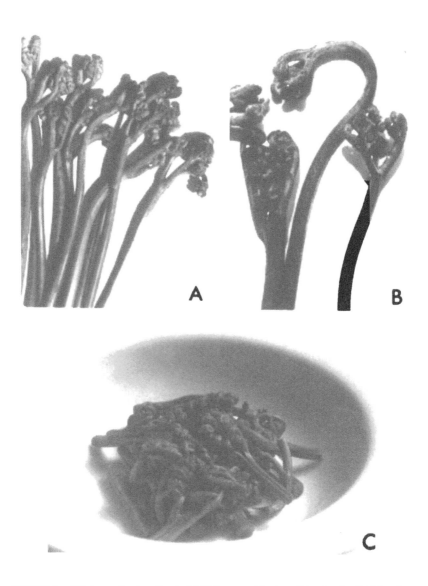

FIGURE 1. (A) Young bracken frond; (B) higher magnification of A, showing fiddlehead or crosier stage of growth. Extreme part of frond is still curled; (C) pickled bracken used as human food.

rats.[8] Tumor incidence in rats fed an unprocessed bracken-containing diet was 78.5%. However, it was 25, 10, and 4.7% in rats fed a diet containing processed bracken fern treated with wood ash, sodium bicarbonate, and NaCl, respectively. Although the carcinogenic activity of bracken fern was markedly reduced by such treatment, weak carcinogenic activity was still retained in the bracken fern thus prepared. Furthermore, to study the relationship between the stage of maturation of bracken fern and carcinogenic activity, the carcinogenicity of mature bracken fern was compared with that of immature young fern with curled tops. From the results obtained in this experiment, it was evident that the mature bracken still retained fairly strong carcinogenic activity, although it was weak compared with the young bracken. Female CD rats (Charles River Sprague-Dawley rats) fed a bracken diet developed mammary cancer in high incidence, in addition to ileal and urinary bladder tumors.[9] Carcinogenicity of bracken fern in laboratory animals other than rats and farm

Table 1
CARCINOGENICITY OF BRACKEN
FERN

Animal[a]	Target organs and histological findings
Rat	Ileum, cecum (adenoma, adenocarcinoma, sarcoma) Urinary bladder (papilloma, carcinoma)
	Sprague-Dawley: mammary cancer
Mouse	Swiss, dd: lung adenoma, lymphatic leukemia
	C57BL/6: jejunal adenoma
Quail	Cecum, colon, ileum (adenocarcinoma)
Hamster	Cecum, ileum (adenocarcinoma)
Guinea pig	Small intestine (adenoma, adenocarcinoma) Urinary bladder (carcinoma)
Cattle	Urinary bladder (papilloma, carcinoma, hemangioendothelioma)

[a] All animals were fed a diet containing bracken powder, except quail which was given ethanol extract of bracken.

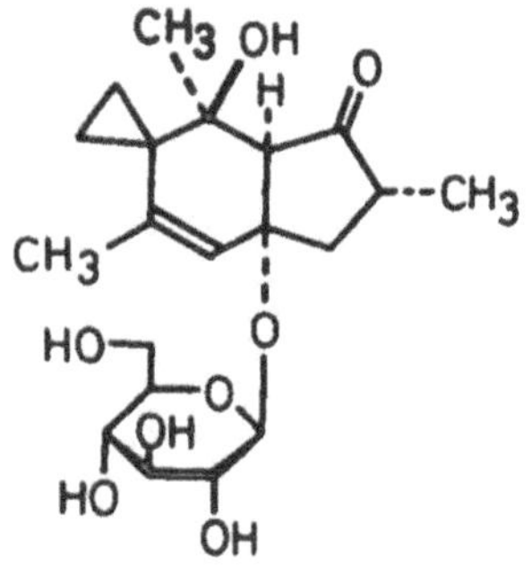

Ptaquiloside

FIGURE 2.   Chemical structure of ptaquiloside.

livestock is summarized in Table 1. Hirono et al.[10] recently reported that rats fed a bracken diet developed papillomas of the tongue, pharynx, esophagus, and forestomach, and squamous cell carcinoma of the pharynx.

IV. ISOLATION OF BRACKEN CARCINOGEN PTAQUILOSIDE

Based on the carcinogenicity in rats, fractionation of the aqueous extracts of bracken fern was performed, resulting in the isolation of a bracken carcinogen, ptaquiloside (Figure 2). The isolation procedure of ptaquiloside is shown in Figure 3.[11,12]

The dried powdered bracken was extracted with boiling water with vigorous agitation for 10 min, and this extraction process was repeated three times. A resin, Amberlite® XAD-2, was added to the combined aqueous extract (I-A) and the mixture stirred and filtered. The resin adsorbate was eluted with methanol and the resulting methanol solution concentrated to give the eluate residue (II-B). The results of the carcinogenicity assay on the aqueous filtrate residue (II-A) and the eluate residue (II-B) clearly revealed that the latter fraction (II-B) was carcinogenic, as shown in Table 2. The eluate residue (II-B) was dissolved in *n*-

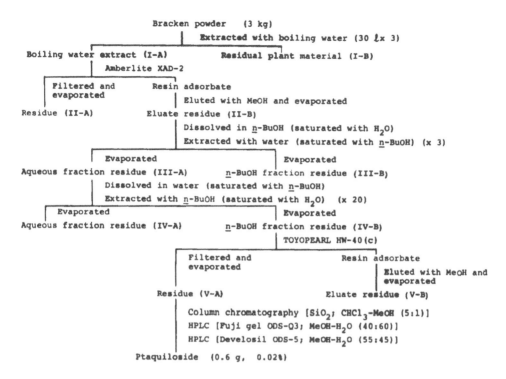

FIGURE 3.   Isolation of a bracken carcinogen, ptaquiloside.

butanol saturated with water and extracted with water saturated with *n*-butanol three times. The carcinogenicity of the two fractions thus obtained, the aqueous fraction (III-A) and the *n*-butanol fraction (III-B), was examined: the former fraction (III-A) was found to show carcinogenic activity which was slightly stronger than the latter fraction (III-B) (Table 2). For further fractionation the aqueous fraction (III-A) was dissolved in water saturated with *n*-butanol and extracted with *n*-butanol saturated with water 20 times. Again in this stage, studies on the carcinogenicity of the resulting two fractions, the aqueous fraction (IV-A) and the *n*-butanol fraction (IV-B), were carried out, the results indicating that the latter fraction (IV-B) was markedly carcinogenic (Table 2). Subsequently, the *n*-butanol fraction (IV-B) dissolved in water was treated with a resin, Toyopearl® HW-40 (c), in order to adsorb flavonoids on the resin, and the mixture was stirred and filtered. The residue (V-A) obtained on concentration of the filtrate showed strong carcinogenicity, whereas the eluate residue (V-B) obtainable from the resin adsorbate was not carcinogenic (Table 2). Thus, separation using the resin adsorption and the solvent partitions described above provided the fraction (V-A) exhibiting strong carcinogenicity.

The fraction (V-A) was analyzed by means of high-performance liquid chromatography (HPLC), and a new compound, ptaquiloside, was isolated as one of the major components of the fraction (V-A). To isolate ptaquiloside on a large scale, the fraction (V-A) was chromatographed on silica gel with chloroform-methanol (5:1). Fractions containing ptaquiloside obtained after silica gel chromatography were further separated and purified by repeating reversed-phase HPLC (i, Fuji® gel ODS-Q3 and methanol-water [40:60]; ii, Develosil® ODS-5 and methanol-water [55:45]) to afford ptaquiloside which was at least 99% pure and obtained in 0.02% yield based on the dried powdered bracken. Ptaquiloside is a colorless amorphous compound with the following physical and spectral properties:[12] molecular formula, $C_{20}H_{30}O_8$; $[\alpha]^{22}D - 188°$ (c 1.00, CH_3OH); IR (KBr) 3400 (broad), 1724, 1640 (weak) cm^{-1} SIMS m/z 421 (M + Na)$^+$.

The original procedure for isolating ptaquiloside was inefficient and complicated and led

Table 2
CARCINOGENICITY OF BRACKEN FRACTION ADMINISTERED TO RATS

	Rats[a]			Bracken diet		Amount of fraction ingested per rat (g)	Experimental period (d)	Incidence of tumor			No. of rats with hyperplastic nodules of the liver
Fraction	No.	Strain	Age (weeks)	Concentration over usual bracken content	Administration period (d)			Mammary gland	Intestine	Urinary bladder	
I-A[b]											
I-B											
II-A	5	ACI	6	× 5	174	416.9	174	0/5	0/5	0/5	
II-B	5	ACI	6	× 5	174	42.9	174	1/5	4/5	4/5	
III-A	7	CD	4	× 5	133	27.0	192	7/7 (83)[c]	7/7	4/7	2
III-B	7	CD	4	× 5	133	16.0	176	7/7	5/7	4/7	3
IV-A	7	CD	4	× 7.5 (7 d) subsequently × 6	120	14.1	171	0/7	3/7	0/7	0
IV-B	7	CD	4		127	19.1	164	7/7 (105)	7/7	6/7	4
V-A	7	CD	4	× 5 (35 d) subsequently × 3	133	9.4	218	7/7 (93)	7/7	5/7	5[d]
V-B	7	CD	4		119	5.1	213	0/7	0/7	0/7	0

[a] All rats used were female.
[b] Carcinogenicity of boiling water extract was reported previously.[25]
[c] Appearance of the first mammary tumor (in days).
[d] vs. V-B; $p < 0.05$

to only 0.02% yield based on the dry weight of bracken. Recently, an efficient and convenient method for the isolation of ptaquiloside was developed.[13] The yield was five times greater than the original procedure.

Extraction procedure and preparation of the *n*-butanol extract — The dried, finely powdered, plant materials (3.0 kg) were stirred in water (30 l) at room temperature for 2 h and filtered. After the resin Amberlite® XAD-2 (wet volume, 12 l) was added to the filtrate, the mixture was stirred at room temperature for 1 h and filtered. The filtrate was again treated with fresh Amberlite® XAD-2 (wet volume, 12 l). Methanol (30 l) was added to the combined resin XAD-2, and the mixture was stirred for 3 h at room temperature. The mixture was filtered, and the resin was washed with methanol (30 l). The combined methanol solution was concentrated under reduced pressure, and the residue (90 g) was dissolved in water saturated with *n*-butanol (1 l). The aqueous solution was extracted with *n*-butanol saturated with water (5 × 500 ml). Concentration of the combined organic solution under reduced pressure afforded the *n*-butanol extract as a dark-brown, amorphous solid (33 g).

Isolation of ptaquiloside — The *n*-butanol extract (33 g) was chromatographed on silica gel (600 g) with EtOAc, EtOAc-MeOH (92:8), and MeOH, successively. Concentration of the fractions eluted with EtOAc-MeOH (92:8) under reduced pressure gave a residue (9 g), which was chromatographed on ODS-silica gel (180 g) with H_2O-MeOH (60:40). The resulting crude ptaquiloside was further purified by HPLC with H_2O-MeOH (50:50) and freeze-dried to give pure ptaquiloside (3 g, 0.1%) as a colorless amorphous powder.

V. CHEMICAL PROPERTIES OF PTAQUILOSIDE

Ptaquiloside is unstable at room temperature under both acidic and basic conditions and undergoes aromatization to give 1-indanone derivatives such as pterosin B and pterosin O, depending on the solvent used; the half-life of ptaquiloside in 0.01 *M* sulfuric acid-methanol at 22°C was about 2 h. Under the particular alkaline conditions (0.01 *M* Na_2CO_3 solution, 22°C, 20 min), ptaquiloside was converted with concomitant elimination of D-(+)glucose into an unstable conjugated dienone, colorless oil. The dienone was extremely unstable in a weakly acidic aqueous solution at room temperature and immediately converted to pterosin B (Figure 4). From the detailed analysis of ^{1}H- and ^{13}C-NMR spectra coupled with chemical evidence, the planar structure of ptaquiloside was established.[12] Subsequently, the absolute stereostructure of ptaquiloside was elucidated by the ^{1}H-NMR spectral method and an X-ray crystallographic analysis of ptaquiloside tetraacetate.[14]

VI. ISOLATION OF *P*-HYDROXYSTYRENE GLYCOSIDES

Two new *p*-hydroxystyrene glycosides isolated from the carcinogenic fraction of aqueous extracts of bracken fern were named ptelatoside-A and ptelatoside-B (Figure 5). Their structures were established to the *p*-β-primeverosyloxystyrene and *p*-β-neohesperidosyloxy-styrene, by chemical and spectral means.[15] A synthesis of ptelatoside-A was achieved.[15] Hyperplastic nodules of the liver were infrequently encountered in rats fed a bracken diet (Table 3). Such lesions occurred also in rats fed a diet containing the carcinogenic fractions of bracken water extract and more frequently with an increase in concentration of the fraction[16] (Table 2). In the case of fractions IV and V, hyperplastic nodules were induced only in rats fed subfractions IV-B and V-A, which had proved to be strongly carcinogenic. These hyperplastic nodules were frequently multiple and detectable even by macroscopic observation. From these results, it was evident that the causative principles of these hyperplastic nodules are present in the carcinogenic fractions of bracken fern. Styrene and styrene oxide are known to be mutagenic in *Salmonella typhimurium*.[17-19] It was strongly suggested that not only ptaquiloside, but also *p*-hydroxystyrene glycosides are the causative principle of these hyperplastic nodules of the liver.

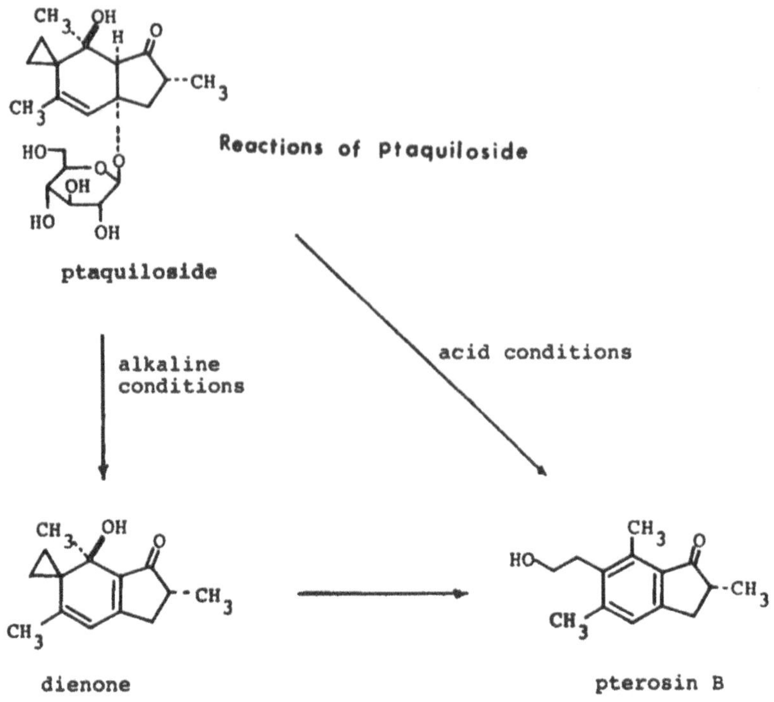

FIGURE 4.   Reaction of ptaquiloside.

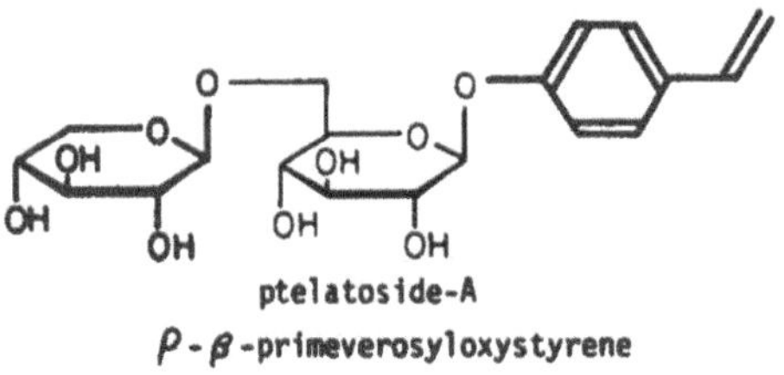

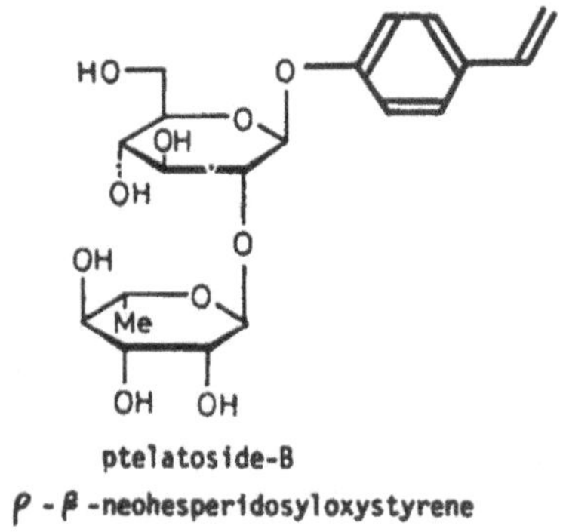

FIGURE 5.   Chemical structures of *p*-hydroxystyrene glycosides; ptela-toside-A and ptelatoside-B.

Table 3
HYPERPLASTIC NODULES OF THE
LIVER IN RATS FED BRACKEN
DIET

Group	Strain	Sex	No.	No. of rats with hyperplastic nodules
Experimental	CD	M	15	1
		F	14	5[b]
Control	CD	M	16	0
		F	16	0
Experimental	ACI	M	19	4[c]
		F	11	0
Control	ACI	M	25	0
		F	25	0

[a] Rats received 30% bracken diet for 260 d until the termination of experiment.
[b] vs. Control F; $p < 0.05$
[c] vs. Control M; $p < 0.01$

VII. CARCINOGENICITY TEST OF PTAQUILOSIDE AND PTELATOSIDE-A

A. Carcinogenicity Test of Ptaquiloside[20]

Female CD rats were used for carcinogenicity testing ptaquiloside. Group 1 consisted of 12 female CD rats, each of which was given an intragastric administration of 780 mg of ptaquiloside per kilogram body weight on day 25 after birth. They were then administered 100 mg ptaquiloside per kilogram body weight 17 d after the first administration. Subsequently, they were given consecutive intragastric administration once a week for 7 weeks as shown in Table 4. Group 2, consisting of 12 female CD rats, 28 d old, received intragastric administration of ptaquiloside twice a week for 8 weeks and only a single administration in week 9. The administration schedule of ptaquiloside in experiment group 2 is also shown in Table 4. The average total doses of ptaquiloside per rat in groups 1 and 2 were 300 mg and 339 mg, respectively. Ptaquiloside was freshly dissolved in 1 ml of physiological saline each time and administered by means of a metal gastric tube. Since ptaquiloside is unstable at room temperature under both acidic and basic conditions, it was kept at −20°C in a freezer. The experiment was terminated 300 d after the start of administration of ptaquiloside.

All rats of group 1 had severe hematuria and urinary incontinence after the first intragastric administration of 780 mg ptaquiloside per kilogram body weight and 2 of 12 rats died 3 d after the administration. Five rats in group 1 died within 83 d after the start of the experiment. The remaining seven rats survived more than 190 d, and all had mammary cancer (Table 5). The earliest mammary tumor was detected by palpation on day 82. The average number of mammary tumors per rat was 4.8. Histologically, the tumors were adenocarcinoma, papillary carcinoma, and anaplastic carcinoma, i.e., the same as those induced in CD rats fed a bracken diet. Mammary cancer induced in a rat which was sacrificed 265 d after the start of experiment showed metastasis in the lung. Furthermore, four of these seven rats (57%) also had multiple ileal adenocarcinomas. No urinary bladder tumors were induced. However, squamous metaplasia or preneoplastic hyperplasia of the urinary bladder mucosa was present in eight of ten rats which survived more than 40 d. In group 2, all rats, except one which died 60 d after the start of experiment, survived more than 165 d, and eight rats

Table 4
ADMINISTRATION SCHEDULE OF PTAQUILOSIDE

	Dose of ptaquiloside by intragastric administration (mg/kg body weight)[a]	
Week	Group 1	Group 2
1	780	100 (× 2)
2	—	100 (× 2)
3	100	150 (× 2)
4	200	150 (× 2)
5	200	100 (× 2)
6	200	100 (× 2)
7	150	100 (× 2)
8	100	100 (× 2)
9	100	100
10	100	

[a] Group 1 was given weekly and group 2 was twice weekly, except a single administration was made on week 9.

Table 5
INCIDENCE OF TUMOR INDUCED BY PTAQUILOSIDE IN FEMALE CD RATS

Group	Initial No. of rats	Effective No. of rats[a]	No. of rats with tumors of		
			Mammary gland	Ileum	Urinary bladder
Group 1	12	7	7 (4.8)[b]	4	0
Group 2	12	11	10 (3.9)	10	1
Control	15	15	0	0	0

[a] No. surviving for more than 165 d after the start of the experiment.
[b] Figures in parentheses are the average number of tumors per rat.

survived beyond 270 d. Mammary cancers were induced in 10 of 11 (effective number) rats (91%). The earliest mammary tumor was detected on day 94 after the start of ptaquiloside administration. The average number of mammary tumors per rat was 3.9. The histological types and incidence were similar to those in group 1. One animal which died 276 d after the start of the experiment showed lung metastasis of the mammary cancer. Multiple ileal adenocarcinomas were also observed in 10 of 11 rats (91%). Although urinary bladder papilloma was induced only in one rat, hyperplasia of the bladder mucosa was encountered in 7 of 11 rats. No tumors were found in rats of the control group. The low incidence of urinary bladder tumor in this experiment was considered to be due to the short period of ptaquiloside administration and relatively early death of the animals from mammary cancer. The incidence of mammary cancer was 100 and 91% in groups 1 and 2, respectively. Ileal tumor was also induced in high incidence, 57 and 91%, and the terminal 20 cm of the ileum was the most common site, as in the case of rats fed bracken diet. From these results, it is evident that ptaquiloside is one of the carcinogenic principles of bracken fern.

In our recent experiment, 15 female ACI rats of 5-weeks initial age were given a diet containing 0.027 to 0.08% ptaquiloside throughout the 210-d experimental period. Both ileal and urinary bladder tumors developed in all rats in the experimental group. These results demonstrated that, like the bracken diet, ptaquiloside induced tumors in both the ileum and urinary bladder.[21] Van der Hoeven et al.[22] isolated from bracken fern a new mutagenic compound which has the same planar structure as ptaquiloside and named it aquilide A.

A previous experiment by our group indicated no significant difference in the incidence of intestinal tumors between germ-free and conventional rats fed a diet containing bracken, suggesting that gut microflora does not play a definite role in bracken tumorigenesis.[23] Since it is assumed that ptaquiloside is converted with concomitant elimination of glucose into an unstable dienone under particular alkaline conditions,[12] it is logical that no significant difference was observed in the incidence of intestinal tumors between germ-free and conventional rats fed a bracken diet.

B. Carcinogenicity Test of Ptelatoside-A

Ptelatoside-A, one of the new p-hydroxystyrene glycosides isolated from the carcinogenic fraction of aqueous extracts of bracken, was successfully synthesized, and the carcinogenicity was tested in CD rats. A total of 23 rats of both sexes were fed a diet containing 0.13% ptelatoside-A for 109 and 125 d in males and females, respectively. The experiment was terminated 520 d after the start of feeding. No significant difference was observed in the incidence of tumors and hyperplastic nodules of the liver in the experimental group compared with the corresponding control group. Such a negative result may be due to either the small dose of ptelatoside-A used in this experiment or the lack of the carcinogenicity in ptelatoside-A.

VIII. REPRODUCTION OF ACUTE BRACKEN POISONING WITH PTAQUILOSIDE[24]

A Holstein-Friesian, 6-month-old female calf was given ptaquiloside. Ptaquiloside was dissolved in 500 ml of 0.9% saline and administered by drench, once in the morning before feeding, for 6 out of 7 d at the following rates: 400 mg/d for the first 24 d, 800 mg/d for 14 d, and 1600 mg/d for 4 d. The total amount of ptaquiloside administered was 27.2 g. Leukocyte counts increased greatly after the beginning of ptaquiloside administration and reached a maximum ($25.3 \times 10^3/mm^3$) on day 17 after the start of administration, following large fluctuations. Subsequently, they decreased and reached a level of $4.7 \times 10^3/mm^3$ on day 64 after the start of the experiment. Neutrophilic granulocyte numbers followed a similar pattern to the leukocytes. They began to decrease rapidly about 50 d after the start of administration, to a minimum level of $0.1 \times 10^3/mm^3$. The granulocytopenia continued for about 35 d until the autopsy, despite cessation of ptaquiloside administration. The erythrocyte level remained between 6 and $8 \times 10^6/mm^3$ during the course of the study. Thrombocyte levels showed a relatively slow depression accompanied by mild fluctuations and reached a minimum level of $1 \times 10^5/mm^3$ 50 d after the beginning of the administration of ptaquiloside (Figure 6). After the ptaquiloside administration was stopped, the thrombocyte count gradually began to recover, but still remained less than $2 \times 10^5/mm^3$ 1 month later. The calf was autopsied 86 d after the start of the administration of ptaquiloside. Sternal bone marrow was found to be mostly replaced with fat marrow, and only small foci of erythropoietic cells and a small number of megakaryocytes remained. Thus, it was demonstrated that the causative principle of cattle bracken poisoning is also ptaquiloside.

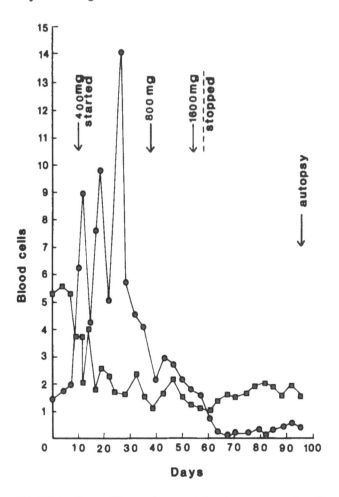

FIGURE 6.   Neutrophilic granulocyte and thrombocyte counts of a calf receiving drench of ptaquiloside. ● Neutrophilic granulocytes ($\times$ 10^3/mm^3); ■ thrombocytes ($\times$ 10^5/mm^3)

REFERENCES

1. **Evans, W. C., Patel, M. C., and Koohy, Y.**, Acute bracken poisoning in homogastric and ruminant animals, *Proc. R. Soc. Edinburgh Sect. B*, 81, 29, 1982.
2. **Evans, I. A.**, Bracken carcinogenicity, in *Chemical Carcinogens*, Vol. 2, 2nd ed., ACS Monogr. 182, Searle, C. E., Ed., American Chemical Society, Washington, D.C., 1984, 1171.
3. **Evans, I. A. and Mason, J.**, Carcinogenic activity of bracken, *Nature (London)*, 208, 913, 1965.
4. **Pamukcu, A. M. and Price, J. M.**, Induction of intestinal and urinary bladder cancer in rats by feeding bracken fern *(Pteris aquilina)*, *J. Natl. Cancer Inst.*, 43, 275, 1969.
5. **Hirono, I., Shibuya, C., Fushimi, K., and Haga, M.**, Studies on carcinogenic properties of bracken, *Pteridium aquilinum*, *J. Natl. Cancer Inst.*, 45, 179, 1970.
6. **Hodge, W. H.**, Fern foods of Japan and the problem of toxicity, *Am. Fern J.*, 63, 77, 1973.
7. **Ogino, H., Fujimoto, M., and Hirono, I.**, Reexamination of histological findings of ileal sarcomas induced in rats given diet containing bracken fern, *J. Cancer Res. Clin. Oncol.*, 112, 6, 1986.
8. **Hirono, I., Shibuya, C., Shimizu, M., and Fushimi, K.**, Carcinogenic activity of processed bracken used as human food, *J. Natl. Cancer Inst.*, 48, 1245, 1972.
9. **Hirono, I., Aiso, S., Hosaka, S., Yamaji, T., and Haga, M.**, Induction of mammary cancer in CD rats fed bracken diet, *Carcinogenesis*, 4, 885, 1983.

10. Hirono, I., Hosaka, S., and Kuhara, K., Enhancement by bracken of induction of tumors of the upper alimentary tract by *N*-propyl-*N*-nitrosourethan, *Br. J. Cancer*, 46, 423, 1982.
11. Hirono, I., Yamada, K., Niwa, H., Shizuri, Y., Ojika, M., Hosaka, S., Yamaji, T., Wakamatsu, K., Kigoshi, H., Niiyama, K., and Uosaki, Y., Separation of carcinogenic fraction of bracken fern, *Cancer Lett.*, 21, 239, 1984.
12. Niwa, H., Ojika, M., Wakamatsu, K., Yamada, K., Hirono, I., and Matsushita, K., Ptaquiloside, a novel norsesquiterpene glucoside from bracken, *Pteridium aquilinum* var. *latiusculum*, *Tetrahedron Lett.*, 24, 4117, 1983.
13. Ojika, M., Kigoshi, H., Kuyama, H., Niwa, H., and Yamada, K., Studies on *pteridium aquilinum* var. latiusculum. IV. Isolation of three *p*-hydroxystyrene glycosides and an efficient method for the isolation of ptaquiloside, an unstable bracken carcinogen, *J. Nat. Prod.*, 48, 634, 1985.
14. Niwa, H., Ojika, M., Wakamatsu, K., Yamada, K., Ohba, S., Saito, Y., Hirono, I., and Matsushita, K., Stereochemistry of ptaquiloside, a novel norsesquiterpene glucoside from bracken, *Pteridium aquilinum* var. *latiusculum*, *Tetrahedron Lett.*, 24, 5371, 1983.
15. Ojika, M., Wakamatsu, K., Niwa, H., Yamada, K., and Hirono, I., Isolation and structures of two new *p*-hydroxystyrene glycosides, ptelatoside-A and ptelatoside-B from bracken, *Pteridium aquilinum* var. *latiusculum*, and synthesis of ptelatoside-A, *Chem. Lett.*, p. 397, 1984.
16. Hirono, I., Aiso, S., Yamaji, T., Niwa, H., Ojika, M., Wakamatsu, K., and Yamada, K., Hyperplastic nodules of the liver induced in rats fed bracken diet, *Cancer Lett.*, 22, 151, 1984.
17. Milvy, P. and Garro, A. J., Mutagenic activity of styrene oxide (1,2-epoxyethylbenzene), a presumed styrene metabolite, *Mutat. Res.*, 40, 15, 1976.
18. Vainio, H., Pääkonen, R., Rönnholm, K., Raunio, V., and Pelkonen, O., A study on the mutagenic activity of styrene and styrene oxide, *Scand. J. Work Environ. Health*, 3, 147, 1976.
19. De Meester, C., Poncelet, F., Roberfroid, M., Rondelet, J., and Mercier, M., Mutagenicity of styrene and styrene oxide, *Mutat. Res.*, 56, 147, 1977.
20. Hirono, I., Aiso, S., Yamaji, T., Mori, H., Yamada, K., Niwa, H., Ojika, M,. Wakamatsu, K., Kigoshi, H., Niiyama, K., and Uosaki, Y., Carcinogenicity in rats of ptaquiloside isolated from bracken, *Gann*, 75, 833, 1984.
21. Hirono, I., Ogino, H., Fujimoto, M., Yamada, K., Yoshida, Y., Ikagawa, M., and Okumura, M., Induction of tumors in ACI rats given a diet containing ptaquiloside, a bracken carcinogen, *J. Natl. Cancer Inst.*, 79, 1143, 1987.
22. Van der Hoeven, J. C. M., Lagerweij, W. J., Posthumus, M. A., van Veldhuizen, A., and Holterman, H. A. J., Aquilide A, a new mutagenic compound isolated from bracken fern (*Pteridium aquilinum* [L.] Kuhn), *Carcinogenesis*, 4, 1587, 1983.
23. Sumi, Y., Hirono, I., Hosaka, S., Ueno, I., and Miyakawa, M., Tumor induction in germ-free rats fed bracken (*Pteridium aquilinum*), *Cancer Res.*, 41, 250, 1981.
24. Hirono, I., Kono, Y., Takahashi, K., Yamada, K., Niwa, H., Ojika, M., Kigoshi, H., Niiyama, K., and Uosaki, Y., Reproduction of acute bracken poisoning in a calf with ptaquiloside, a bracken constituent, *Vet. Rec.*, 115, 375, 1984.
25. Hirono, I., Ushimaru, Y., Kato, K., Mori, H., and Sasaoka, I., Carcinogenicity of boiling water extract of bracken, *Pteridium aquilinum*, *Gann*, 69, 383, 1978.

Chapter 9

CARBOXYATRACTYLOSIDE

Richard J. Cole, Horace G. Cutler, and Barry P. Stuart

TABLE OF CONTENTS

I. INTRODUCTION

The therapeutic and toxic properties of the root of the birdlime thistle *Atractylis gummifera* (L.) have been reported as far back in history as about 300 B.C. when Theophrastus noted that this plant known then as "Chamaeleon" was lethal to animals.[1] Later reports by Dioscorides of Anazarbus (First Century, A.D.) in his "De Materia Medica,"[2] Pliny, the Elder in "De Historia Plantarium",[1] and Mesue the Younger, about 1000 A.D., reflected its potency.[3] The thistle *A. gummifera* is found primarily in the Mediterranean region, but also occurs in Morocco, Algeria, Tunisia, Asia Minor, Greece, Spain, Portugal, and Italy.[4] Ancient naturalists familiar with the toxicity of the plant gave it such common names as "Chamaeleon", reflecting the changing color of its flowers,[1,2] and "Mezereon" or "Lion of the Earth".[3] Throughout history the rhizomes of this thistle were recommended for treatment of a wide array of diseases including bubonic plague, drowsiness, melancholy, toothache, bleeding gingiva, intestinal parasites, ulcers, poisonous snakebite, and hydropsy, and have been associated with numerous accidental, fortuitous, or criminal poisonings.[4] In his *Species Plantarum*,[5] Linneus classified the plant in the Compositae family and gave it the scientific name, *A. gummifera*. More recently, Fiori and Paoletti[6,7] have named the plant *Carlina gummifera Less.* and distinguish the species as two varieties: the *α-typical* and the *β-Fontanesii*. The *α-typical* is found in southern Italy and Sicily whereas the *β-Fontanesii* is found in Sardinia and Corsica.[4]

An excellent review on the history, chemistry, biochemistry, and toxicity of atractyloside and carboxyatractyloside has been published.[4]

II. DISTRIBUTION OF CARBOXYATRACTYLOSIDE IN PLANTS

A. *Atractylis gummifera L.*

The toxic principle of *A. gummifera* was originally thought to be primarily due to atractyloside (AT). In 1964 Stanislas and Vignais[8] isolated a glucoside from *A. gummifera* that they named gummiferin. The identification of gummiferin as being identical to carboxyatractyloside (CAT) was the result of studies by Luciani et al.[9] and Defaye et al.[10,11] Luciani et al.[9] suggested this possibility from the similar behavior of the substances in inhibition of mitochondrial oxidative phosphorylation, while Defaye et al.[10,11] provided chemical evidence when they pyrolytically converted gummiferin into AT, and biological evidence from studies on competitive inhibition of ADP-stimulated respiration and binding properties. It was found that the proportions of AT and the considerably more toxic CAT varied greatly according to the freshness of the rhizomes. The amount of CAT was greater in fresh rhizomes and decreased dramatically with aging, as CAT was converted to AT via decarboxylation.[12]

B. *Xanthium* spp.

Toxicosis in domestic animals associated with eating young sprouts or burrs of cocklebur (*Xanthium strumarium*) is characterized by nausea, depression, ataxia, and death. The toxic principle of *X. canadense* was originally reported to be hydroquinone.[13] Kupiecki et al.[14] isolated a crystalline chemical from seeds of *X. strumarium* that exhibited potent hypoglycemic activity in the rat. In subsequent studies, Craig et al.[15] identified the hypoglycemic agent in *X. strumarium* as CAT. In addition, studies by Cole et al.[16] and Stuart et al.[17] redefined the toxic principle in *X. strumarium* responsible for the often reported toxicosis in domestic animals as CAT. The CAT was found only in the seeds, and young sprouts were toxic only as long as the cotyledons were present on the seedling. It is assumed that the toxic principle in other species of *Xanthium* is also due to CAT. No AT was detected in *X. strumarium*.

C. *Coffea arabica*

Obermann and Spiteller[18] isolated and identified atractyligenin, the aglycone of AT, as a glycoside in coffee beans (*Coffea arabica* L.) Obermann et al.[19] had previously isolated a metabolic product (2-β-hydroxy-15-oxyatractylan-4α carbonic acid) of this glycoside in the urine of coffee drinkers.

D. *Callilepsis laureola*

A plant, *Callilepsis laureola*, has been known to cause fatalities among the indigenous population of South Africa. AT was isolated and identified from dried, pulverized rhizomes.[20] Although AT was identified as the toxic principle responsible for the hypoglycemic and nephrotoxic activity, the presence of CAT in fresh rhizomes would not be surprising in view of the relationship of AT and CAT in fresh and dried rhizomes of *A. gummifera*.[12]

E. *Wedelia asperrima*

Wedelia asperrima Benth, a member of the family Compositae, is responsible for fatalities among sheep in northwestern Queensland. Oelrichs et al.[21] have recently isolated and identified the chemical structure of the toxic principle (wedeloside). Although wedeloside is an amino sugar, its aglycone is nearly identical to the aglycone of CAT. The difference between the two is the presence of an additional hydroxyl group on the wedeloside aglycone.

III. CHEMISTRY

CAT (I) is a glucoside that contains a molecule of D-glucose linked by a β-glycosidic bond to the hydroxyl group at position C2 of the aglycone moiety, carboxyatractyligenin (II). The D-glucose moiety contains sulfuric acid groups at positions C'3 and C'4 and an isovalaric group at C'2. CAT can be extracted from plant tissues with water or aqueous acetone. Purification usually involves concentration and defatting of the extract, followed by salting out of the potassium salts of AT and CAT. The precipitate is treated with 85% alcohol, and the insoluble material contains reasonably pure CAT. Pure CAT, $C_{31}H_{44}O_{18}S_2K_2 \cdot H_2O$, m.p. 280 to 282°C, $[\alpha]_D = -45.8$ (C = 1, water) as the dipotassium salt can be obtained after crystallization from water solution.[12]

Alkaline hydrolysis (20% KOH) of CAT yields carboxyatractyligen (II) [$C_{20}H_{28}O_6$, m.p. 235 to 238°C (dec.), $[\alpha]_D = 79.6°$ (C = 1, MeOH)] in addition to a sugar, sulfate, and isovalerate.

Treatment of carboxyatractyligen (II) with diazomethane gave the dimethylester [crystals from 50% aqueous methanol, m.p. 85°C, $[\alpha]_D = -68.8°$ (C = 0.675, MeOH)]. Acetylation of the dimethylester gave the diacetate dimethylester [crystals from acetone, 149°C, $[\alpha]_D = 21.6°$ (C = 0.565, MeOH)]. Heating of carboxyatractyligenin ether in solid state or solution yields atractyligenin.

The structure of CAT was primarily established by transformation of CAT into AT (III) by heating in glycol solution. A detailed discussion on the chemistry of AT (III) has been presented previously.[4]

IV. TOXICOSIS IN LIVESTOCK

The case histories, clinical signs, and lesions of the early documented cases of naturally occurring or experimentally induced livestock intoxication with cocklebur (*Xanthium* sp.) were reported in detail by Marsh et al.[22] Until recently the toxic principle of *X. strumarium* was generally thought to be hydroquinone.[13,23] Although the symptoms and lesions were reportedly reproduced by authentic hydroquinone administration, the latter was not present in glycosidic combination in the plant.[13] Early reports, however, had suggested the toxin to be a glycoside called xanthostrumarin.[13] More recently a hypoglycemic agent was isolated from *X. strumarium* by Murphy and Craig.[24] Craig et al.[15] identified the agent in crystalline form as CAT and showed it to induce hypoglycemia in rats, rabbits, and dogs. Cole et al.[16] isolated CAT from aqueous extracts of *X. strumarium* cotyledons and burrs, and showed that toxic signs produced in swine were similar for both natural cocklebur extracts and authentic CAT. In addition, these investigators were unable to recover hydroquinone from the plant or burr of *X. strumarium*. Authentic hydroquinone produced hyperglycemia and failed to produce the typical lesions of CAT intoxication which are comparable to those of spontaneous cocklebur poisoning in swine.[17] Clinical cocklebur poisoning in calves or pigs is associated with animals eating the plant during its palatable cotyledonary stage, which is prevalent in flood plains or in fields where herbicides are not properly used. Clinical signs in swine of cocklebur or CAT intoxication are acute depression, weakness, convulsions, and death.[17,22,23] Signs are similar in calves; however, calves may also become belligerent.[25] Marked serofibrinous ascites, edema of the gallbladder wall, and lobular accentuation of the liver are principal gross lesions.[17,22] Acute to subacute centrilobular hepatic necrosis is the primary histologic change; however, occasionally acute nephrosis or ischemic neuronal degeneration is present.[17] There is a marked hypoglycemia with elevated serum glutamic oxalacetic transaminase (SGOT), isocitric dehydrogenase (ICD), and sorbitol dehydrogenase compatible with the liver necrosis.[17] Authentic CAT administered intravenously to 20-kg pigs at 10 to 100 mg IV produced acute toxicity, whereas the amount of cotyledonary-stage cocklebur (*X. strumarium*) seedlings required for toxicity was 0.75 to 3% of body weight orally.[16,17]

In a recent report by Martin et al.[25] of naturally occurring cocklebur poisoning in calves, the authors described acute centrilobular hepatic necrosis and associated elevation of serum (gamma) glutamyltransferase typical of experimental intoxication in swine. There was no mention, however, of the prominent gallbladder edema which is a characteristic lesion of cocklebur intoxication in swine and to some extent in cattle.[17,22]

The clinical signs and lesions seen in cattle and swine from cocklebur intoxication are similar to that described for Noogoora burr (*X. pungens*).[26] A common plant in Queensland, Noogoora burr is poisonous to livestock, particularly in the dicotyledonary stage. In calves and pigs the lethal dose is 1.8 to 2% body weight with death occurring in 12 to 48 h. Clinical signs observed include excitement, nervousness, trembling, tetany, abdominal pain, muscular spasms, recumbency, and death. Lesions are characterized by reddening or hemorrhagic inflammation of the stomach and intestines as well as endocardial or epicardial hemorrhage, liver swelling, gallbladder edema, and distended urinary bladder.[26] Although the toxic principle has not been definitively identified, it is presumed to be a glycoside, and in view of the plant's close relationship to *X. strumarium*, the involvement of CAT would not be surprising.

V. TOXICOSIS IN LABORATORY ANIMALS

Seawright et al.[27] showed that active extracts from cotyledons of *Xanthium* spp. produced hypoglycemia and acute centrilobular necrosis of the liver in mice and that CAT produced similar hepatic necrosis in rats and mice. Hatch et al.[28] demonstrated prominent hepatic lobular accentuation in rats given CAT intraperitoneally; however, the necrosis observed microscopically was not marked or associated with a particular hepatic lobular structure as in other reports. Carpenedo et al.[29] in contrast to Seawright et al.[27] and Hatch et al.[28] observed no hepatic necrosis in rats administered CAT. The i.p. LD$_{50}$ for CAT in the studies of Seawright et al.[27] was 10.6 (7.5 to 15.1) and 2.9 (1.5 to 5.8) mg/kg for the mouse and rat, respectively, while the s.c. LD$_{50}$ was 5.4 (3.8 to 7.7) and 2.9 (2.1 to 4.1) mg/kg which were in general agreement with those reported by Luciani et al.[15,16] and Carpenedo et al.[29]

The 24-h LD$_{50}$ of CAT given i.p. to male rats by Hatch et al.[28] was 13.5 mg/kg. In mice both active and aqueous extracts were more toxic by the s.c. route.[27]

The i.p. LD$_{50}$ of AT in male rats was 143 mg/kg (153.01 to 133.64 mg/kg).[29] Carpenedo et al.[29] have shown that AT given to male rats (50 mg/kg i.p.) produces selective nephrosis in the distal portion of the renal proximal convoluted tubule. This portion of the renal nephron tubular epithelium is laden with mitochondria. There was an associated increase in urinary volume and excretion of potassium, glucose, ketones, and protein.[29] The distal tubule was unaffected by AT either morphologically or functionally (as measured by a lack of sodium loss).[30] After i.p. injection of AT, the corticomedullary junction of the kidneys became pale. Microscopically there was desquamation of the necrotic proximal tubular epithelium with granular casts formed in the distal convoluted tubules. Rats which survived showed regenerative attempts characterized by flattened tubular epithelium and vesicular hyperchromatic nuclei with occasional mitotic figures. The glomeruli were not affected. According to Carpenedo et al.[29] mice developed similar alterations as rats; however, rabbits and guinea pigs were unaffected. No morphological hepatocellular injury occurred in these studies. The authors reported normal measurements of SGPT, SGOT, and total serum bilirubin, although the attached tables in the article suggested mild to moderate elevation of SGOT at 100 mg/kg AT when compared to controls.[29] There was marked elevation of serum urea, but serum creatinine was only slightly increased. As in livestock there was marked dose-related hypoglycemia in these studies. Extracts of *C. laureola* which contain AT produced nephrotoxicity in rats similar to that seen by Carpenedo et al.[29]

The ambient temperature has been shown to directly influence the toxicity of AT in rats and mice; the toxicity increases with decrease in temperature.[30]

In contrast to AT in rats, Carpenedo et al.[29] found that CAT administered at lethal or sublethal doses produced no morphological renal changes or altered urea and creatine clearances. Isolated kidney mitochondria were, however, highly sensitive to the effects of CAT.[30] The authors considered the second carboxylic group of CAT to enhance the polarity of the compound and thus reduce reabsorption by the tubules and subsequent proximal tubular cell damage. Altered liver function was also not observed in rats given CAT in these studies[29,30] in direct contrast to that in livestock. Although both AT and CAT inhibit oxidative phosphorylation of heart mitochondria *in vitro*, the heart is not affected in animals.[29] Santi similarly observed no glycogen depletion in the heart of dosed animals.[31]

VI. TOXICOSIS IN HUMANS

As was alluded to in the introduction, numerous cases of accidental or purposeful poisonings in humans have been reported for *A. gummifera,* including an accidental poisoning involving a class of Italian school children in which three girls died.[4,32] Poisoning by *A. gummifera L.* has also occurred in humans in North Africa where the plant grows in dry conditions and apparently contains both AT (III) and CAT.[30,33] Both compounds are presumed to be highly toxic to humans based on the relative toxicities in mice, rats, and dogs.[30]

The marked hypoglycemia, convulsions, cerebral edema, neuronal necrosis, gastric irritation and hemorrhage, acute nephrosis, and hepatocellular necrosis seen in human cases are similar observations to those seen in animal toxicoses.[33,34] In South Africa the rhizome of *C. laureola* (contains AT) has been associated with zonal liver necrosis and renal tubular necrosis.[35] Mortality results from hypoglycemic coma, uremia, or hepatic failure.[35] In general, the prognosis for recovery from AT or CAT intoxication is poor unless supportive treatment such as pumping the stomach and appropriate fluid therapy is instituted.

VII. ATRACTYLOSIDE/CARBOXYATRACTYLOSIDE MECHANISM OF ACTION

AT or CAT administered to animals resulted in an initial hyperglycemia followed by a severe hypoglycemia.[29,30] During the hypoglycemic phase both serum lactate and nonesterified fatty acids increased, while both hepatic and skeletal muscle glycogen stores decreased.[30] The toxicity of AT may relate to tissue inhibition of oxidative processes, decrease in oxygen utilization, with a shift to anaerobic metabolism and glycogen breakdown. Luciani et al.[30] have shown decreased oxidative phosphorylation in homogenates of rat liver and kidney, but not heart.

Santi[36] was the first to demonstrate that AT inhibits mitochondrial oxidative phosphorylation and the citric acid cycle. CAT is more effective than AT in this inhibition of oxidative phosphorylation and is reported to be ten times more toxic *in vivo* due to the presence of a second carboxylic group in position four of the atractyligenin ring.[37,38] In other studies, Luciani et al.[30] showed the LD_{50} of CAT to be 30 to 40 times lower than that of AT. Inhibition of phosphate uptake by CAT is not influenced by varying the ADP concentration, whereas with AT the adenine nucleotides can competitively overcome inhibition.[27,38] Both AT and CAT inhibit energy transfer by preventing translocation of adenine nucleotide.[30] The difference in the toxicity of kinetics may be related to the tightness of binding to the mitochondrial membrane.[30]

The mechanism of specific ADP/ATP transport through the inner mitochondrial membrane is catalyzed by the hydrophobic protein adenine nucleotide translocase (ANT).[39] This polypeptide translocase controls the overall mechanism of oxidative phosphorylation such as nucleotide specificity, kinetics, temperature, and pH dependency of ATP synthesis or hydrolysis.[39] AT (a competitive inhibitor) and CAT (a noncompetitive inhibitor) have been extensively used to investigate mitochondrial ADP/ATP transport.[27,30,40] This nucleotide transport is essential for mitochondrial calcium uptake, oxidative phosphorylation, cell respiration, ATP reserves, glycolysis, glycogenolysis, gluconeogenesis, amino acid synthesis, blood glucose concentration, and fatty acid oxidation.[27,28] AT and CAT inhibit the adenonucleotide transport from the outside of the inner mitochondrial membrane[39] where they inhibit membrane aggregation induced by cations such as magnesium (Mg^{2+}) and thus increase the membrane negative charge.[41] The mitochondrial transport polypeptide translocase has multiple fractions or affinity sites which bind nucleoside (di-, tri-) phosphates, acyl-CoA and its esters, or AT/CAT.[40,42] This affinity of CAT for the translocase has allowed investigators to study the control of mitochondrial respiration. These studies have shown that the major limiting factor in oxidative phosphorylation in liver and heart mitochondria is the respiratory chain capacity and membrane ATPase.[42,43,44] Luciani et al.[30] have also shown that AT or CAT are not interfering with the cellular respiration mechanism. The CAT inhibitor mechanism permits access of exogenous ATP from the cellular cytosol to the mitochondrial matrix via the CAT-sensitive translocase with net ATP uptake inhibition.[45]

Following the investigation of spontaneous cocklebur intoxication in swine, Hatch et al.[28] attempted to determine the mechanism of action of CAT in male rats. These investigators pretreated rats with either cytochrome P-450 dependent enzymes, a P-450 inhibitor, a hemoprotein synthesis inhibitor, a cytochrome P-448 inducer, or a glutathione precursor or inhibitor. Following pretreatment, the rats were given a calculated i.p. LD_{50} dose (13.5 mg/kg body weight) of CAT. The studies showed that CAT detoxification may occur through a hemoprotein independent, phenylbutazone-inducible enzyme, and partly through a P-448-dependent enzyme. CAT detoxification apparently was not reported to be P-450 or glutathione dependent in these preliminary studies.[28]

Luciani et al.[30] have shown that both *in vivo* and *in vitro* effects of AT are characterized by inhibited oxidative phosphorylation, impaired respiratory rate, accelerated anaerobic glycolysis, lactate production, and glycogenolysis through inhibition of the mitochondrial translocase. The oxidative phosphorylation dysfunction with diminished ATP energy production results in inhibition of respiratory electron flow and interferes with the extra mitochondrial cytosolic redox systems. There is evidence to suggest that AT depresses cytosolic phosphorylation in liver and kidney, with accelerated glucose uptake.[30] Since the first steps in gluconeogenesis require ATP and acetyl CoA, this pathway was inhibited by AT with subsequent elevation of blood nonesterified fatty acids. Luciani et al.[30] have suggested that hormonal influence, as with glucagon or insulin, or neural adrenergic receptors may play an as yet unidentified role in glycoside toxicity. It is apparent that AT and/or CAT intoxication interferes with many cellular metabolic pathways by initially blocking the mitochondrial energy translocase mechanism with subsequent cellular and organ dysfunction. Hypoglycemia, hepatic necrosis, and nephrosis are the clinicopathologic hallmarks of AT and/or CAT intoxication in man and animals.

VIII. EFFECTS OF CARBOXYATRACTYLOSIDE ON PLANTS

In 1895, Arthur published his observations on the dormancy of *Xanthium* species[49] and publications by other authors followed until 1957.[46,47,48] It had been established that the genus *Xanthium* produces fruiting bodies (burrs) that are biloculate and that a seed is contained in each locule. The seeds are located next to each other; one is superior and the other slightly, but not fully, inferior. When the burr shatters and the seeds disperse, it is the superior one that germinates rapidly whereas the inferior one remains dormant for several years before germination occurs.[50] The dormancy is characteristically prolonged, even under ideal growing conditions, and it was this botanical idiosyncrasy that attracted attention for 90 years. It was believed that a self-inhibitor of germination was present in the seed that accounted for the delayed germination. Recent evidence strongly implicates CAT as the chemical agent responsible for that self-inhibition.

In 1957, Wareing and Foda[47] demonstrated the presence of two plant growth inhibitors in cocklebur seed. They extracted seeds with water, separated the inhibitory components by paper chromatography using isopropanol-1% NH_4OH in water (4:1, v/v), divided the developed chromatograms into ten equal parts, and bioassayed each segment in the etiolated wheat coleoptile "straight growth" test. Two inhibitory zones were noted. The first, which ran from R_f 0.1 to 0.3, was designated inhibitor A, and the second, with R_f 0.4 to 0.5, was called inhibitor B. Inhibitor A was approximately 33% more inhibitory than inhibitor B. The work was discontinued, and no further publications on the isolation and chemical identification of either inhibitor A or B were forthcoming.[59] At that time, Wareing was in scientific pursuit of abscisic acid, a potent plant-growth inhibitor since discovered in many plants and plant parts including ash, birch, sycamore, willow, young cotton bolls, seeds of ash and pear,[51] and the fungus *Cercospora cruenta*.[52,53] Neither of the inhibitors in cocklebur seed had the chemical characteristics of abscisic acid because they were both considerably more polar.

The discovery of CAT and AT by Danieli and co-workers in 1972 from the Mediterranean thistle, *Atractylis gummifera* L.[37] and the later discovery of CAT from *X. strumarium* L. appeared to be only of pharmacological interest because of its hypoglycemic properties.[15] Because of the chemical features of CAT, its behavior in early experiments with paper chromatograms, and its inhibitory effect on etiolated wheat coleoptiles, it became obvious that Wareing's inhibitor A and CAT had strikingly similar properties.[54]

When authentic CAT was chromatographed on paper in isopropanol-1% NH_4OH in water (4:1, v/v), in the ascending mode, dried, sprayed with anisaldehyde reagent, and gently heated, a bright magenta area developed at R_f 0.00 to 0.28. Another salmon pink area developed at R_f 0.46 to 0.56. When a complementary, unsprayed part of the developed chromatogram was cut into ten equal parts (R_f 0.00 to 0.10; 0.10 to 0.20; etc.), added to phosphate-citrate buffer supplemented with 2% sucrose pH 5.6,[55] and bioassayed with etiolated wheat coleoptiles, inhibitory activity was seen from R_f 0.00 to 0.30 and was recorded as 88, 76, and 86%, respectively, relative to control segments from chromatograms developed in solvent and bioassayed. When ($\pm$) abscisic acid was developed identically, it was observed to migrate between R_f 0.6 to 0.90 on subsequent chromogenic reaction with anisaldehyde; and these same areas, upon elution, inhibited wheat coleoptiles 100%. Hence, the identity of inhibitor A appears to be CAT, while that of inhibitor B is still unresolved. But it is certain that neither inhibitor A nor B is abscisic acid. Further indirect evidence for inhibitor A being CAT was the choice of chromatographic developing solvent by Wareing. Purified CAT is particularly unstable in aqueous solution as the free acid. The species that he had in his mildly basic solvent was, perhaps, the most stable.

In carefully controlled experiments CAT was shown to have marked plant-growth inhibitory activity. Wheat seed (*Triticum aestivum L.*, cv. Wakeland) was grown on moist sand in the dark for 4 d at 22 $\pm$ 1°C,[56] at which time the shoots were removed and the apical 2 mm were cut off in a Van der Weij guillotine and discarded. The next 4 mm were cut and retained for bioassay. Solutions of the potassium salt of CAT were formulated in phosphate-citrate buffer[12] at 10^{-3}, 10^{-4}, 10^{-5}, and 10^{-6} M. Ten 4-mm coleoptile sections were placed in each test tube that contained a specific molar concentration of CAT. Tubes were rotated at 0.25 rpm for 24 h at 22°C in the dark. Coleoptiles were then measured with a photographic enlarger at $\times$ 3 magnification, and data were recorded and statistically analyzed. CAT significantly ($p < 0.01$) inhibited coleoptiles at 10^{-3}, 10^{-4}, and 10^{-5} M, 100, 91, and 48% relative to controls, respectively.[54]

Higher plants were also treated with CAT, and the responses were selective. Week-old bean plants (*Phaseolus vulgaris L.*, cv. Black Valentine) were treated with aerosol sprays of 1 ml of the metabolite at 10^{-2}, 10^{-3}, and 10^{-4} M (8470, 847, and 84.7 μg, respectively) per pot: four plants per pot in a triplicated experiment. There were no apparent effects. However, when 6-week-old tobacco (*Nicotiana tabacum L.*, cv. Hicks) was treated with CAT at the same molar concentrations (1 ml per pot; 1 plant per pot) there were unusual responses. Within 24 h all treatment (10^{-2}, 10^{-3}, and 10^{-4} M) showed cupping of leaves and resembled tobacco that had received cold night temperatures (2 to 7°C), even though the ambient night temperature was 18.5°C. During the day, when temperatures reached 35°C, the leaves flattened and resembled normal tobacco plants. This odd effect lasted more than 72 h and the plants appeared to resume normal growth habits except that at 10^{-2} and 10^{-3} M treatments the leaves were slightly distorted 9 d following treatment. There were no necrotic lesions, and the response suggested that selective cells were subtly influenced by the metabolite.

The case for corn was different. When 15-day-old plants (*Zea mays L.*, cv. Norfolk Market White) were treated by pipetting 100 μl of each concentration (10^{-2}, 10^{-3}, and 10^{-4} M) into the leaf sheaths, there was a response within 72 h. At 10^{-2} M (847 μg per plant) there was severe chlorosis within the leaf sheaths; 1 week later, there was massive necrosis of the leaves and stunting at 10^{-2} M so that plants were inhibited approximately 50% relative to controls. At 10^{-3} M treatment, plants exhibited chlorosis inside the leaf sheaths, and there was chlorosis of the leaf margins. Also, plants were inhibited 25% relative to controls. There was some slight chlorosis of the 10^{-4} M-treated plants, but there was no growth inhibition; 14 d following treatment with 10^{-2}, 10^{-3}, and 10^{-4} M CAT, corn was inhibited 43, 14, and 0%, respectively, relative to controls.[54] The specificity of CAT for corn, a monocotyledonous plant, suggests that it should be evaluated among the other members of the grass family for herbicidal properties.

Congeners of CAT have also been assayed. AT, which differs from CAT by the lack of one COOH group, is inactive in the wheat coleoptile bioassay.[54] The aglycone, carboxy-

(1) Carboxyatractyloside, R_1 = D-glucose; R_2 = COOH; (2) Carboxy-atractyligen, R_1 = H; R_2 = COOH; (3) Atractyloside, R_1 = D-glucose; R_2 = H.

atractyligenin, appears to be active at concentrations $>10^{-3} M$ in preliminary studies[56] with wheat coleoptiles.

Fungi have also been assayed against CAT, and it has selective fungistatic properties. In repeated disk assays, various concentrations of the metabolite have consistently inhibited cultures of *Chaetomium cochlioides* 189 ($>$15-mm inhibition zones with 4-mm impregnated disks) grown on nutrient agar in petri dishes. But neither *C. cochlioides* 195, *Aspergillus flavus*, nor *Curvularia lunata* 49 were inhibited.[57] Recent studies have demonstrated that pollen of *Xanthium strumarium L.*, in aqueous suspension, is variably fungitoxic (genus dependent) to *Absidia ramosa*, *Aspergillus awamori*, *A. flavus*, *A. fumigatus*, *A. nidulans*, *A. niger*, *A. sydowi*, *A. terreus*, *Chaetomium globosum*, *Colletotrichum capsici*, *Curvularia lunata*, *Fusarium moniliforme*, and *Trichophyton mentagrophytes*.[58] Whether the inhibitor in the pollen is CAT or a related compound, or compounds, remains to be seen.

IX. CONCLUSION

The glucoside CAT has been at least partly responsible for the toxicity in the rhizomes of *Atractylis gummifera*, in the seeds of *Xanthium strumarium*, and may occur in rhizomes of *Callilepsis laureola*. The aglycone of AT was found in *Caffea arabica*, and a chemical closely resembling the aglycone of CAT was part of the wedeloside molecule isolated from *Wedelia asperrima*. CAT has been implicated in both human and animal toxicosis and is characterized by its pronounced hypoglycemic activity. Its site of action involves inhibition of mitochondrial oxidative phosphorylation as a noncompetitive inhibitor of the ADP transport system. This is in contrast to AT which acts as a competitive inhibitor of the ADP transport system.

CAT also acts as an inhibitor of plant growth. It is thought that CAT is the self-inhibitor in *X. strumarium* described by Wareing and Foda.[47] It has also shown some selective fungistatic activity.

REFERENCES

1. **Theophrastus, E.,** *De Historia plantarum*, Cap. XIII, *Libro nono*, 1552, Lungduni, Amstelodami, 1644.
2. **Dioscorides, P. A.,** *De Materia Medica Libri quinque III*, 93, Max Wellman Ed., Berolini, 1906.
3. **Mesue, J.,** *De Simplicibus*, Cap. XXII, G. Costa, Apud Luntas, Venetiis, 1623, 71.
4. **Santi, R. and Luciani, S.,** *Atractyloside: Chemistry, Biochemistry, and Toxicology*, Santi, R. and Luciani, S., Eds., Piccin Medical, Padova, Italy, 1978, 135.

5. **Linneus, C.,** *Species Plantarum,* (1797—1810), Part III, 4th ed., Wildenow, C. L., Tomus III, Berolini, 1960.

6. **Fiori, A.,** *Nouva Flora Analitica d'Italia,* Vol. 2, 701, Firenze, 1925—1929.

7. **Fiori, A. and Paoletti, C.,** *Flora Analitica d'Italia,* Vol. 3, 309, Padova, 1925—1929.

8. **Stanislas, E. and Vignais, P. M.,** Sur les principes toxiques d'*Atractylis gummifera* L., *C. R. Acad. Sci.,* 259, 4872, 1964.

9. **Luciani, S., Martini, N., and Santi, R.,** Effects of carboxyatractyloside a structural analogue of atractyloside on mitochondrial oxidative phosphorylation, *Life Sci.,* 10, 961, 1971.

10. **Defaye, G. and Vignais, P. V.,** Evidence experimentale pour l'identification de la gummiferine au carboxyatractyloside, *C. R. Acad. Sci.,* 273, 2671, 1971.

11. **Defaye, G., Vignais, P. M., and Vignais, P. V.,** Pyrolytic conversion of gummiferin into atractyloside, chemical and biological evidence, *FEBS Lett.,* 25, 325, 1972.

12. **Danieli, B., Bombardelli, E., Bonati, A., and Gabetta, B.,** Carboxyatractyloside, a new glycoside from *Atractylis gummifera* L., *Fitoterapia,* 42, 91, 1971.

13. **Kuzel, N. R. and Miller, C. E.,** A phytochemical study of *Xanthium canadense, J. Am. Pharm. Assoc.,* 39, 202, 1950.

14. **Kupiecki, F. P., Ogzewalla, C. D., and Schell, F. M.,** Isolation and characterization of a hypoglycemic agent from *Xanthium strumarium, J. Pharm. Sci.,* 63, 1166, 1974.

15. **Craig, J. C., Mole, M. L., Billets, S., and El-Feraly, F.,** Isolation and identification of the hypoglycemic agent, carboxyatractylate, from *Xanthium strumarium, Phytochemistry,* 15, 1178, 1976.

16. **Cole, R. J., Stuart, B. P., Lansden, J. A., and Cox, R. H.,** Isolation and redefinition of the toxic agent from cocklebur *Xanthium strumarium), J. Agric. Food Chem.,* 28, 1330, 1980.

17. **Stuart, B. P., Cole, R. J., and Gosser, H. S.,** Cocklebur (*Xanthium strumarium,* L. var. *strumarium*) intoxication in swine: review and redefinition of the toxic principle, *Vet. Pathol.,* 18, 368, 1981.

18. **Obermann, H. and Spiteller, G.,** Die strukturen der "Kaffee-Atractyloside", *Chem. Ber.,* 109, 3450, 1976.

19. **Obermann, H., Spiteller, G., and Hoyer, G. A.,** Struktur eines aus menschenharn isolierten C_{19}-terpenoids-2β-hydroxy-15-oroatractylan-4α-carbonsame, *Chem. Ber.,* 106, 3506, 1973.

20. **Candy, H. A., Pegel, K. H., Brookes, B., and Rodwell, M.,** The occurrence of atractyloside in *Callilepsis laureola, Phytochemistry,* 16, 1308, 1977.

21. **Oelrichs, P. B., Vallely, P. J., McLeod, J. K., and Lewis, I. A. S.,** Isolation of a new potential antitumor agent from *Wedeliza asperrima, J. Nat. Prod.,* 43, 414, 1980.

22. **Marsh, C. D., Roe, G. C., and Clawson, A. B.,** Cockleburs (species of *Xanthium*) as poisonous plants, *U.S. Dep. Agric. Bull.,* 1274, 1, 1924.

23. **Kingsbury, J. M.,** *Poisonous Plants of the United States and Canada,* Prentice-Hall, Englewood Cliffs, NJ, 1964, 440.

24. **Murphy, J. C. and Craig, J. C.,** Pharmacological effects of a new toxin is dated from cocklebur, *J. Miss. Acad. Sci.,* 20, 26, 1975.

25. **Martin, T., Stair, E. L., and Dawson, L.,** Cocklebur poisoning in cattle, *J. Am. Vet. Med. Assoc.,* 189, 562, 1986.

26. **Kenny, G. C., Everist, S. L., and Sutherland, A. K.,** Noogoora burr poisoning of cattle, *Queensl. Agric. J.,* 70, 172, 1950.

27. **Seawright, A. A., Hrdlicka, J., Lee, J. A., and Ogunsan, E. A.,** Toxic substances in the food of animals: Some recent findings of Australian poisonous plant investigations, *J. Appl. Toxicol.,* 2, 75, 1982.

28. **Hatch, R. C., Jain, A. V., Weiss, R., and Clark, D. J.,** Toxicologic study of carboxyatractyloside (active principle in cocklebur — *Xanthium strumarium*) in rats treated with enzyme inducers and inhibitors and gluthathione precursor and depletor, *Am. J. Vet. Res.,* 43, 111, 1982.

29. **Carpenedo, F., Luciani, S., Searavilli, F., Palatini, P., and Santi, R.,** Neptorotoxic effect of atractyloside in rats, *Arch. Toxicol.,* 32, 169, 1974.

30. **Luciani, S., Carpenedo, F., and Tarjan, E. M.,** Effects of atractyloside and carboxyatractyloside in the whole animal, in *Atractyloside: Chemistry, Biochemistry and Toxicology,* Santi, R. and Luciani, S., Eds., Piccin Medical, Padova, Italy, 1978, 109.

31. **Santi, R.,** Pharmacological properties and mechanism of action of atractyloside, *J. Pharm. Pharmacol.,* 16, 437, 1964.

32. **Santi, R. and Cascio, G.,** Ricerche farmacologiche sul principie attivo dell'*Atryactylis gummifera.* I. Azione generale, *Arch. Ital. Sci. Farmacol.,* 5, 534, 1955.

33. **Capdevielle, P. and Darracq, R.,** L'intoxication par le chardon a glu (*Atractylis gummifera* L.), *Med. Trop.,* 40, 137, 1980.

34. **Catanzano, G., Delons, S., and Benyahia, T. D.,** A propos de 2 cas d'intoxication par le chadon a glu, *Maroc. Med.,* 49, 651, 1969.

35. **Debetto, P.,** Plants recently found to contain atractylosides, in *Atractyloside: Chemistry, Biochemistry and Toxicology,* Santi, R. and Luciani, S., Eds., Piccin Medical, Padova, Italy, 1978, 126.

36. **Santi, R.,** Potassium atractylate, a new inhibitor of the tricarboxylic acid cycle, *Nature (London)*, 182, 257, 1958.
37. **Danieli, B., Bombardelli, E., Bonati, A., and Gabetta, B.,** Structure of the deterpenoid carboxyatractyloside, *Phytochemistry*, 11, 3501, 1972.
38. **Luciani, S., Martini, N., and Santi, R.,** Effects of carboxyatractyloside a structural analogue of atractyloside on mitochondrial oxidative phosphorylation, *Life Sci.*, 10, 961, 1971.
39. **Panov, A., Filippova, S., and Lyakhovich, V.,** Adenine nucleotide translocase as a site of regulation by ADP of the rat liver mitochondria permeability to hydrogen and potassium ions, *Arch. Biochem. Biophys.*, 199, 420, 1980.
40. **Wohlrab, H.,** Purification of a reconstitutively active mitochondrial phosphate transport protein, *J. Biol. Chem.*, 235, 8170, 1980.
41. **Stoner, C. D., Sirak, H. D., and Richardson, M.,** Magnesium induced inner membrane aggregation in heart mitochondria prevention and reversal by carboxyatractyloside and bongherkic-acid, *J. Cell Biol.*, 77, 417, 1978.
42. **Pande, S. V., Goswami, T., and Parvin, R.,** Protective role of adenine nucleotide translocase in oxygen deficient hearts, *Am. J. Physiol.*, 247, H25, 1984.
43. **Davis, E. J. and Davis-Van Thienen, W. I. A.,** Rate control of phosphorylation-coupled respiration by rat liver mitochondria, *Arch. Biochem. Biophys.*, 233, 573, 1984.
44. **Forman, N. G. and Wilson, D. F.,** Dependence of mitochondrial oxidative phosphorylation on activity of the adenine nucleotide translocase, *J. Biol. Chem.*, 258, 8649, 1983.
45. **Aprille, J. R.,** Net uptake of adenine nucleotides by newborn rat liver mitochondria, *Arch. Biochem. Biophys.*, 207, 157, 1981.
46. **Thornton, N. C.,** Factors influencing germination and development of dormancy in cocklebur seeds, *Contrib. Boyce Thompson Inst.*, 7, 477, 1935.
47. **Wareing, P. F. and Foda, H. A.,** Growth inhibitors and dormancy in *Xanthium* seed, *Physiol. Plant.*, 10, 266, 1957.
48. **Esashi, Y., Komatsu, H., Ishihara, N., and Ishizawa, K.,** Dormancy and impotency of cocklebur seeds. VIII. Lack of germination responsiveness in primarily dormant seeds to cyanide, azide, anoxia and chilling, *Plant Cell Physiol.*, 23, 41, 1982.
49. **Arthur, J. C.,** Delayed germination in cocklebur and other paired seeds, *Proc. Soc. Prom. Agric. Sci.*, 16, 70, 1895.
50. **King, K. J.,** *Weeds of the World, Biology and Control*, Leonard Hill Books, London, 1966, chap. 6.
51. **Wareing, P. F. and Phillips, I. D. J.,** *The Control of Growth and Differentiation in Plants*, Pergamon Press, Oxford, 1970, chap. 11.
52. **Ichimura, M., Oritani, T., and Yamashita, K.,** The metabolism of (2Z,4E)-α-ionylideneacetic acid in *Cercospora cruenta*, a fungus producing (+)-abscisic acid, *Agric. Biol. Chem.*, 47, 1895, 1983.
53. **Oritani, T., Ichimura, M., and Yamashita, K.,** The metabolism of analogs of abscisic acid in *Cercospora cruenta*, *Agric. Biol. Chem.*, 46, 1959, 1982.
54. **Cutler, H. G. and Cole, R. J.,** Carboxyatractyloside: a compound from *Xanthium strumarium* and *Atractylis gummifera* with plant growth inhibiting properties. The probable "inhibitor" A, *J. Nat. Prod.*, 46, 690, 1983.
55. **Nitsch, J. P. and Nitsch, C.,** Studies on the growth of coleoptile and first internode sections. A new, sensitive straight-growth test for auxins, *Plant Physiol.*, 31, 94, 1956.
56. **Cutler, H. G.,** Unpublished data.
57. **Hancock, C. R., Barlow, H. W., and Lacey, H. J.,** The East Malling coleoptile straight growth test method, *J. Exp. Bot.*, 15, 166, 1964.
58. **Cutler, H. G.,** Secondary metabolites from plants and their allelochemic effects, in *Bioregulators for Pest Control*, A.C.S. Symp. Ser. No. 276, Hedin, P. A., Ed., American Chemical Society, Washington, D.C., 1985, chap. 32.
59. **Tripathi, R. N., Dubey, N. K., and Dixit, S. N.,** Fungitoxicity of pollen grains with special reference to *Xanthium strumarium* (compositae), *Grana*, 24, 61, 1985.
60. **Wareing, P. F.,** Personal communication.

INDEX

G

H

L

M

occurrence, 173—175

properties, structure, and chemical reactions, 163—170

quantitation, 170—172

synthesis of divicine and isouramil, 172—173

Vigna angularis, 102, 106

5-Vinyloxazolidine-2-thione, 28, 29

Viola odorata, 146

Vitamin A, calcinogenic glycoside effects, 225

Vitamin C, 167—169, 184, 185, 190—191

Vitamin D, see Calcinogenic glycosides

Vitamin E, 177, 188, 190—192

T

Wasabia japonica, 4, 17

Water balance, saponins and, 126

Watercress, 4

Water snail, 123

Wedelia asperrima, 256

Wedeloside, 256

Weed-seed meal, 20

Wheat, 261

White mustard, 2, 4

White sweet clover, 103, 113

X

Xanthine dehydrogenase. 51

Xanthine oxidase, 51

Xanthium, 255, 257

   pungens, 257

   trumarium, 255, 257, 260, see also Carboxyatractyloside

Xanthostrumarin, 257

Xysmalobium undulatum, 70

Y

Yamogenin, 108

Yams, saponins, 100

Yellow oat grass, 212

Yucca, 130

   mohavensis, 127, 130

   schidigen, 128

   schottii, 128

Z

Zea mays, 211, 261

Zierin, 48

Zinc, cyanide and, 52

Zygaenid moths, 47